Thermus Species

BIOTECHNOLOGY HANDBOOKS

Series Editors: Tony Atkinson and Roger F. Sherwood

Centre for Applied Microbiology and Research
Division of Biotechnology
Salisbury, Wiltshire, England

<table>
<tr><td>Volume 1</td><td>PENICILLIUM AND ACREMONIUM
Edited by John F. Peberdy</td></tr>
<tr><td>Volume 2</td><td>BACILLUS
Edited by Colin R. Harwood</td></tr>
<tr><td>Volume 3</td><td>CLOSTRIDIA
Edited by Nigel P. Minton and David J. Clarke</td></tr>
<tr><td>Volume 4</td><td>SACCHAROMYCES
Edited by Michael F. Tuite and Stephen G. Oliver</td></tr>
<tr><td>Volume 5</td><td>METHANE AND METHANOL UTILIZERS
Edited by J. Colin Murrell and Howard Dalton</td></tr>
<tr><td>Volume 6</td><td>PHOTOSYNTHETIC PROKARYOTES
Edited by Nicolas H. Mann and Noel G. Carr</td></tr>
<tr><td>Volume 7</td><td>ASPERGILLUS
Edited by J. E. Smith</td></tr>
<tr><td>Volume 8</td><td>SULFATE-REDUCING BACTERIA
Edited by Larry L. Barton</td></tr>
<tr><td>Volume 9</td><td>THERMUS SPECIES
Edited by Richard Sharp and Ralph Williams</td></tr>
</table>

A Continuation Order Plan is available for this series. A continuation order will bring delivery of each new volume immediately upon publication. Volumes are billed only upon actual shipment. For further information please contact the publisher.

Thermus Species

Edited by

Richard Sharp

The Centre for Applied Microbiology and Research
Porton Down, Salisbury, England

and

Ralph Williams

Queen Mary and Westfield College
London, England

Plenum Press • New York and London

Library of Congress Cataloging-in-Publication Data

Thermus species / edited by Richard Sharp and Ralph Williams.
    p.   cm. -- (Biotechnology handbooks ; v. 9)
  Includes bibliographical references and index.
  ISBN 0-306-44925-0
  1. Thermophilic bacteria. I. Sharp, Richard (Richard J.)
II. Williams, R. A. D. (Ralph Anthony David), 1938- .
III. Series.
QR84.8.T46 1995
589.9'2--dc20
                                       95-12394
                                          CIP

ISBN 0-306-44925-0

© 1995 Plenum Press, New York
A Division of Plenum Publishing Corporation
233 Spring Street, New York, N. Y. 10013

10 9 8 7 6 5 4 3 2 1

Contributors

Gudni A. Alfredsson • University of Iceland, Institute of Biology, Microbiology Laboratory, Ármúli 1A, IS-108 Reykjavik, Iceland

Peter L. Bergquist • Bacterial Genetics and Microbiology, School of Biological Sciences, University of Auckland, New Zealand

Doug Cossar • Cangene Corporation, 3403 American Drive, Mississauga, Ontario, Canada, L4V 1T4

Milton S. da Costa • Departmento de Zoologia, Universidade de Coimbra, 3049 Coimbra Cedex, Portugal

Melanie L. Duffield • Centre for Applied Microbiology and Research, Porton Down, Salisbury, Wiltshire SP4 0JG, United Kingdom

Jakob K. Kristjansson • University of Iceland, Institute of Biology, Microbiology Laboratory, Ármúli 1A, IS-108 Reykjavik, Iceland

Hugh W. Morgan • Thermophile Research Unit, University of Waikato, Private Bag 3105, Hamilton, New Zealand

Tairo Oshima • Department of Life Science, Tokyo Institute of Technology, Nagatsuta, Yokohama, Japan

Neil D. H. Raven • Centre for Applied Microbiology and Research, Porton Down, Salisbury, Wiltshire SP4 0JG, United Kingdom

Richard Sharp • Centre for Applied Microbiology and Research, Porton Down, Salisbury, Wiltshire SP4 0JG, United Kingdom

Ralph Williams • Queen Mary & Westfield College, Biochemistry Department, Faculty of Basic Medical Sciences, Mile End Road, London E14 NS, United Kingdom

Preface

There is considerable interest in thermophile microorganisms, in their environments, their ability to survive at temperatures which normally denature proteins, but more importantly, as a valuable resource for biotechnology.

The first reported isolation of *Thermus* by Tom Brock was in 1969. This initiated the present era of thermophilic research with the realization that where liquid water is available, there may be no limits to the temperature at which microorganisms can grow. Considerable research into the ecology, physiology, metabolism, and thermostable enzymes of thermophiles has led to their evaluation for a range of industrial and commercial processes. The past fifteen years have been an explosive period of discovery of many new genera and species, including the descriptions of a new fundamental kingdom—the Archaea.

Much of the current research has been focused on the Archaea; but it is significant that during this period, the original type strain YT-1 of *Thermus aquaticus* described by Brock has provided a major step forward in molecular biology. DNA polymerase from strain YT-1 has proved to be the major success in the commercialization of enzymes from thermophilic microorganisms to date.

The ease with which *Thermus* strains can be handled in laboratories without specialized equipment, together with the large investment in describing their structure, metabolism, and genetics, should ensure a continuing effort in *Thermus* research.

This book brings together many groups that have isolated and contributed to our knowledge of *Thermus*. It covers aspects of ecology and isolation, taxonomy, physiology, molecular biology and genetics, cell structure and biotechnology. This volume aims primarily at established researchers in universities, research institutes, and industry, but it introduces this important group of thermophilic bacteria to undergraduates and postgraduates who are interested in thermophilic microorganisms as potential biotechnological tools and as fascinating laboratory curiosities.

Richard Sharp
Ralph Williams

Contents

Chapter 3

Physiology and Metabolism of *Thermus* 67

Richard Sharp, Doug Cossar, and Ralph Williams

Chapter 4

Enzymes of *Thermus* and Their Properties 93

Melanie L. Duffield and Doug Cossar

Chapter 5

The Cell Walls and Lipids of *Thermus*   143

Milton S. da Costa

Chapter 6

**Genetics of *Thermus*: Plasmids, Bacteriophage,
Potential Vectors, Gene Transfer Systems**   157

Neil D. H. Raven

Chapter 7

Genes and Genetic Manipulation in *Thermus thermophilus* .....

Tairo Oshima

Chapter 8

The Taxonomy and Identification of *Thermus*

RALPH WILLIAMS and RICHARD SHARP

1. INTRODUCTION AND GENERAL PROPERTIES OF *THERMUS*

Bacteria of the genus *Thermus* were first isolated in the late 1960s from hot springs in Japan and the USA. *Thermus* strains also appear to be widespread in neutral, hot, aqueous artificial environments such as domestic and industrial hot water systems and thermally polluted streams. The reference strains that have been used in taxonomic studies are listed in Table I. Isolates from sources in Yellowstone National Park were originally allocated to the type species, *Thermus aquaticus* (Brock and Freeze, 1969), of this new genus. Subsequently, sites in Yellowstone National Park were sampled for *Thermus*, but isolates were only obtained from sources between 55 °C and 80 °C and pH 6.0 to 10.5 (Munster *et al.*, 1986). In Japan, early isolates were named "*Flavobacterium thermophilum*" (Oshima and Imahori, 1971), but validly redescribed as "*Thermus thermophilus*" (Oshima and Imahori, 1974). Other Japanese strains were given the invalid species names "*Thermus flavus*" (Saiki *et al.*, 1972), and "*Thermus caldophilus*" (Taguchi *et al.*, 1982), but these strains seem to belong to the *T. thermophilus* genospecies (see sections 5.2. and 6.1). Other terrestrial hot springs in Iceland were sampled (Cometta *et al.*, 1982b; Hudson *et al.*, 1987a; Kristjansson and Alfredsson, 1983; Pask-Hughes and Williams, 1977). Extensive studies were also made of many New Zealand isolates, (Hudson *et al.*, 1987a) and the validly named strain *Thermus filiformis* (Hudson *et al.*, 1987b). *Thermus*

RALPH WILLIAMS • Queen Mary & Westfield College, Biochemistry Department, Faculty of Basic Medical Sciences, Mile End Road, London E14 NS, United Kingdom.
RICHARD SHARP • Centre for Applied Microbiology and Research, Porton Down, Salisbury, Wiltshire SP4 0JG, United Kingdom.

Thermus Species, edited by Richard Sharp and Ralph Williams. Plenum Press, New York, 1995.

Table I. Sources of *Thermus* Isolates Used in Taxonomic Studies

Species	Strain	ATCC	NCIB	DSM	Originator	Source
T. aquaticus	YT1^T	25104	11243	625	Brock, T. D.	U.S.A.
T. filiformis	Wai33A1^T	43280	12588		Hudson, J. A.	New Zealand
	T351	31674			Hudson, J. A.	New Zealand
T. ruber	BKMB1258	35948	11269	1279	Loginova, L. G.	Kamchatka
	16210	51134			Sharp, R. J.	Iceland
	16234	51135			Sharp, R. J.	Iceland
"*T. rubens*"	KY12265	31556			Ado, Y.	Japan
"*T. brockianus*"	YS38^T		12676		Sharp, R. J.	U.S.A.
"*T. thermophilus*"	HB8^T	27624	11244	579	Oshima, T.	Japan
	RQ1				da Costa, M. S.	Azores
	B		11247		Williams, R. A. D.	Iceland
	ZK1				Horikoshi, K.	Japan
	IB21	43815			Alfredsson, G.	Iceland
"*T. flavus*"	AT62	33923		674	Saiki, T.	Japan
"*T. caldophilus*"	GK24				Ohta, T.	Japan
"*T. lacteus*"	KY12264	31557			Ado, Y.	Japan
"*T. anaerobicus*"	M10				Baldursson, S.	Iceland
"*T. oshimai*"	SPS17				da Costa, M. S.	Portugal
"*T. imahorii*"	Vi7				da Costa, M. S.	Portugal

were purified from hot springs in the Czech Republic (Peckova, 1990) and Continental Portugal and the Azores Islands (Prado *et al.*, 1988; Santos *et al.*, 1989), New Mexico (Hudson *et al.*, 1989), Belgium (Degryse *et al.*, 1978), England (Pask-Hughes and Williams, 1975), Kenya (Williams, unpublished), and the Australian Artesian Basin (Denman *et al.*, 1991). Halotolerant strains of *Thermus* were also isolated from shallow marine thermal vents off Iceland (Hudson *et al.*, 1989), and hot springs in the Azores and Japan. These cultures were similar to their terrestrial counterparts except in their halotolerance. The halotolerant Icelandic strain IB21 (Kristjansson *et al.*, 1986) has high DNA:DNA homology with *T. thermophilus* (Williams, 1989).

Red-pigmented strains of *Thermus ruber*, with a temperature optimum of 60 °C, were isolated from hot springs in the Kamchatka Peninsula, USSR (Loginova and Egorova, 1976; Loginova *et al.*, 1984), Iceland, the island of Sao Miguel in the Azores, and from an aerated fermenter fed with yeast wastes (Hensel *et al.*, 1986). The patent strain "*Thermus rubens*" ATCC 35948 (Ado *et al.*, 1982) is indistinguishable from *T. ruber*. The strains tested by DNA:DNA homology were single genospecies (Sharp and Williams, 1988; Ruffett, 1992).

2. CELL WALL STRUCTURE

The bacteria of the genus *Thermus* stain as Gram-negative have a cell wall that resembles that of true Gram-negative bacteria. There are many features of *Thermus* that appear to be more consistent with Gram-positivity, and it seems likely that *Thermus* belongs to an ancient line that predates the development of distinct Gram-positive and Gram-negative types. Upon initial isolation, many strains form filaments, but after repeated transfers in laboratory media, most of them grow as pleomorphic rod-shaped cells and short filaments, with the exception of the *T. filiformis* type strain (Hudson *et al.*, 1987b), which always produces very long intertwined filaments.

The definition of *Thermus* as Gram-negative is open to doubt for a number of reasons, despite the multilayered cell wall structure and the mode of cell division by invagination rather than Gram-positive septum formation. The ornithine type of peptidoglycan has not been reported except in Gram-positive bacteria; although rather rare, this type is found in *Dienococcus*. Branched chain fatty acids in the lipids are also Gram-positive, as is the use of menaquinones rather than ubiquinones as respiratory chain components. Strains of *Thermus* show a high sensitivity to the penicillins, actinomycin D, and to novobiocin, which are all characteristics of the Gram-positive bacteria.

Endospores are not seen in *Thermus* cultures grown on any medium, including starch agar. Electron microscopy shows that the cell envelope, external to the plasma membrane, is composed of a thin dense layer presumed to be peptidoglycan, surrounded by a highly corrugated outer layer that is connected to the peptidoglycan layer only at the indentations (Brock and Edwards, 1970). The surface of the cells resembles annelid worms because of the regularity of the indentations of this outer layer (Williams, 1975). Unusual morphological structures, such as "rotund bodies" that are sometimes visible with phase contrast microscopy, are of two types. The 'aggregation' structure consists of several cells bound together by the external layer of the cell enclosing a number of cells and some intercellular space; the 'vesicular' type develops from the surface of a single cell (Brock and Edwards, 1970; Kraepelin and Gravenstein, 1980; Becker and Starzyk, 1984). By contrast with other strains of *Thermus*, *T. filiformis* has a stable filamentous morphology. An extra cell wall coat surrounds the corrugated layer and runs uninterrupted over zones of septum formation (Hudson *et al.*, 1987b). A calcium-protein complex (Berenguer *et al.*, 1988) is a major component of the cell wall of *T. thermophilus* HB8. Morphological features described above are not valuable in distinguishing species within the genus *Thermus*.

2.1. Peptidoglycan

All strains of *Thermus* examined have a peptidoglycan containing the dibasic amino acid ornithine (Pask-Hughes and Williams, 1978; Hensel *et al.*, 1986). This relatively rare peptidoglycan type is not found in Gram-negative bacteria, but is present in strains of the Gram-positive genus *Deinococcus*, which is remotely related to *Thermus* (Hensel *et al.*, 1986, Embley *et al.*, 1993). This feature is of value because a simple test can be carried out to distinguish isolates of *Thermus* from thermophilic members of the genus *Bacillus*, which may be yellow but have a peptidoglycan containing diaminopimelic acid (DAP). This amino acid can be identified by qualitative thin-layer chromatography of whole cell hydrolysates because neither ornithine or DAP occurs in proteins. Partial purification of peptidoglycan by extraction with hot 4% sodium dodecyl sulfate followed by hot 10% trichloracetic acid before acid hydrolysis of the residue helps the clarity of the chromatograms (Williams, 1989; Georganta *et al.*, 1993). This process, done on cellulose plates, developed with: methanol, 80, pyridine, 10, 1.25M HCl, 20, then dried in air, treated with 0.1% ninhydrin in acetone, and heated to 100 °C for 5 min. Olive green spots fading to yellow indicate DAP. Ornithine may be detected by two-dimensional chromatography: first—ethanol, 100; butan-1-ol, 100, water, 50, propionic acid, 20, and second—with butan-1-ol, 100, acetone, 100, water, 50, dicyclohexylamine, 20. Color is developed with 0.2% ninhydrin in acetone containing 7% glacial acetic acid. The identification of amino acids is helped by some specific color reactions: aspartate is turquoise; glycine is claret; tyrosine and phenylalanine are grey-brown; histidine is grey; and proline is yellow.

3. PIGMENTS AND LIPIDS

The lipids, including the pigments, of *Thermus* are described in detail in Chapter 5. Research on lipids has been directed towards two objectives: determine whether these components are taxonomic; and discover whether their composition changes with growth conditions, and variations in temperature.

3.1. Pigments

Most strains have yellow, orange, or red carotenoids, but some strains are almost colorless, showing a pale pink-fawn color due to cytochromes. The orange strains contain phytoene and carotene, but the pigments of the red isolates (*T. ruber*) resemble neurosporoxanthin and retrodehydro-γ-carotene (Loginova *et al.*, 1984). Cultures from natural habitats that are exposed to sunlight are generally pigmented, while those from man-made thermal environments lacking illumination are frequently nonpigmented. In some strains pigmentation has been said to be unstable. Cometta reported the frequent isolation of colorless "mutants" from *T. aquaticus* YT1 grown in continuous culture (Cometta *et al.*, 1982a). By contrast, pigmentation of strain YS45 varied directly with the intensity of illumination in chemostat cultures, but was never completely lost, and colorless "mutants" were never detected (Cossar and Sharp, 1989).

3.2. Menaquinones

Menaquinones [MK] with 7(minor), 8(major) and 9(minor) isoprene groups were detected among the lipids of *T. aquaticus* YT1 (Collins and Jones, 1981; Hensel *et al.*, 1986). MK-8 is the predominant quinone detected in all strains investigated (Williams, 1989), rather than the ubiquinones found in typical Gram-negative bacteria. The identification of MK-8 is an important test used to distinguish *Thermus* from the thermophilic bacilli, in which the predominant component is MK-7. To test for quinone type, cell paste is freeze-dried and extracted with chloroform and methanol (2:1) in the dark overnight (Collins and Jones, 1981), filtered and concentrated under reduced pressure, applied to a preparative Keiselgel 60F TLC plate, and developed with 15% diethyl ether in light petroleum (60–80). The band absorbing UV light (254 nm) at Rf 0.8, corresponding to the menaquinones, is redissolved in chloroform, concentrated under a stream of nitrogen, applied to a reverse phase TLC (RP18F254, Merck) and developed with acetone. The Rf values of quinones are observed under UV light (254 nm). Ultraviolet absorption spectra of extracts of the preparative plates in iso-octane may also be determined between 230 and 350 nm (Collins and Jones, 1981).

3.3. Fatty Acids

Iso- and anteiso-branched fatty acids (predominantly iso-C15 and iso-C17) are the principal aliphatic components of *Thermus* lipids (Hensel *et al.*, 1986; Oshima, 1978; Pask-Hughes and Shaw, 1982; Prado *et al.*, 1988; Donato *et al.*, 1990; Nordstrom and Laakso, 1992). The predominance of branched fatty acids has been suggested to confirm a strain's membership of the genus *Thermus* (Donato *et al.*, 1990). In *T. aquaticus* and *T. thermophilus* the major fatty acid is iso-C17, while for most Portuguese isolates iso-C15 is the most abundant. *T. filiformis* is unusual as the anteiso-

C17 is the main constituent. It would be interesting to know the fatty acid analysis of other New Zealand isolates. For *T. ruber* the iso-C15 predominates over the iso-C17 fatty acid and also increases as the temperature rises (Donato *et al.*, 1991). Branched-chain fatty acids with even numbers of carbon atoms, unsaturated fatty acids and cyclopentane fatty acids are either present in traces, or have not been detected (see Chapter 5).

3.4. Complex Lipids

The polar fraction of the lipids of several yellow-pigmented and colorless strains includes a major glycolipid which is a diglycosyl-(N-acyl)-glycosaminyl-glucosyl-diacylglycerol containing glucose or galactose, and glucosamine or galactosamine (Oshima, 1978; Pask-Hughes and Shaw, 1982; Prado *et al.*, 1988). Such abundant membrane components, rich in carbohydrate, appear to be characteristic of *Thermus*. In *T. thermophilus* and *T. flavus*, the major glycolipid represents 70% of the total lipids (Oshima and Yamakawa, 1974; Pask-Hughes and Shaw, 1982). The glycolipid of *T. aquaticus* YT1 contains 3 glucose (1 galactosamine but Portuguese isolates contained no galactosamine); it had constituents approximately 3 glucose; 1 glucosamine, 1 glycerol, 3 fatty acids (Prado *et al.*, 1988); and it represented 87%–97% of the total glycolipid.

The major phospholipid of *Thermus* has not been completely characterized. The *T. aquaticus* YT1 phospholipid comprised 1 glycerol, 1 phosphate, 1 long chain amine, 3 fatty acids (Ray *et al.*, 1971a), but five strains, including YT1, were found to contain a major glycophospholipid including N-acylglucosamine (Pask-Hughes and Shaw, 1982). Materials extracted from *Thermus* by classical preparation methods for lipopolysaccharide (LPS) were found to contain neither heptose or ketodeoxyoctulosonate and were therefore not considered to be LPS (Pask-Hughes and Williams, 1978).

4. PHENOTYPIC PROPERTIES

4.1. Physiological and Biochemical Tests

4.1.1. High Temperature or Yellow-Pigmented Strains

Colonies of *Thermus* sp. on agar plates appear as orange, bright yellow to pale cream or fawn colonies. In general strains isolated from man-made environments tend to lack pigment. Morphology varies slightly but colonies are predominantly circular, raised and may be either rough or smooth. Munster *et al.* (1986) distinguished two groups of strains isolated from Yellowstone by spreading and nonspreading colony morphologies, and this morphological difference correlates with the two species that occur in this site.

The pH and temperature ranges for growth were between pH 6.0–7.0 and pH 9.0–10.0 and from 40–45 °C to 70–75 °C. *T. thermophilus* and related strains were typified by maximum growth temperatures of 80 °C.

All strains are Gram-negative, and with the exception of *T. filiformis* form pleomorphic rod-shaped cells and short filaments. *T. aquaticus* YT1 was isolated in a complex medium consisting of synthetic salts base supplemented with 0.1% yeast extract and 0.1% tryptone (Brock and Freeze, 1969). The species grows less with ammonium ions as the nitrogen source, and with acetate, sucrose, citrate, succinate or glucose as sole carbon sources. No growth was observed when the concentration of the organic components was increased to 1%, and this was considered typical of most *Thermus* sp. However, *T. thermophilus* and related strains are routinely cultured on media containing more than 1% organic matter.

A number of strains will grow in basal salts medium, with ammonium as nitrogen source and an appropriate carbon source. Many strains, nevertheless, require vitamins for growth on minimal medium with single carbon sources (Alfredsson *et al.*, 1985). Nitrate is not reduced by *T. aquaticus*, *T. filiformis*, nor most of the red-pigmented strains examined, but serves as a terminal electron acceptor for some high temperature strains, several of which also reduce nitrite (Brock, 1978; Hudson *et al.*, 1987b). Fourteen of the strains isolated by Munster *et al.* (1986) reduced nitrate, and most strains except the Vizela isolates of cluster C grew anaerobically in the presence of nitrate (Santos *et al.*, 1989). Such strains have been enriched and isolated for the production of formate-linked nitrate reductase (Baldursson and Kristjansson, 1990) and tentatively named "*T. anaerobicus.*" No *Thermus* strain has been shown to be capable of fermentation of carbohydrates.

Thermus sp. often utilize diverse carbohydrates, carboxylic acids, amino acids, peptides, and proteins. Many studies have found a nutritional diversity amongst *Thermus* strains (Alfredsson *et al.*, 1985; Pask-Hughes and Williams, 1977; Santos *et al.*, 1989). This is suggested to correlate with the geographical source of the isolates (Hudson *et al.*, 1989). However, strains with different physiological properties, and shown to comprise different genospecies (Williams, 1989), coexist in geothermal areas in the U.S.A. (Munster *et al.*, 1986), and Europe (Santos *et al.*, 1989). Studies that involved similar media and test conditions are in broad agreement about the properties of key strains. Significant discrepancies found with particular tests may be due to the use of different test procedures. Since some *Thermus* strains may be inhibited by low concentrations of organic substrates, the utilization of single carbon source compounds is tested with 2–4 gL^{-1} of the substrate. Incomplete substrate utilization and growth inhibition have been reported for a number of strains, but no explanation for this phenomenon has been proposed (Sonnleitner et al., 1982).

Many strains utilize acetate, pyruvate, proline, and glutamate (Alfredsson et al., 1985; Hudson et al., 1989) while polyols are used by a minority of strains. Several monosaccharides are used as single carbon sources by strains of Thermus, but pentoses are not usually metabolized at all (Alfredsson et al., 1985; Hudson et al., 1986; Hudson et al., 1987a; Munster et al., 1986). Disaccharides, but not necessarily their constituent monosaccharides, may be used. Most strains from Iceland utilized sucrose and maltose, while only two used glucose, and none grew on fructose (Alfredsson et al., 1985). Both β-galactosidase and α-galactosidase are produced by many strains and are characteristic of several taxonomic groups (Santos et al., 1989).

Proteins such as elastin, fibrin, and casein are hydrolyzed by many isolates, but some strains are incapable of hydrolyzing each substrate (Munster et al., 1986; Hudson et al., 1989; Santos et al., 1989). Most strains hydrolyze Tween 80, arbutin, and esculin, and reduction of tellurite has been shown to discriminate between several phenotypic groups (Hudson et al., 1987a). Sensitivities to fosfomycin, chloramphenicol, spectinomycin, kanamycin, and cycloserine differ among phenetic taxonomic groups (Santos et al., 1989).

4.1.2. Low Temperature Red-Pigmented Strains

The red-pigmented strains related to T. ruber form orange or red colonies. Spectroscopic studies (Sharp and Williams, 1988) of the extracted carotenoid pigments showed similar spectral features and the observed color differences probably depend on the concentration of the pigment. The colony morphology varies from circular or irregular to punctiform with raised, flat, or convex elevation. The colony diameter is generally 1 mm or less (Ruffett, 1992). The only large study of red-pigmented strains isolated from widely differing geographical locations showed no growth at 30 °C or 75 °C although all strains grew at 45 °C and 65 °C with 30% able to grow at 70 °C (Ruffett et al., 1992). All strains grew at pH 6.0 and pH 9.0 with variable growth at pH 5.0 and 10.0.

While some taxonomic studies have shown red-pigmented isolates to cluster with yellow-pigmented strains, the two groups have striking phenotypic and genetic differences. A study of eighteen isolates from Iceland indicated most strains required a vitamin supplement for growth and used fructose, maltose, mannose, or sucrose as a sole carbon source. Unlike the yellow-pigmented strains they did not grow on acetate or pyruvate. Pro-

Table II. Alignment of 16s rDNA Sequences of *Thermus* Strains[a]

```
                            (SIGNATURE for THERMUS/DEINOCOCCUS)
                                      gcytaag
                                           |----------
                                           1111     222

T.fil        1  GCTCAGGGTGAACGCTGGCGGCGTGCCTAAGACATGCAAGTCGTGCGGGC
Tok20A1      1  GCTCAGGGTGAACGCTGGCGGCGTGCCTAAGACATGCAAGTCGTGCGGGC
OK6A1        1  GCTCAGGGTGAACGCTGGCGGCGTGCCTAAGACATGCAAGTCGTGCGGGC
RT41A        1  GCTCAGGGTGAACGCTGGCGGCGTGCCTAAGACATGCAAGTCGTGCGGGC
T351         1  GCTCAGGGTGAACGCTGGCGGCGTGCCTAAGACATGCAAGTCGTGCGGGC
Tok8A1       1  GCTCAGGGTGAACGCTGGCGGCGTGCCTAAGACATGCAAGTCGTGCGGGC


T.br YS38    1  NNNNNNNNNNNNNNNNNNNNNNNCGTGCCTAAGACATGCAAGTCGGGCGGGC
ZHGIA1       1  GCTCAGGGTGAACGCTGGCGGCGTGCCTAAGACATGCAAGTCGGGCGGGC
OSRAM4       1  NNNNNNNNNNNNNNNNNNNNNNNNNNNNNNNNNNNNNNNNNNNNNNNNNNNN
ZFIA2        1  GCTCAGGGTGAACGCTGGCGGCGTGCCTAAGACATGCAAGTCGrGCGGGC


Vi7          1  NNNNNNNNNNNNNNNNNNNNNNNNNNNNNNNTAAGACATGCAAGTCGAGCGGGG
NMX2A1       1  GCTCAGGGTGAACGCTGGCGGCGTGCCTAAGACATGCAAGTCGAGCGGGG
             ---------------------------------------------------

T.aq YT1     1  GCTCAGGGTGAACGCTGGCGGCGTGCCTAAGACATGCAAGTCGTGCGGGC-
YS025        1  GCTCAGGGTGAACGCTGGCGGCGTGCCTAAGACATGCAAGTCGTGCGGGCC
YS052        1  GCTCAGGGTGAACGCTGGCGGCGTGCCTAAGACATGCAAGTCGTGCGGGCC
                                                                 *

HSA1         1  GCTCAGGGTGAACGCTGGCGGCGTGCCTAAGACATGCAAGTCGTGCGGGC
T.th HB8*    1  GCTCAGGGTGAACGCTGGCGGCGTGCCTAAGACATGCAAGTCGTGCGGGC
T.th HB8     1  GCTCAGGGTGAACGCTGGCGGCGTGCCTAAGACATGCAAGTCGTGCGGGC
T.th HB27    1  GCTCAGGGTGAACGCTGGCGGCGTGCCTAAGACATGCAAGTCGTGCGGGC
T.flavus     1  GCTCAGGGTGAACGCTGGCGGCGTGCCTAAGACATGCAAGTCGTGCGGGC


SPS14        1  NNNNNNNNNNNNNNNNNNNNNNNNNNNNNNNNCTAAGACATGCAAGTCGTGCGGGG

T.ruber      1  GCTCAGGGTGAACGCTGGCGGTATGCCTAAGACATGCAAGTCGAACGGGC

D.rad        1  GCTCAGGGTGAACGCTGGCGGCGTGCTTAAGACATGCAAGTCGAACGCGG
```

teolysis of casein and gelatin was observed in most strains and hydrolysis of hide powder by some 40% of strains. Like the yellow-pigmented strains, most red strains do not grow in 2.5% tryptone. All the isolates studied by Ruffett (1992) were catalase positive and oxidase negative, all reduced tellurite and hydrolyzed arbutin, and most did not reduce nitrate.

4.2. Numerical Analysis

A number of taxonomic studies of *Thermus* species have been reported, including numerical analyses (Munster *et al.*, 1986; Hudson *et al.*, 1986, 1987a, 1989; Santos *et al.*, 1989; and Ruffett *et al.*, 1992) and char-

Table II. (*Continued*)

```
                    HELIX 6 REGION                    BULGE
            ---V1--------------------|
            2222222    22222222222 1111

T.fil       51 TGCGGGGTTTTACTCCGTGGTCAGCGGCGGACGGGTGAGTAACGCGTGGG
Tok20A1     51 CGCGGGGTTTTACTTCGCGGTCAGCGGCGGACGGGTGAGTAACGCGTGGG
OK6A1       51 TGCGGGGTTTTACTTCGCGGTCAGCGGCGGACGGGTGAGTAACGCGTGGG
RT41A       51 TGCGGGGTTTTACTTCGCGGTCAGCGGCGGACGGGTGAGTAACGCGTGGG
T351        51 TGCGGGGTTTTACTTCGCGGTCAGCGGCGGACGGGTGAGTAACGCGTGGG
Tok8A1      51 CGCGGGGTTTTACTTCGCGGTCAGCGGCGGACGGGTGAGTAACGCGTGGG
                       *

T.br YS38   51 CATGGGGTTTTACTCCGTGGTCAGCGGCGGACGGGTGAGTAACGCGTGGG
ZHGIA1      51 CATGGGGTTTTACTCCGTGGTCAGCGGCGGACGGGTGAGTAACGCGTGGG
OSRAM4      51 NNNNNNNNNNNNNNNNNNNNNNNNNNNNNNNNNNNNNNNNNNNNNNNNNNN
ZFIA2       51 CGTGGGGTTT--CT-CACGGTCAGCGGCGGACGGGTGAGTAACGCGTGGG
                       *       **  *  **

Vi7         51 CA---GG-TTTATGCCGTGTCCAGCGGCGGACGGGTGAGTAACGCGTGGG
NMX2A1      51 CA---GG-TTTATACC-TGTTCAGCGGCGGACGGGTGAGTAACGCGTGGG
                            *   *     *

T.aq YT1    51 CGTGGGG---TATCTCACGGTCAGCGGCGGACGGGTGAGTAACGCGTGGG
YS025       51 CGTGGGG---TATCTCACGGTCAGCGGCGGACGGGTGAGTAACGCGTGGG
YS052       51 CGTGGGG---TATCTCACGGTCAGCGGCGGACGGGTGAGTAACGCGTGGG

HSA1        51 CGCGGGGTTTTACTCTGCGGTCAGCGGCGGACGGGTGAGTAACGCGTGGG
T.th HB8*   51 CGCGGGGTTTTACTCCGTGGTCAGCGGCGGACGGGTGAGTAACGCGTGGG
T.th HB8    51 CGCGGGGTTTTACTCCGTGGTCAGCGGCGGACGGGTGAGTAACGCGTGGG
T.th HB27   51 CGCGGGGTTTTACTCCGTGGTCAGCGGCGGACGGGTGAGTAACGCGTGGG
T.flavus    51 CGCGGGGTTTTACTCCGTGGTCAGCGGCGGACGGGTGAGTAACGCGTGGG
                              *  *

SPS14       51 --TGG---TT--CGCCAC--CCAGCGGCGGACGGGTGAGTAACGCGTGGG

T.ruber     51 TG------TTTA-TGCA-GC-CAGTGGCGGACGGGTGAGTAACACGTAGG

D.rad       51 T-----------CTTCGGACCGAGTGGCGCACGGGTGAGTAACACGTAAC
```

acterization studies (Cometta *et al.*, 1982b; Sharp and Williams, 1988). Attempts have been made by a number of groups to attribute phenotypic differences to geographical location and to the temperature and pH profiles of the source.

A study of twenty-eight Icelandic isolates from one geographical location, using 33 character states, indicated considerable phenotypic variation (Cometta *et al.*, 1982b). It is interesting to note that even in this early comparative study, the close phenetic similarity of *T. thermophilus* and *T. flavus* was observed.

```
                    STEM 8                    HELIX 9
                    111111---222222----222222----11111

T.fil        101  TGACCTACCTGGAAGAGGGGGACAACCTGGGGAAACTCGGGCTAATCCCC
Tok20A1      101  TGACCTACCCGGAAGAGGGGGACAACCTGGGGAAACTCGGGCTAATCCCC
OK6A1        101  TGACCTACCCGGAAGAGGGGGACAACCTGGGGAAACTCGGGCTAATCCCC
RT41A        101  TGACCTACCCGGAAGAGGGGGACAACCTGGGGAAACTCGGGCTAATCCCC
T351         101  TGACCTACCCGGAAGAGGGGGACAACCTGGGGAAACTCGGGCTAATCCCC
Tok8A1       101  TGACCTACCCGGAAGAGGGGGACAACCTGGGGAAACTCGGGCTAATCCCC
                       *


T.br YS38    101  TGACCTACCCGGAAGTGTGGGACAACCCGGGGAAACTCGGGCTAATCCCG
ZHGIA1       101  TGACCTACCCGGAAGTGTGGGACAACCCGGGGAAACTCGGGCTAATCCCG
OSRAM4       101  NNNNNNNNNNNNNNNNNNNNNNNNNNNNNNNNNNNNNNNNNNNNNNNNNNN
ZFIA2        101  TGACCTACCCGGAAGAGGGGGACAACCTGGGGAAACCTAGGCTAATCCCC
                       *  *        *      * * *           *


Vi7          101  TGACCTACCCGGAAGAGGCGGACAACCTGGGGAAACCCAGGCTAATCCGC
NMX2A1       101  TGACCTACCCGGAAGAGGCGGACAACCTGGGGAAACCCAGGCTAATCCGC


T.aq YT1     101  TGACCTACCCGGAAGAGGGGGACAACATGGGGAAACCCAGGCTAATCCCC
YS025        101  TGACCTACCCGGAAGAGGGGGACAACCTGGGGAAACCCAGGCTAATCCCC
YS052        101  TGACCTACCCGGAAGAGGGGGACAACCTGGGGAAACCCAGGCTAATCCCC
                       *


HSA1         101  TGACCTACCCGGAAGAGGGGGACAACCCGGGGAAACTCGGGCTAATCCCC
T.th HB8*    101  TGACCTACCCGGAAGAGGGGGACAACCCGGGGAAACTCGGGCTAATCCCC
T.th HB8     101  TGACCTACCCGGAAGAGGGGGACAACCCGGGGAAACTCGGGCTAATCCCC
T.th HB27    101  TGACCTACCCGGAAGAGGGGGACAACCCGGGGAAACTCGGGCTAATCCCC
T.flavus     101  TGACCTACCCGGAAGAGGGGGACAACCCGGGGAAACTCGGGCTAATCCCC


SPS14        101  TGACCTGCCCGGAAGTGGGGGACAACCCGGGGAAACCCGGGCTAATCCCC


T.ruber      101  TGACCTACCCCAAAGTCTGGGACAACTAGGAGAAATCTTAGCTAATCCTG


D.rad        101  TGACCTACCCAGAAGTCATGAATAACTGGCCGAAAGGTCAGCTAATACGT
```

(continued)

A study of 48 isolates from several hot springs in Yellowstone National Park using 71 phenetic characteristics distinguished two major clusters at the level of 80% similarity (Munster *et al.*, 1986). Group 1 was considered to be representative of *T. aquaticus* although the type strain YT1 only clustered at 83% similarity with the group. Four isolates in a small sub-cluster were distinguished by their lack of proteolytic activity and anae-robic growth on nitrate. Strains in group 2 were typified by a spreading colony morphology on agar plates. This group included several reference strains, including *T. thermophilus* strains HB8, B and *T. flavus*. Subsequent DNA:DNA hybridization analysis (Williams, 1989) confirmed that the

Table II. (*Continued*)

```
                HELIX 10 POSSIBLE SPECIES PROBE REGION
                111111111----111111111
                |-------------------V2-------------------|

T.fil       151 CATGTGGTCGTGCCCTTTGGGGTGCGATTAAAGGGTGAAGAGCCCGCTTC
Tok20Al     151 CATGTGGTCGTGCCCTTTGGGGTGCGATTAAAGGGTGAAGAGCCCGCTTC
OK6Al       151 CATGTGGTCGTGCCCTTTGGGGTGCGATTAAAGGGTGAAGAGCCCGCTTC
RT41A       151 CATGTGGTCGTGCCCTTTGGGGTGCGATTAAAGGGTGAAGAGCCCGCTTC
T351        151 CATGTGGTCGTGCCCTTTGGGGTGCGATTAAAGGGTGAAGAGCCCGCTTC
Tok8Al      151 CATGTGGTCGTGCCCTTTGGGGTGCGATTAAAGGGTGAAGAGCCCGCTTC


T.br YS38   151 CATGTGGTCATGTCCTGTGGGGCATGATTAAAGGGCGAG-AGTCCGCTTC
ZHGIAl      151 CATGTGGTCATGTCCTGTGGGGCATGATTAAAGGGCGAG-AGTCCGCTTC
OSRAM4      151 NNNNNNNNNNNNNNNNNNNNNNNNNNNNNNNNNNNNNNNN-NNNNNNNNNN
ZFIA2       151 CATGTGGACACGTCCTGTGGGGCGTGTTTAAAGGGTGACGAGCCCGCTTC
                   *    *           *    *          *   ** *


Vi7         151 CATGTGGTCCTGTCCTGTGGGGTAGGACTAAAGGGTGAATAGCCCGCTTC
NMX2Al      151 CATGTGGTCCTGTCCTGTGGGGCAGGACTAAAGGGTGGATAGCCCGCTTC
                            *                         *


T.aq YT1    151 CATGTGGACACATCCTGTGGGGTGTGTTTAAAGGGTTTT--GCCCGCTTC
YS025       151 CATGTGGACACATCCTGTGGGGTGTGTTTAAAGGGTTTT--GCCCGCTTC
YS052       151 CATGTGGACACATCCTGTGGGGTGTGTTTAAAGGGTTTT--GCCCGCTTC


HSAl        151 CATGTGGACCCGCCCCTTGGGGCGTGTCCAAAGGGCTTT--GCCCGCTTC
T.th HB8*   151 CATGTGGACCCGCCCCTTGGGGTGTGTCCAAAGGGCTTT--GCCCGCTTC
T.th HB8    151 CATGTGGACCCGCCCCTTGGGGTGTGTCCAAAGGGCTTT--GCCCGCTTC
T.th HB27   151 CATGTGGACCCGCCCCTTGGGGTGTGTCCAAAGGGCTTT--GCCCGCTTC
T.flavus    151 CATGTGGACCCGCCCCTTGGGGTGTGTCCAAAGGGCTTT--GCCCGCTTC


SPS14       151 CATGTGGTCCGGCCCC-TGGGCCGTGACTAAAG-GCCAAAAGCC-GCTTC


T.ruber     151 GATGTGGACATCTACTTTGGTGGATGTTTAAA--GCTT---CGGCGCTTT


D.rad       151 GATGTGGTGATTTGCCGTGGCAAATCACTAAA--GATT-TAT--CGCTTC
```

group 1 strains tested were *T. aquaticus* and indicated that the Yellowstone isolates (but not, of course, the *T. thermophilus* strains) in group 2 formed a second distinct DNA homology group to be called *T. brockianus*.

A taxonomic study of 45 Thermus isolates from New Zealand hot springs showed evidence of seven clusters at or above 73% similarity (Hudson *et al.*, 1986). The majority of the New Zealand strains appeared distinct from the six reference strains examined, only two of which, *T. aquaticus* YT1 and *Thermus* strain X1, clustered with any of the New Zealand isolates. There appeared to be little correlation between clusters and the geographical area of isolation, although there was evidence of correlation with the

```
                OPPOSITE                              HELIX 12
                STEM  8        1111111111112222222-------22222222

T.fil       201 CGGATGGGCCCGCGTCCCATCAGCTAGTTGGTGGGGTAATGGCCTACCAA
Tok20A1     201 CGGATGGGCCCGCGTCCCATCAGCCAGTTGGTAGGGTAATGGCCTACCAA
OK6A1       201 CGGATGGGCCCGCGTCCCATCAGCCAGTTGGTAGGGTAATGGCCTACCAA
RT41A       201 CGGATGGGCCCGCGTCCCATCAGCCAGTTGGTAGGGTAATGGCCTACCAA
T351        201 CGGATGGGCCCGCGTCCCATCAGCCAGTTGGTAGGGTAATGGCCTACCAA
Tok8A       201 CGGATGGGCCCGCGTCCCATCAGCTAGTTGGTAGGGTAATGGCCTACCAA
                                        *


T.br YS38   201 CGGATGGGCCCGCGTCCCATCAGCTAGTTGGTGGGGTAAAGGCCCACCAA
ZHGIA1      201 CGGATGGGCCCGCGTCCCATCAGCTAGTTGGTGGGGTAAAGGCCCACCAA
OSRAM4      201 NNNNNNNNNNNNNNNTCCCATCAGCTAGTTGGTGGGGTNNNGGCCCACCAA
ZFIA2       201 CGGATGGGCCCGCGTCCCATCAGCTAGTTGGTGGGGTAAGAGCCCACCAA
                                                     **


Vi7         201 CGGATGGGCCCGCGTCCCATCAGCTAGTTGGTGGGGTAAAGGCCCACCAA
NMX2A1      201 CCCATGGGCCCGCGTCCCATCAGCTAGTTGGTGGGGTAAAGGCCCACCAA
                **


T.aq YT1    201 CGGATGGGCCCGCGTCCCATCAGCTAGTTGGTGGGGTAAGAGCCCACCAA
YSO25       201 CGGATGGGCCCGCGTCCCATCAGCTAGTTGGTGGGGTAAGAGCCCACCAA
YSO52       201 CGGATGGGCCCGCGTCCCATCAGCTAGTTGGTGGGGTAAGAGCCCACCAA


HSA1        201 CGGATGGGCCCGCGTCCCATCAGCTAGTTGGTGGGGTAATGGCCCACCAA
T.th HB8*   201 CGGATGGGCCCGCGTCCCATCAGCTAGTTGGTGGGGTAATGGCCCACCAA
T.th HB8    201 CGGATGGGCCCGCGTCCCATCAGCTAGTTGGTGGGGTAATGGCCCACCAA
T.th HB27   201 CGGATGGGCCCGCGTCCCATCAGCTAGTTGGTGGGGTAATGGCCCACCAA
T.flavus    201 CGGATGGGCCCGCGTCCCATCAGCTAGTTGGTGGGGTAATGGCCCACCAA


SPS14       201 CGGATGGGCCCGCGTCCCATCAGCTAGTTGGTGGGGTAACGGCCCACCAA

T.ruber     201 GGGATGGGCCTGCGGCGCATCAGGTAGTTGGTGGGGTAATGGCCTACCAA

D.rad       201 TGGATGGGGTTGCGTTCCATCAGCTGGTTGGTGGGGTAAAGGCCTACCAA
```

(continued)

source temperatures and pH. The key distinguishing characteristics among the seven groups were reduction of tellurite, degradation of elastin and fibrin, hydrolysis of p-nitrophenyl laurate, and growth at 45 °C. A further study of 51 Icelandic isolates (Hudson *et al.*, 1987a) from various locations distinguished four key cluster groups at 73% similarity. The two largest groupings of 25 (cluster A) and 17 (cluster B) strains were composed entirely of Icelandic isolates. Cluster B of seven strains contained three Icelandic and two New Zealand isolates together with *T. flavus* and *T. ruber*. A further small cluster C was composed of two Icelandic and one New Zealand isolates. *T. aquaticus* YT1, *T. lacteus, T. rubens,* and *T. thermo-*

Table II. (*Continued*)

		HELIX 12 111111111111	HELIX 13 1111111---------1111111	HELIX 14 1111111--
T.fil	251	GGCGACGACGGGTAGCCGGTCTGAGAGGATGGCCGGCCACAGGGGCACTG		
Tok20Al	251	GGCGACGACGGGTAGCCGGTCTGAGAGGATGGCCGGCCACAGGGGCACTG		
OK6Al	251	GGCGACGACGGGTAGCCGGTCTGAGAGGATGGCCGGCCACAGGGGCACTG		
RT41A	251	GGCGACGACGGGTAGCCGGTCTGAGAGGATGGCCGGCCACAGGGGCACTG		
T351	251	GGCGACGACGGGTAGCCGGTCTGAGAGGATGGCCGGCCACAGGGGCACTG		
Tok8Al	251	GGCGACGACGGGTAGCCGGTCTGAGAGGATGGCCGGCCACAGGGGCACTG		
T.br YS38	251	GGCGACGACGGGTAGCCGGTCTGAGAGGATGGCCGGCCACAGGGGCACTG		
ZHGIAl	251	GGCGACGACGGGTAGCCGGTCTGAGAGGACGGCCGGCCACAGGGGCACTG		
OSRAM4	251	GGCGACGACGGGTAGCCGGTCTGAGAGGATGGCCGGCCACAGGGGCACTG		
ZFIA2	251	GGCGACGACGGGTAGCCGGTCTGAGAGGACGGCCGGCCACAGGGGCACTG		
		` *`		
Vi7	251	GGCGACGACGGGTAGCCGGTCTGAGAGGATGGCCGGCCACAGGGGCACTG		
NMX2Al	251	GGCGACGACGGGTAGCCGGTCTGAGAGGATGGCCGGCCACAGGGGCACTG		
T.aq YT1	251	GGCGACGACGGGTAGCCGGTCTGAGAGGACGGCCGGCCACAGGGGCACTG		
YS025	251	GGCGACGACGGGTAGCCGGTCTGAGAGGACGGCCGGCCACAGGGGCACTG		
YSO52	251	GGCGACGACGGGTAGCCGGTCTGAGAGGACGGCCGGCCACAGGGGCACTG		
HSA1	251	GGCGACGACGGGTAGCCGGCCTGAGAGGGTGGCCGGCCACAGGGGCACTG		
T.th HB8*	251	GGCGACGACGGGTAGCCGGTCTGAGAGGATGGCCGGCCACAGGGGCACTG		
T.th HB8	251	GGCGACGACGGGTAGCCGGTCTGAGAGGATGGCCGGCCACAGGGGCACTG		
T.th HB27	251	GGCGACGACGGGTAGCCGGTCTGAGAGGATGGCCGGCCACAGGGGCACTG		
T.flavus	251	GGCGACGACGGGTAGCCGGTCTGAGAGGATGGCCGGCCACAGGGGCACTG		
		` *       *`		
SPS14	251	GGCAACGACGGGTAGCCGGTCTGAGAGGATGGCCGGCCACAGGGGCACTG		
T.ruber	251	GCCGACGACGCGTAGCTGGTCTGAGAGGACGATCAGCCACAGGGGTACTG		
D.rad	251	GGCGACGACGGATAGCCGGCCTGAGAGGGTGGCCGGCCACAGGGGCACTG		

philus all failed to cluster with the Icelandic isolates. The source temperature did not appear to correlate with the cluster groups but there was evidence that source pH might be a determinant in defining the cluster group. The key distinguishing characteristics were the reduction of tellurite, nitrate and methylene blue, and the hydrolysis of elastin.

Hudson *et al.* (1989) carried out a further numerical taxonomic analysis of 131 isolates from several geothermal regions: New Zealand, Iceland, Yellowstone, and New Mexico (USA), Japan, Fiji, USSR, and UK; it in-

```
                            HELIX 15
            HELIX 14        357R primer          HELIX 16
            ------1111111   atgccctccgttcgtc      1111---
                            11111--------------11111
```

```
T.fil        301 AGACACGGGCCCCACTCCTACGGGAGGCAGCAGTTAGGAATCTTCCGCAA
Tok20A1      301 AGACACGGGCCCCACTCCTACGGGAGGCAGCAGTTAGGAATCTTCCGCAA
OK6A1        301 AGACACGGGCCCCACTCCTACGGAAGGCAGCAGTTAGGAATCTTCCGCAA
RT41A        301 AGACACGGGCCCCACTCCTACGGGAGGCAGCAGTTAGGAATCTTCCGCAA
T351         301 AGACACGGGCCCCACTCCTACGGGAGGCAGCAGTTAGGAATCTTCCGCAA
Tok8A1       301 AGACACGGGCCCCACTCCTACGGGAGGCAGCAGTTAGGAATCTTCCGCAA


T.br YS38    301 AGACACGGGCCCCACTCCTACGGGAGGCAGCAGTTAGGAATCTTCCGCAA
ZHGIA1       301 AGACACGGGCCCCACTCCTACGGGAGGCAGCAGTTAGGAATCTTCCGCAA
OSRAM4       301 AGACWCGGGCCNNNCTCCTACGGGAGGCAGCAGTTAGGAATCTTCCGCAA
ZFIA2        301 AGACACGGGCCCCACTCCTACGGGAGGCAGCAGTTAGGAATCTTCCGCAA


Vi7          301 AGACACGGGCCCCACTCCTACGGGAGGCAGCAGTTAGGAATCTTCCGCAA
NMX2A1       301 AGACACGGGCCCCACTCCTACGGGAGGCAGCAGTTAGGAATCTTCCGCAA


T.aq YT1     301 AGACACGGGCCCCACTCCTACGGGAGGCAGCAGTTAGGAATCTTCCGCAA
YS025        301 AGACACGGGCCCCACTCCTACGGGAGGCAGCAGTTAGGAATCTTCCGCAA
YS052        301 AGACACGGGCCCCACTCCTACGGGAGGCAGCAGTTAGGAATCTTCCGCAA


HSA1         301 AGACACGGGCCCCACTCCTACGGGAGGCAGCAGTTAGGAATCTTCCGCAA
T.th HB8*    301 AGACACGGGCCCCACTCCTACGGGAGGCAGCAGTTAGGAATCTTCCGCAA
T.th HB8     301 AGACACGGGCCCCACTCCTACGGGAGGCAGCAGTTAGGAATCTTCCGCAA
T.th HB27    301 AGACACGGGCCCCACTCCTACGGGAGGCAGCAGTTAGGAATCTTCCGCAA
T.flavus     301 AGACACGGGCCCCACTCCTACGGGAGGCAGCAGTTAGGAATCTTCCGCAA


SPS14        301 AGACACGGGCCCCACTCCTACGGGAGGCAGCAGTTAGGAATCTTCCGCAA

T.ruber      301 AGACACGGACCCCACTCCTACGGGAGGCAGCAGTTAGGAATCTTCCGCAA

D.rad        301 AGACACGGGTCCCACTCCTACGGGAGGCAGCAGTTAGGAATCTTCCACAA
```

(continued)

cluded many of the strains used in the two earlier studies. Fourteen clusters were formed at the level of 73% similarity. Ten of these groups segregated on geographical origin and three others contained only 1 strain from a different geographical source to the remainder of the cluster. Two of the ten strains of group A, NMX2A1 and HS1A1 segregate according to 16SrDNA sequences (Table II). *T. aquaticus* YT1 clustered with some other Yellowstone strains in group I. *T. thermophilus* clustered with two Icelandic isolates in group J, but *T. flavus* and *T. ruber* were unclustered but close together as were *T. rubens* and *T. lacteus. T. filiformis* remained unclustered although strains with high DNA homology with it occur in groups E

Table II. (*Continued*)

```
           HELIX 16                              HELIX 17
      --222----222---1111                   1111------2222----2
                                            ¦--------V3-
```

T.fil	351	TGGGCGAAAGCCTGACGGAGCGACGCCGCTTGGAGGAGGAAGCCCTTCGG
Tok20A1	351	TGGGCGCAAGCCTGACGGAGCGACGCCGCTTGGAGGAGGAAGCCCTTCGG
OK6A1	351	TGGGCGAAAGCCTGACGGAGCGACGCCGCTTGGAGGAGGAAGCCCTTCGG
RT41A	351	TGGGCGAAAGCTTGACGGAGCGACGCCGCTTGGAGGAGGAAGCCCTTCGG
T351	351	TGGGCGAAAGCCTGACGGAGCGACGCCGCTTGGAGGAGGAAGCCCTTCGG
Tok8A1	351	TGGGCGAAAGCCTGACGGAGCGACGCCGCTTGGAGGAGGAAGCCCTTCGG
		*

T.br YS38	351	TGGGCGCAAGCCTGACGGAGCGACGCCGCTTGGAGGAGGAAGCCCTTCGG
ZHGIA1	351	TGGGCGCAAGCCTGACGGAGCGACGCCGCTTGGAGGAGGAAGCCCTTCGG
OSRAM4	351	TGGGCGCAAGCCTGACGGAGCGACGCMNHWTGGAGGAGGAAAG-CTTCGG
ZFIA2	351	TGGACGGAAGTCTGACGGAGCGACGCCGCTTGGAGGAGGAAGCCCTTCGG
		* * * ***

Vi7	351	TGGACGGAAGTCTGACGGAGCGACGCCGCTTGGAGGAGGAAGCCCTTCGG
NMX2A1	351	TGGACGGAAGTCTGACGGAGCGACGCCGCTTGGAGGAGGAAGCCCTTCGG

T.aq YT1	351	TGGGCGCAAGCCTGACGGAGCGACGCCGCTTGGAGGAGGAAGCCCTTCGG
YS025	351	TGGGCGCAAGCCTGACGGAGCGACGCCGCTTGGAGGAGGAAGCCCTTCGG
YS052	351	TGGGCGCAAGCCTGACGGAGCGACGCCGCTTGGAGGAGGAAGCCCTTCGG

HSA1	351	TGGGCGCAAGCCTGACGGAGCGACGCCGCTTGGAGGAAGAAGCCCTTCGG
T.th HB8*	351	TGGGCGCAAGCCTGACGGAGCGACGCCGCTTGGAGGAAGAAGCCCTTCGG
T.th HB8	351	TGGGCGCAAGCCTGACGGAGCGACGCCGCTTGGAGGAAGAAGCCCTTCGG
T.th HB27	351	TGGGCGCAAGCCTGACGGAGCGACGCCGCTTGGAGGAAGAAGCCCTTCGG
T.flavus	351	TGGGCGCAAGCCTGACGGAGCGACGCCGCTTGGAGGAAGAAGCCCTTCGG

SPS14	351	TGGGCGAAAGCCTGACGGAGCGACGCCGCTTGCGGGACGAAGCCCCTCGG
T.ruber	351	TGGGCGAAAGCCTGACGGAGCGATACCGCTTGAAGGACGAAGCCCTTCGG
D.rad	351	TGGGCGCAAGCCTGATGGAGCGACGCCGCGTGAGGGATGAAGGTTTTCGG

(RT358), G (TOK22), and H (T351, RT6A1). Based on the level of 65% similarity, Hudson proposed that there was evidence of eight major taxonomic groups (species) of *Thermus*. At this level of discrimination the geographical segregation was lost. The largest cluster (1) included 61 strains from every country included in the study. The three validly published species represented three of these groups, *T. aquaticus* in 5, *T. filiformis* in 6 with *T. thermophilus,* and two other New Zealand but mostly Icelandic strains, and *T. ruber* in 8 with *T. flavus.* Confirmation of the remaining five as species was required. Differential characterization of the eight clusters included reduction of tellurite, nitrate, and methylene blue, hydrolysis of

```
                HELIX 17                    HELIX 18
                222-----1111     11111--------2222----2222---------
                -------¦         ¦-------------------V4------

T.fil      401 GGTGTAAACTCCTGAACCCAGGACGAAATCCCTGATGAGGGGGATGACGG
Tok20A1    401 GGTGTAAACTCCTGAACCCAGGACGAAATCCCTGATGAGGGGGATGACGG
OK6A1      401 GGTGTAAACTCCTGAACCCAGGACGAAATCCCTGATGAGGGGGATGACGG
RT41A      401 GGTGTAAACTCCTGAACCCAGGACGAAATCCCTGATGAGGGGGATGACGG
T351       401 GGTGTAAACTCCTGAACCCAGGACGAAATCCCTGATGAGGGGGATGACGG
Tok8A1     401 GGTGTAAACTCCTGAACCCAGGACGAAATCCCTGATGAGGGGGATGACGG


T.br YS38  401 GGTGTAAACTCCTGAACTGGGGACGAAAGCCCCGATGAGGGGGATGACGG
ZHGIA1     401 GGTGTAAACTCCTGAACTGGGGACGAAAGCCCCGATGAGGGGGATGACGG
OSRAM4     401 GGTGTAAACTCCTGAACTGGGGACGAAAGCCCCGNNGANGGGGATGACGG
ZFIA2      401 GGTGTAAACTCCTGAACTGGGGACGAAAGCCCTGATGAGGGGGATGACGG
                                                *


Vi7        401 GGTGTAAACTCCTGAACWGGGGACGAAAGCCCCGTATAGGGGGATGACGG
NMX2A1     401 GGTGTAAACTCCTGAACTGGGGACGAAAGCCCCGTGTAGGGGGATGACGG
                                                *


T.aq YT1   401 GGTGTAAACTCCTGAACCCGGGACGAAACCCCCGATGAGGGGACTGACGG
YS025      401 GGTGTAAACTCCTGAACCCGGGACGAAACCCCCGATGAGGGGACTGACGG
YS052      401 GGTGTAAACTCCTGAACCCGGGACGAAACCCCCGATGAGGGGACTGACGG


HSA1       401 GGTGTAAACTCCTGAACCCGGGACGAAACCCCCGATGAGGGGACTGACGG
T.th HB8*  401 GGTGTAAACTCCTGAACCCGGGACGAAACCCCCGACGAGGGGACTGACGG
T.th HB8   401 GGTGTAAACTCCTGAACCCGGGACGAAACCCCCGACGAGGGGACTGACGG
T.th HB27  401 GGTGTAAACTCCTGAACCCGGGACGAAACCCCCGAGGAGGGGACTGACGG
T.flavus   401 GGTGTAAACTCCTGAACCCGGGACGAAACCCCCGAGGAGGGGACTGACGG
                                                *


SPS14      401 GGTGTAAACCGCTGAACCTGGGACGAAAACCCCCACAAGGGGACTGACGG

                                          T.ruber deletion
T.ruber    401 GGTGTAAACTTCTGAACTCGGGACGATAA---------------TGACGG

D.rad      401 ATCGTAAACCTCTGAATCTGGGACGAAAGAGCCTTCGGGCA-GATGACGG
```

(continued)

hide powder azure, production of β-galactosidases, hydrolysis of p-nitro-phenyl fatty acids, sensitivity to chloramphenicol and spectinomycin, and the reddening of triple sugar iron medium.

Forty-two strains isolated from the Azores and mainland Portugal were examined using 48 phenetic characters (Santos *et al.*, 1989). Six clusters were defined at the 75% similarity level. Five clusters were homogenous, formed of isolates from one locations, and the sixth (cluster E) from strains isolated from Furnas, Sao Miguel in the Azores, and Sao Pedro do Sul in mainland Portugal. *T. aquaticus* YT1 and *T. thermophilus* failed to

Table II. (*Continued*)

```
              HELIX 18                        gtcgtcggcgccattatg 530rev
              --11111                         gcagccgcggtaatacgg 530f
              -¦
```

T.fil	451	TACTGGGGTAATAGCGCCGGCCAACTCCGTGCCAGCAGCCGCGGTAATAC
Tok20A1	451	TACTGGGGTAATAGCGCCGGCCAACTCCGTGCCAGCAGCCGCGGTAATAC
OK6A1	451	TACTGGGGTAATAGCGCCGGCCAACTCCGTGCCAGCAGCCGCGGTAATAC
RT41A	451	TACTGGGGTAATAGCGCCGGCCAACTCCGTGCCAGCAGCCGCGGTAATAC
T351	451	TACTGGGGTAATAGCGCCGGCCAACTCCGTGCCAGCAGCCGCGGTAATAC
Tok8A1	451	TACTGGGGTAATAGCGCCGGCCAACTCCGTGCCAGCAGCCGCGGTAATAC

T.br YS38	451	TACCCAGGTAATAGCGCCGGCCAACTCCGTGCCAGCAGCCGCGGTAATAC
ZHGIA1	451	TACCCAGGTAATAGCGCCGGCCAACTCCGTGCCAGCAGCCGCGGTAATAC
OSRAM4	451	TATCCAGGTWATAGCGCCGGCNNNNNNNNNNNNNNNNNNNNNNNNNNNNNN
ZFIA2	451	TACCCAGGTAATAGCACCGGCCAACTCCGTGCCAGCAGCCGCGGTAATAC
		* *

Vi7	451	TACCCAGGTAATAGCGCCGGCCAACTCCGTGCCAGCAGCCGCGGTAATAC
NMX2A1	451	TACCCAGGTAATAGCGCCGGCCAACTCCGTGCCAGCAGCCGCGGTAATAC

T.aq YT1	451	TACCGGGGTAATAGCGCCGGCCAACTCCGTGCCAGCAGCCGCGGTAATAC
YS025	451	TACCGGGGTAATAGCGCCGGCCAACTCCGTGCCAGCAGCCGCGGTAATAC
YS052	451	TACCGGGGTAATAGCGCCGGCCAACTCCGTGCCAGCAGCCGCGGTAATAC

HSA1	451	TACCGGGGTAATAGCGCCGGCCAACTCCGTGCCAGCAGCCGCGGTAATAC
T.th HB8*	451	TACCGGGGTAATAGCGCCGGCCAACTCCGTGCCAGCAGCCGCGGTAATAC
T.th HB8	451	TACCGGGGTAATAGCGCCGGCCAACTCCGTGCCAGCAGCCGCGGTAATAC
T.th HB27	451	TACCGGGGTAATAGCGCCGGCCAACTCCGTGCCAGCAGCCGCGGTAATAC
T.flavus	451	TACCGGGGTAATAGCGCCGGCCAACTCCGTGCCAGCAGCCGCGGTAATAC

SPS14	451	TACCAGGGTAATANNNNNNNNNNNNNNNNNNNNNNNNNNNNNNNNNNNNNN
T.ruber	451	TACCGAGGTAATAGCACCGGCTAACTCCGTGCCAGCAGCCGCGGTAATAC
D.rad	451	TACCAGAGTAATAGCACCGGCTAACTCCGTGCCAGCAGCCGCGGTAATAC

cluster at this level, the later clustering at 73% with two strains from the Azores. These three strains were the only group to grow in 3% NaCl and also at 80 °C. This study indicated that geographical location and separation were reflected in the phenotypic grouping of strains although the source pH and temperature appeared to have little significance. The key distinguishing characteristics were yellow or no pigmentation, growth in 2.5% tryptone, fibrin hydrolysis, and production of α-galactosidase, anaerobic growth with nitrate, and utilization of succinate. At a lower level of %S_{SM} the study resolves only three clusters comprising A, B, C, and D together, clusters E and F and the three halotolerant strains that grow at

STEM 20

T.fil	501	GGAGGGCGCGAGCGTTACCCGGATTTACTGGGCGTAAAGGGCGTGTAGGC
Tok20A1	501	GGAGGGCGCGAGCGTTACCCGGATTTACTGGGCGTAAAGGGCGTGTAGGC
OK6A1	501	GGAGGGCGCGAGCGTTACCCGGATTTACTGGGCGTAAAGGGCGTGTAGGC
RT41A	501	GGAGGGCGCGAGCGTTACCCGGATTTACTGGGCGTAAAGGGCGTGTAGGC
T351	501	GGAGGGCGCGAGCGTTACCCGGATTTACTGGGCGTAAAGGGCGTGTAGGC
Tok8A1	501	GGAGGGCGCGAGCGTTACCCGGATTTACTGGGCGTAAAGGGCGTGTAGGC

T.br YS38	501	GGAGGGCGCGAGCGTTACCCGGATTTACTGGGCGTAAAGGGCGTGTAGGC
ZHGIA1	501	GGAGGGCGCGAGCGTTACCCGGATTTACTGGGCGTAAAGGGCGTGTAGGC
OSRAM4	501	NNNNNNNNNNNNNNNNNNNNNNNNNNNNNNNNNNNNNNNNNNNNNNNNNNN
ZFIA2	501	GGAGGGTGCGAGCGTTACCCGGATTTACTGGGCGTAAAGGGCGTGTAGGC
		*

Vi7	501	GGAGGGCGCGAGCGTTACCCGGATTTACTGGGCGTAAAGGGCGTGTAGGC
NMX2A1	501	GGAGGGCGCGAGCGTTACCCGGATTTACTGGGCGTAAAGGGCGTGTAGGC

T.aq YT1	501	GGAGGGCGCGAGCGTTACCCGGATTTACTGGGCGTAAAGGGCGTGTAGGC
YS025	501	GGAGGGCGCGAGCGTTACCCGGATTTACTGGGCGTAAAGGGCGTGTAGGC
YS052	501	GGAGGGCGCGAGCGTTACCCGGATTTACTGGGCGTAAAGGGCGTGTAGGC

HSA1	501	GGAGGGCGCGAGCGTTACCCGGATTCACTGGGCGTAAAGGGCGTGTAGGC
T.th HB8*	501	GGAGGGCGCGAGCGTTACCCGGATTCACTGGGCGTAAAGGGCGTGTAGGC
T.th HB8	501	GGAGGGCGCGAGCGTTACCCGGATTCACTGGGCGTAAAGGGCGTGTAGGC
T.th HB27	501	GGAGGGCGCGAGCGTTACCCGGATTCACTGGGCGTAAAGGGCGTGTAGGC
T.flavus	501	GGAGGGCGCGAGCGTTACCCGGATTCACTGGGCGTAAAGGGCGTGTAGGC

SPS14	501	NNNNNNNNNNNNNNNNNNNNNNNNNNNNNNNNNNNNNNNNNNNNNNNNNNN
T.ruber	501	GGAGGGTGCGAGCGTTACCCGGATTTACTGGGTGTAAAGGGCGTGTAGGC
D.rad	501	GGAGGGTGCAAGCGTTACCCGGAATCACTGGGCGTAAAGGGCGTGTAGGC

(continued)

80 °C. These are confirmed as genospecies by DNA:DNA homology (Williams, 1989 (see 5.1.).

The red pigmented strains with maximum growth temperatures in the region of 65–70 °C, some 10 °C lower then the yellow pigmented strains have received little taxonomic attention and have provoked little interest as a source of thermostable enzymes. Sharp and Williams (1988) studied seventeen strains isolated from pools at Hveragerthi in Iceland with pH values of 7.6–8.8 and 50–80 °C. Optimum growth temperature was 60 °C, no growth was observed at 30 °C or 70 °C. Most strains were able to grow on fructose, maltose, mannose, and sucrose as sole carbon source and the % G+C ranged from 59–63. The strains appeared relatively homogeneous, showing little significant differentiation.

Table II. (*Continued*)

```
                               HELIX
                   2222222-------111111-----111111-----222

T.fil      551 GGCCTGGTGCGTCTGGCGTTAAAGACCGCGGCTCAACCGCGGGGGTGCGC
Tok20Al    551 GGCCTGGTGCGTCTGGCGTTAAAGACCGCGGCTCAACCGCGGGGGTGCGC
OK6Al      551 GGCCTGGTGCGTCTGGCGTTAAAGACCGCGGCTCAACCGCGGGGGTGCGC
RT41A      551 GGCCTGGTGCGTCTGGCGTTAAAGACCGCGGCTCAACCGCGGGGGTGCGC
T351       551 GGCCTGGTGCGTCTGGCGTTAAAGACCGCGGCTCAACCGCGGGGGTGCGC
Tok8Al     551 GGCCTGGTGCGTCTGGCGTTAAAGACCGCGGCTCAACCGCGGGGGTGCGC


T.br YS38  551 GGCTTGGGGCGTCCCATGTGAAAGACCACGGCTCAACCGTGGGGGAGCGT
ZHGIAl     551 GGCTTGGGGCGTCCCATGTGAAAGACCACGGCTCAACCGTGGGGGAGCGT
OSRAM4     551 NNNNNNNNNNNNNNNNNNNNNNNNNNNNNNNNNNNNNNNNNNNNNNNNNNN
ZFIA2      551 GGCCTGGGGCGTCCCATGTGAAAGACCACGGCTCAACCGTGGGGGAGCGT
                  *


Vi7        551 GGCTTGGGGCGTCCCATGTGAAAGGCCACGGCTCAACCGTGGAGGAGCGT
NMX2Al     551 GGCCTGGGGCGTCCCATGTGAAAGGCCACGGCTCAACCGTGGAGGAGCGT
                  *


T.aq YT1   551 GGCTTGGGGCGTCCCATGTGAAAGGCCACGGCTCAACCGTGGAGGAGCGT
YS025      551 GGCTTGGGGCGTCCCATGTGAAAGGCCACGGCTCAACCGTGGAGGAGCGT
YS052      551 GGCTTGGGGCGTCCCATGTGAAAGGCCACGGCTCAACCGTGGAGGAGCGT


HSAl       551 GGCCTGGGGCGTCCCATGTGAAAGACCACGGCTCAACCGTGGGGGAGCGT
T.th HB8*  551 GGCCTGGGGCGTCCCATGTGAAAGACCACGGCTCAACCGTGGGGGAGCGT
T.th HB8   551 GGCCTGGGGCGTCCCATGTGAAAGACCACGGCTCAACCGTGGGGGAGCGT
T.th HB27  551 GGCCTGGGGCGTCCCATGTGAAAGACCACGGCTCAACCGTGGGGGAGCGT
T.flavus   551 GGCCTGGGGCGTCCCATGTGAAAGACCACGGCTCAACCGTGGGGGAGCGT

SPS14      551 NNNNNNNNNNNNNNNNNNNNNNNNNNNCCACGGCTCAACCGTGGAACCGCGC

T.ruber    551 GGTCTCTCAAGTCCGATGCTAAAGACCGAAGCTCAACTTCGGGGGTGCGT

D.rad      551 GGATATTTAAGTCTGGTTTTAAAGACCGAGGCTCAACCTCGGGAGTGGAC
```

Seventy seven strains of *T. ruber* and six strains of *Rhodothermus* isolated from Iceland, Yellowstone National Park, Portugal, Sao Miguel in the Azores, and New Zealand were examined using sixty five different biochemical and physiological parameters (Ruffett, 1992). The data was analyzed using unweighted average linkage analysis and indicated the formation of three clusters. A small cluster of six strains was composed of *Rhodothermus marinus* (Alfredsson *et al.*, 1988) and five environmental isolates from Portugal and New Zealand. This small group were halotolerant growing in 2% salt and producing cellobiase and pullulanase. The two large groups of forty-five and thirty-two strains were differentiated on the ability to use a wide range of single carbon sources. Analysis of %G+C and

```
                                      HELIX 22                 HELIX 23
                 2222
T.fil        601 TGGATACGGCCGGGCTAGACGGTGGGAGAGGGTGGTGGAATTCCCGGAGT
Tok20A1      601 TGGATACGGCCGGGCTAGACGGTGGGAGAGGGTGGTGGAATTCCCGGAGT
OK6A1        601 TGGATACGGCCGGGCTAGACGGTGGGAGAGGGTGGTGGAATTCCCGGAGT
RT41A        601 TGGATACGGCCGGGCTAGACGGTGGGAGAGrGTGGTGGAATTCCCGGAGT
T351         601 TGGATACGGCCGGGCTAGACGGTGGGAGAGGGTGGTGGAATTCCCGGAGT
Tok8A1       601 TGGATACGGCTGGGCTAGACGGTGGGAGAGGGTGGTGGAATTCCCGGAGT
                           *

T.br YS38    601 GGGATACGCTCAGGCTAGACGGCGGGAGGGGGTGGTGGAATTCCCGGAGT
ZHGIA1       601 GGGATACGCTCAGGCTAGACGGCGGGAGGGGGTGGTGGAATTCCCGGAGT
OSRAM4       601 NNNNNNNNNNNNNNNNNNNNNNNNNNNNNNNNNNNNNNNNNNATTCCCGGAGT
ZFIA2        601 GGGATACGCTCAGGCTAGACGGCGGGAGGGGGTGGTGGAATTCCCGGAGT

Vi7          601 GGGATACGCTCAGGCTAGAGGGTGGGAGAGGGTGGTGGAATTCCCGGAGT
NMX2A1       601 GGGATACGCTCAGGCTAGAGGGTGGGAGAGGGTGGTGGAATTCCCGGAGT

T.aq YT1     601 GGGATACGCTCAGGCTAGACGGTGGGAGAGGGTGGTGGAATTCCCGGAGT
YS025        601 GGGATACGCTCAGGCTAGACGGTGGGAGAGGGTGGTGGAATTCCCGGAGT
YS052        601 GGGATACGCTCAGGCTAGACGGTGGGAGAGGGTGGTGGAATTCCCGGAGT

HSA1         601 GGGATACGCTCAGGCTAGACGGTGGGAGAGGGTGGTGGAATTCCCGGAGT
T.th HB8*    601 GGGATACGCTCAGGCTAGACGGTGGGAGAGGGTGGTGGAATTCCCGGAGT
T.th HB8     601 GGGATACGCTCAGGCTAGACGGTGGGAGAGGGTGGTGGAATTCCCGGAGT
T.th HB27    601 GGGATACGCTCAGGCTAGACGGTGGGAGAGGGTGGTGGAATTCCCGGAGT
T.flavus     601 GGGATACGCTCAGGCTAGACGGTGGGAGAGGGTGGTGGAATTCCCGGAGT

SPS14        601 CGNATACGCCCGGGCTAGACGGCGGGAGAGGGTGGTGGAATTCCCGGAGT

T.ruber      601 TGGATACTGTGAGGCTAGACGGTCGGAGAGGGTAGCGGAATTTCCGGAGT

D.rad        601 TGGATACTGGATGTCTTGACCTCTGGAGAGGTAACTGGAATTCCTGGTGT
```

(continued)

DNA:DNA homology indicated little genetic difference between these groups but significant difference from *Rhodothermus*.

5. NUCLEIC ACID STUDIES

5.1. DNA Base Composition

The mean base composition reported for the DNA of the strains examined varies between 57–65% G+C, but analyses of most strains give results above 60% G+C. Most analyses have been done by comparative methods such as buoyant density in caesium chloride gradients and spec-

Table II. (*Continued*)

HELIX 23   BULGE 24

T.fil	651	AGCGGTGAAATGCGCAGATACCGGGAGGAACGCCGATGGCGAAGGCAGCC
Tok20Al	651	AGCGGTGAAATGCGCAGATACCGGGAGGAACGCCGATGGCGAAGGCAGCC
OK6Al	651	AGCGGTGAAATGCGCAGATACCGGGAGGAACGCCGATGGCGAAGGCAGCC
RT41A	651	AGCGGTGAAATGCGCAGATACCGGGAGGAACGCCGATGGCGAAGGCAGCC
T351	651	AGCGGTGAAATGCGCAGATACCGGGAGGAACGCCGATGGCGAAGGCAGCC
Tok8Al	651	AGCGGTGAAATGCGCAGATACCGGGAGGAACGCCGATGGCGAAGGCAGCC
T.br YS38	651	AGCGGTGAAATGCGCAGATACCGGGAGGAACGCCGATAGCGAAGGCAGCC
ZHGIAl	651	AGCGGTGAAATGCGCAGATACCGGGAGGAACGCCGATAGCGAAGGCAGCC
OSRAM4	651	AGCGGTGAAATGCGCAGATACCGGGAGGAACGCCNATNGCGAAGGCAGCC
ZFIA2	651	AGCGGTGAAATGCGCAGATACCGGGAGGAACGCCGATGGCGAAGGCAGCC
Vi7	651	AGCGGTGAAATGCGCAGATACCGGGAGGAACGCCGATGGCGAAGGCAGCC
NMX2Al	651	AGCGGTGAAATGCGCAGATACCGGGAGGAACGCCGATGGCGAAGGCAGCC
T.aq YT1	651	AGCGGTGAAATGCGCAGATACCGGGAGGAACGCCGATGGCGAAGGCAGCC
YS025	651	AGCGGTGAAATGCGCAGATACCGGGAGGAACGCCGATGGCGAAGGCAGCC
YS052	651	AGCGGTGAAATGCGCAGATACCGGGAGGAACGCCGATGGCGAAGGCAGCC
HSAl	651	AGCGGTGAAATGCGCAGATACCGGGAGGAACGCCGATGGCGAAGGCAGCC
T.th HB8*	651	AGCGGTGAAATGCGCAGATACCGGGAGGAACGCCGATGGCGAAGGCAGCC
T.th HB8	651	AGCGGTGAAATGCGCAGATACCGGGAGGAACGCCGATGGCGAAGGCAGCC
T.th HB27	651	AGCGGTGAAATGCGCAGATACCGGGAGGAACGCCGATGGCGAAGGCAGCC
T.flavus	651	AGCGGTGAAATGCGCAGATACCGGGAGGAACGCCGATGGCGAAGGCAGCC
SPS14	651	AGCGGTGAAATGCGCAGATACCGGGAGGAACGCCAATGGCGAAGGCAGCC
T.ruber	651	AGCGGTGAAATGCGCAGATACCGGAAGGAACGCCAATAGCGAAAGCAGCT
D.rad	651	AGCGGTGGAATGCGTAGATACCAGGAGGAACACCAATGGCGAAGGCAAGT

troscopic analysis of melting temperature (Owen and Hill, 1979). Direct analyses can be carried out on enzymically hydrolyzed DNA by the separation of nucleosides or nucleotides by high performance liquid chromatography (Breter and Zahn, 1973; Zillig *et al.*, 1989; Ip *et al.*, 1985). The bases of DNA are not modified to a great extent, except for small amounts of N^4-methylcytosine and N^6-methyladenine due to restriction-modification systems (see Chapter 4).

5.2. DNA:DNA Homology

The studies of DNA:DNA homology with *Thermus* have used two of the available techniques for validation of species. The nitrocellulose filter

HELIX 25

T.fil	701	ACCTGGTCCACCCGTGACGCTGAGGCGCGAAAGCGTGGGGAGCAAACCGG
Tok20A1	701	ACCTGGTCCACTCGTGACGCTGAGGCGCGAAAGCGTGGGGAGCAAACCGG
OK6A1	701	ACCTGGTCCACCCGTGACGCTGAGGCGCGAAAGCGTGGGGAGCAAACCGG
RT41A	701	ACCTGGTCCACCCGTGACGCTGAGGCGCGAAAGCGTGGGGAGCAAGCCGG
T351	701	ACCTGGTCCACCCGTGACGCTGAGGCGCGAAAGCGTGGGGAGCAAACCGG
Tok8A1	701	ACCTGGTCCACTCGTGACGCTGAGGCGCGAAAGCGTGGGGAGCAAACCGG
		* *
T.br YS38	701	ACCTGGCTCGTTCGTGACGCTGAGGCGCGAAAGCGTGGGGAGCAAACCGG
ZHGIA1	701	ACCTGGCTCGTTCGTGACGCTGAGGCGCGAAAGCGTGGGGAGCAAACCGG
OSRAM4	701	ACCTGGCTCGTTCGTGACGCTGAGGCGCGAAAGCRTGGGGAGCAAACCGG
ZFIA2	701	ACCTGGCTCGTTCGTGACGCTGAGGCGCGAAAGCGTGGGGAGCAAACCGG
Vi7	701	ACCTGGTCCACTTCTGACGCTGAGGCGCGAAAGCGTGGGGAGCAAACCGG
NMX2A1	701	ACCTGGTCCACTTCTGACGCTGAGGCGCGAAAGCGTGGGGAGCAAACCGG
T.aq YT1	701	ACCTGGTCCACTCGTGACGCTGAGGCGCGAAAGCGTGGGGAGCAAACCGG
YS025	701	ACCTGGTCCACTCGTGACGCTGAGGCGCGAAAGCGTGGGGAGCAAACCGG
YS052	701	ACCTGGTCCACTCGTGACGCTGAGGCGCGAAAGCGTGGGGAGCAAACCGG
HSA1	701	ACCTGGTCCACCCGTGACGCTGAGGCGCGAAAGCGTGGGGAGCAAACCGG
T.th HB8*	701	ACCTGGTCCACCCGTGACGCTGAGGCGCGAAAGCGTGGGGAGCAAACCGG
T.th HB8	701	ACCTGGTCCACCCGTGACGCTGAGGCGCGAAAGCGTGGGGAGCAAACCGG
T.th HB27	701	ACCTGGTCCACCCGTGACGCTGAGGCGCGAAAGCGTGGGGAGCAAACCGG
T.flavus	701	ACCTGGTCCACCCGTGACGCTGAGGCGCGAAAGCGTGGGGAGCAAACCGG
SPS14	701	ACCTGGCCCGCCCGTGACGCTGAGGCGCGAAAGCGTGGGGAGCAAACCGG
T.ruber	701	ACCTGGACGATTTGTGACGCTGAGGCGCGAAAGCGTGGGGAGCAAACCGG
D.rad	701	TACTGGACAGAAGGTGACGCTGAGGCGCGAAAGTGTGGGGAGCAAACCGG

(continued)

method has been used most frequently followed by the spectrophotometric estimation of the reassociation rate of DNA from pairs of strains. Research, mainly concerned with the relationship of *Thermus* to *Deinococcus* (Hensel *et al.*, 1986), included a small DNA hybridization study indicating that four *T. ruber* strains were closely related and quite distinct from the yellow strains examined (*T. aquaticus, T. thermophilus, T. flavus*). This has been confirmed with seventeen strains of *T. ruber* by nitrocellulose filter hybridization (Sharp and Williams, 1988) and extended to cover two phenotypic clusters comprising strains 15540, 16519, and 16311 (cluster A) on the one hand and *T. ruber, T. rubens,* and strains 16266 and 16068 (cluster B) on the other (Ruffett, 1992).

Table II. (*Continued*)

(SIGNATURE for THERMUS/DEINOCOCCUS)
ccctaaacg

T.fil	751	ATTAGATACCCGGGTAGTCCACGCCCTAAACGATGCGCACTAGGTCTCTG
Tok20A1	751	ATTAGATACCCGGGTAGTCCACGCCCTAAACGATGCGCACTAGGTCTCTG
OK6A1	751	ATTAGATACCCGGGTAGTCCACGCCCTAAACGATGCGCACTAGGTCTCTG
RT41A	751	ATTAGATACCCGGGTAGTCCACGCCCTAAACGATGCGCACTAGGTCTCTG
T351	751	ATTAGATACCCGGGTAGTCCACGCCCTAAACGATGCGCACTAGGTCTCTG
Tok8A1	751	ATTAGATACCCGGGTAGTCCACGCCCTAAACGATGCGCACTAGGTCTCTG

T.br YS38	751	ATTAGATACCCGGGTAGTCCACGCCCTAAACGATGCGCGCTAGGTCTCTG
ZHGIA1	751	ATTAGATACCCGGGTAGTCCACGCCCTAAACGATGCGCGCTAGGTCTCTG
OSRAM4	751	ATTAGATACCMGGGTAGTCCAAGCCCTAAACGATGCGCGCTAGGTCTCTG
ZFIA2	751	ATTAGATACCCGGGTAGTCCACGCCCTAAACGATGCGCGCTAGGTCTCTG
		*

Vi7	751	ATTAGATACCCGGGTAGTCCACGCCCTAAACGATGCGCGCTAGGTCTTTG-
NMX2A1	751	ATTAGATACCCGGGTAGTCCACGCCCTAAACGATGCGCGCTAGGTCTTTGG
		*

T.aq YT1	751	ATTAGATACCCGGGTAGTCCACGCCCTAAACGATGCGCGCTAGGTCTCTG
YS025	751	ATTAGATACCCGGGTAGTCCACGCCCTAAACGATGCGCGCTAGGTCTCTG
YSO52	751	ATTAGATACCCGGGTAGTCCACGCCCTAAACGATGCGCGCTAGGTCTCTG

HSA1	751	ATTAGATACCCGGGTAGTCCACGCCCTAAACGATGCGCGCTAGGTCTCTG
T.th HB8*	751	ATTAGATACCCGGGTAGTCCACGCCCTAAACGATGCGCGCTAGGTCTCTG
T.th HB8	751	ATTAGATACCCGGGTAGTCCACGCCCTAAACGATGCGCGCTAGGTCTCTG
T.th HB27	751	ATTAGATACCCGGGTAGTCCACGCCCTAAACGATGCGCGCTAGGTCTCTG
T.flavus	751	ATTAGATACCCGGGTAGTCCACGCCCTAAACGATGCGCGCTAGGTCTCTG

SPS14	751	ATTAGATACCCGGGTAGTCCACGCCCTAAACGTTGCGCGCTAGGTCTCTG

T.ruber	751	ATTAGATACCCGGGTAGTCCACGCCCTAAACGATGAGTGCTGGGTGTCCG

D.rad	751	ATTAGATACCCGGGTAGTCCACACCCTAAACGATGTACGTTGGCTAAGCG

5.3. Ribosomal RNA Structure

The taxonomic position of *Thermus* relative to most other genera remains uncertain. Early molecular biology studies included a catalogue of oligonucleotides derived by hydrolysis of 16s RNA, which indicated a distinct, if remote, relationship between *Thermus* [represented by *T. aquaticus* and *T. ruber*], and *Dienococcus* (Hensel *et al.*, 1986). More recently the sequence data have become available for the gene for the 16s RNA of *T. thermophilus* HB8 (Murzina *et al.*, 1988; Hartmann *et al.*, 1989) with discrepancies in two areas between the two sets of data. Alignment with other sequences lead to the conclusion that *T. thermophilus* falls in the cluster of

```
                                                    HELIX 29
                      ¦ V5 ¦
```

```
T.fil       801 GGTTATCTGGGGGCCGAAGCCAACGCGTTAAGTGCGCCGCCTGGGGAGTA
Tok20A1     801 GGTTAACTGGGGGCCGAAGCCAACGCGTTAAGTGCGCCGCCTGGGGAGTA
OK6A1       801 GGTTATCTGGGGGCCGAAGCCAACGCGTTAAGTGCGCCGCCTGGGGAGTA
RT41A       801 GGTTATCTGGGGGCCGAAGCCAACGCGTTAAGTGCGCCGCCTGGGGAGTA
T351        801 GGTTATCTGGGGGCCGAAGCCAACGCGTTAAGTGCGCCGCCTGGGGAGTA
Tok8A1      801 GGTTATCTGGGGGCCGAAGCCAACGCGTTAAGTGCGCCGCCTGGGGAGTA


T.br YS38   801 GGTTTTCTGGGGGCCGAAGCCAACGCGTTAAGCGCGCCGCCTGGGGAGTA
ZHGIA1      801 GGTTTTCTGGGGGCCGAAGCCAACGCGTTAAGCGCGCCGCCTGGGGAGTA
OSRAM4      801 GGNNNNNNNNNNNNNNNNNNNNNNNNNNNNNNNNNNNNNNNNNNNNNNNNN
ZFIA2       801 GGGTTTCTGGGGGCCGAAGCCAACGCGTTAAGCGCGCCGCCTGGGGAGTA


Vi7         801 GGTTTACCT-GGGGCCGAAGCCAACGCGTTAAGCGCGCCGCCTGGGGAGTA
NMX2A1      801 GGTTTACCTGGGGGCCGAAGCCAACGCGTTAAGCGCGCCGCCTGGGGAGTA
                         *


T.aq YT1    801 GGTTATCTGGGGGCCGAAGCTAACGCGTTAAGCGCGCCGCCTGGGGAGTA
YS025       801 GGTTATCTGGGGGCCGAAGCTAACGCGTTAAGCGCGCCGCCTGGGGAGTA
YS052       801 GGTTATCTGGGGGCCGAAGCTAACGCGTTAAGCGCGCCGCCTGGGGAGTA


HSA1        801 GGTCTCCTGGGGGCCGAAGCTAACGCGTTAAGCGCGCCGCCTGGGGAGTA
T.th HB8*   801 GGTCTCCTGGGGGCCGAAGCTAACGCGTTAAGCGCGCCGCCTGGGGAGTA
T.th HB8    801 GGTCTCCTGGGGGCCGAAGCTAACGCGTTAAGCGCGCCGCCTGGGGAGTA
T.th HB27   801 GGTCTCCTGGGGGCCGAAGCTAACGCGTTAAGCGCGCCGCCTGGGGAGTA
T.flavus    801 GGTCTCCTGGGGGCCGAAGCTAACGCGTTAAGCGCGCCGCCTGGGGAGTA


SPS14       801 GGTCCTCTGGGGGCCGAAGCTAACGCGTTAAGCGCGCCGCCTGGGGAGTA

T.ruber     801 ACATCTGTTGGGTGCCGTAGCTAACGCGTTAAGCACTCCACCTGGGAAGTA

D.rad       801 GGATGCTGTGCTTGGCGAAGCTAACGCGATAAACGTACCGCCTGGGAAGTA
```

(continued)

the green nonsulphur bacteria along with *Thermomicrobium roseum*, *Chloroflexus aurantiacus*, and *Herpetosiphon aurantiacus*. This is the second deepest eubacterial branchpoint after *Thermotoga maritima*. Many 16s RNA and 16s rDNA sequences are now available (Neefs *et al.*, 1991; Olsen *et al.*, 1991) and there is a tendency to deposit such information in databases without publishing the entire sequence in journals (e.g., Saul *et al.*, 1993a). The sequences of the genes for 16sRNA of *T. ruber*, "*T. brockianus*" YS38, "*T. imahorii*" Vi7, and "*T. oshimai*" SPS14 were determined (Embley *et al.*, 1993), and while the sequences of the yellow pigmented *Thermus* were clearly related (Table II), the lower thermophilic bias evident in the sequence of the *T. ruber* gene was important in confirming the relationship

Table II. (*Continued*)

```
                         HELIX 29
                   aa*gaattgacgg 926f
                            cccgggcgtgttcgcca 943rev

T.fil       851 CGGCCGCAAGGCTGAAACTCAAAGGAATTGACGGGGGCCCGCACAAGCGG
Tok20A1     851 CGGCCGCAAGGCTGAAACTCAAAGGAATTGACGGGGGCCCGCACAAGCGG
OK6A1       851 CGGCCGCAAGGCTGAAACTCAAAGGAATTGACGGGGGCCCGCACAAGCGG
RT41A       851 CGGCCGCAAGGCTGAAACTCAAAGGAATTGACGGGGGCCCGCACAAGCGG
T351        851 CGGCCGCAAGGCTGAAACTCAAAGGAATTGACGGGGGCCCGCACAAGCGG
Tok8A1      851 CGGCCGCAAGGCTGAAACTCAAAGGAATTGACGGGGGCCCGCACAAGCGG


T.br YS38   851 CGGCCGCAAGGCTGAAACTCAAAGGAATTGACGGGGGCCCGCACAAGCGG
ZHGIA1      851 CGGCCGCAAGGCTGAAACTCAAAGGAATTGACGGGGGCCCGCACAAGCGG
OSRAM4      851 NNNNNNNNNNNNNNNNNNNNNNNNNNNNNNNNNNNNNNNNNNNNNNNNNN
ZFIA2       851 CGGCCGCAAGGCTGAAACTCAAAGGAATTGACGGGGGCCCGCACAAGCGG


Vi7         851 CGGCCGCAAGGCTGAAACTCAAAGGAATTGACGGGGGCCCGCACAAGCGG
NMX2A1      851 CGGCCGCAAGGCTGAAACTCAAAGGAATTGACGGGGGCCCGCACAAGCGG


T.aq YT1    851 CGGCCGCAAGGCTGAAACTCAAAGGAATTGACGGGGGCCCGCACAAGCGG
YS025       851 CGGCCGCAAGGCTGAAACTCAAAGGAATTGACGGGGGCCCGCACAAGCGG
YS052       851 CGGCCGCAAGGCTGAAACTCAAAGGAATTGACGGGGGCCCGCACAAGCGG


HSA1        851 CGGCCGCAAGGCTGAAACTCAAAGGAATTGACGGGGGCCCGCACAAGCGG
T.th HB8*   851 CGGCCGCAAGGCTGAAACTCAAAGGAATTGACGGGGGCCCGCACAAGCGG
T.th HB8    851 CGGCCGCAAGGCTGAAACTCAAAGGAATTGACGGGGGCCCGCACAAGCGG
T.th HB27   851 CGGCCGCAAGGCTGAAACTCAAAGGAATTGACGGGGGCCCGCACAAGCGG
T.flavus    851 CGGCCGCAAGGCTGAAACTCAAAGGAATTGACGGGGGCCCGCACAAGCGG


SPS14       851 CGGCCGCAAGGCTGAAACTCAAAGGAATTGACGGGGACCCGCACAAGCGG

T.ruber     851 CGGTCGCAAGACTGAAACTCAAAGGAATTGACGGGGGCCCGCACAAGCGG

D.rad       851 CGGCCGCAAGGTTGAAACTCAAAGGAATTGACGGGGGCCCGCACAAGCGG
```

between *Thermus* and *Deinococcus*. The sequences of *T. aquaticus* YT1, *T. flavus* AT62, *Thermus* sp. HSA.1, *T. ruber, Thermus* sp. NMX2A1, *Thermus* sp. OK6A1, *Thermus* sp. Rt41A, *T. filiformis, Thermus* sp. T351, *Thermus* sp. Tok8A1, *Thermus* sp. Tok20A1, *Thermus* sp. ZFIA2, *Thermus* sp. ZHGIA1, have been determined in the laboratory of Bergquist and deposited in Genbank (Saul *et al.*, 1993b) and analyzed for their taxonomic value. The sequences for YS025 and YS052 were determined by F. Reaney at the Deutsche Sammlung von Mikroorganismen. When these are aligned with the other available sequences shown in Table II the results are most illuminating. The close relationship between the sequences of *T. filiformis* and five other strains isolated in New Zealand is confirmed, fully support-

```
                    --------HELIX 33----        STEM 34
              ----32---                    BULGE

T.fil      901 TGGAGCATGTGGTTTAATTCGAAGCAACGCGAAGAACCTTACCAGGCCTT
Tok20A1    901 TGGAGCATGTGGTTTAATTCGAAGCAACGCGAAGAACCTTACCAGGCCTT
OK6A1      901 TGGAGCATGTGGTTTAATTCGAAGCAACGCGAAGAACCTTACCAGGCCTT
RT41A      901 TGGAGCATGTGGTTTAATTCGAAGCAACGCGAAGAACCTTACCAGGCCTT
T351       901 TGGAGCATGTGGTTTAATTCGAAGCAACGCGAAGAACCTTACCAGGCCTT
Tok8A1     901 TGGAGCATGTGGTTTAATTCGAAGCAACGCGAAGAACCTTACCAGGCCTT


T.br YS38  901 TGGAGCATGTGGTTTAATTCGAAGCAACGCGAAGAACCTTACCAGGCCTT
ZHGIA1     901 TGGAGCATGTGGTTTAATTCGAAGCAACGCGAAGAACCTTACCAGGCCTT
OSRAM4     901 NNNNNNNNNNNNNNNNNNNNNNNNNNNNNNNNNNNNNNNNNNNNNNNNNNN
ZFIA2      901 TGGAGCATGTGGTTTAATTCGAAGCAACGCGAAGAACCTTACCAGGCCTT

Vi7        901 TGGAGCATGTGGTTTAATTCGAAGCAACGCGAAGAACCTTACCAGGCCTT
NMX2A1     901 TGGAGCATGTGGTTTAATTCGAAGCAACGCGAAGAACCTTACCAGGCCTT


T.aq YT1   901 TGGAGCATGTGGTTTAATTCGAAGCAACGCGAAGAACCTTACCAGGCCTT
YS025      901 TGGAGCATGTGGTTTAATTCGAAGCAACGCGAAGAACCTTACCAGGCCTT
YS052      901 TGGAGCATGTGGTTTAATTCGAAGCAACGCGAAGAACCTTACCAGGCCTT


HSA1       901 TGGAGCATGTGGTTTAATTCGAAGCAACGCGAAGAACCTTACCAGGCCTT
T.th HB8*  901 TGGAGCATGTGGTTTAATTCGAAGCAACGCGAAGAACCTTACCAGGCCTT
T.th HB8   901 TGGAGCATGTGGTTTAATTCGAAGCAACGCGAAGAACCTTACCAGGCCTT
T.th HB27  901 TGGAGCATGTGGTTTAATTCGAAGCAACGCGAAGAACCTTACCAGGCCTT
T.flavus   901 TGGAGCATGTGGTTTAATTCGAAGCAACGCGAAGAACCTTACCAGGCCTT


SPS14      901 TGGAGCATGTGGTTTAATTCGAAGCAACGCGAAGAACCTTACCAGGCCTT


T.ruber    901 TGGAGCATGTGGTTTAATTCGAAGCAACGCGCAGAACCTTACCAGGTCTT


D.rad      901 TGGAGCATGTGGTTTAATTCGAAGCAACGCGAAGAACCTTACCAGGTCTT
```

(continued)

ing their membership of one genospecies as indicated by DNA:DNA homology (Georganta *et al.*, 1993), in spite of the phenetic heterogeneity among New Zealand strains (Hudson *et al.*, 1989).

There is also an excellent correlation between the sequences of a strain isolated in New Mexico (NMX2A1) and a strain that was isolated in Portugal (Vi7) and for which the name *T. imahorii* is now proposed. The congruence between these sequences is good, only 18 differences (substitutions at 64, 71, 173, 188, 202, 203, 436, 554, 959, 963, 968, 1003, 1099, 1220, 1230, a deletion at 67 and insertions after 800 and at 810, all in NMX2A1 relative to Vi7) in over 1400 nucleotides, apart from those that involve ambiguities in the determination of sequences indicated by N in

Table II. (*Continued*)

```
                              DOMAIN 35
              11111111--222222------222222---3333----3333--1
              !-------------------------V6---------------
```

T.fil	951	GACATGCTAGGGAACCTGGGTGAAAGCCTGGGGTGCCCCGCGAGGGGAGC
Tok20A1	951	GACATGCTAGGGAACCTGGGTGAAAGCCTGGGGTGCCCCGCGAGGGGAGC
OK6A1	951	GACATGCTAGGGAACCCGGGTGAAAGCCTGGGGTGCCCCGCGAGGGGAGC
RT41A	951	GACATGCTAGGGAACCCGGGTGAAAGCCTGGGGTGCCCCGCGAGGGGAGC
T351	951	GACATGCTAGGGAACCCGGGTGAAAGCCTGGGGTGCCCCGCGAGGGGAGC
Tok8A1	951	GACATGCTAAGGAACCTGGGTGAAAGCCTGGGGTGCCCCGCGAGGGGAGC
		* *

T.br YS38	951	GACATGCTAGGGAACCTGCCTGAAAGGGTGRGGTGCCCCGCGAGGGGAGC
ZHGIA1	951	GACATGCTAGGGAACCTGCCTGAAAGGGTGGGGTGCCCCGCGAGGGGAGC
OSRAM4	951	NNNNNNNNNNNNNNNNNNNNNNNNNNNNNNNNNNNNNNNNNNNNNNNNNNN
ZFIA2	951	GACATGCTAGGGAACCTGGGTGAAAGCCTGGGGTGCCC-GCGAGGG-AGC
		** ** * *

Vi7	951	GACATGCTAGGGGACCTAGGTGAAAGCCTGGGGTGCCC-GCGAGGG-AGC
NMX2A1	951	GACATGCTGGGGAACCTGGGTGAAAGCCTGGGGTGCCC-GCGAGGG-AGC
		* * *

T.aq YT1	951	GACATGCTAGGGAACCTGGGTGAAAGCCTGGGGTGCCCCGCGAGGGGAGC
YS025	951	GACATGCTAGGGAACCTGGGTGAAAGCCTGGGGTGCCCCGCGAGGGGAGC
YSO52	951	GACATGCTAGGGAACCTGGGTGAAAGCCTGGGGTGCCCCGCGAGGGGAGC

HSA1	951	GACATGCTAGGGAACCCGGGTGAAAGCCTGGGGTGCCCCGCGAGGGGAGC
T.th HB8*	951	GACATGCTAGGGAACCCGGGTGAAAGCCTGGGGTGCCC-GCGAGGG-AGC
T.th HB8	951	GACATGCTAGGGAACCCGGGTGAAAGCCTGGGGTGCCC-GCGAGGG-AGC
T.th HB27	951	GACATGCTAGGGAACCCGGGTGAAAGCCTGGGGTGCCCCGCGAGGGGAGC
T.flavus	951	GACATGCTAGGGAACCCGGGTGAAAGCCTGGGGTGCCCCGCGAGGGGAGC
		* *

SPS14	951	GACATGCTGGGGAACCCGGGTGAAAACCCGGGGTGCCC-GCAAGGG-AAC

T.ruber	951	GACATCCACGGAACCCTGG-TGAAAGCCGGGGGTGCCC-GCAAGGG-AGC

D.rad	951	GACATGCTAGGAAGGC-GCTGGAGACAGCGCCGTGCCCTTCGGGGA-ACC

one of the sequences. It seems highly likely that strain NMX2A1 from New Mexico is a strain of *T. imahorii,* all the other isolates of which come from Portugal and the Azores.

There is only one difference at 280 between the sequences of *T. brockianus* YS38 from North America and strain ZHG1A1 from Iceland (Hudson *et al.,* 1989). If strain ZHG1A1 does indeed belong to *T. brockianus* this is further evidence that the geographical spread of *Thermus* species is wider than has often been thought. Although the sequence of OSRAM4, isolated like YS38 from Yellowstone National Park, is incomplete, it is also

```
                1111111   ---------36----------      --------38----
                ------|

T.fil      1001  CTTAGCACAGGTGCTGCATGGCCGTCGTCAGCTCGTGCCGTGAGGTGTTG
Tok20A1    1001  CTTAGCACAGGTGCTGCATGGCCGTCGTCAGCTCGTGCCGTGAGGTGTTG
OK6A1      1001  CTTAGCACAGGTGCTGCATGGCCGTCGTCAGCTCGTGCCGTGAGGTGTTG
RT41A      1001  CTTAGCACAGGTGCTGCATGGCCGTCGTCAGCTCGTGCCGTGAGGTGTTG
T351       1001  CTTAGCACAGGTGCTGCATGGCCGTCGTCAGCTCGTGCCGTGAGGTGTTG
Tok8A1     1001  CTTAGCACAGGTGCTGCATGGCCGTCGTCAGCTCGTGCCGTGAGGTGTTG


T.br YS38  1001  CCTAGCACAGGTGCTGCATGGCCGTCGTCAGCTCGTGTCGTGAGATGTTG
ZHGIA1     1001  CCTAGCACAGGTGCTGCATGGCCGTCGTCAGCTCGTGTCGTGAGATGTTG
OSRAM4     1001  NNNNNNNNNNNNNNNNNNNNNNNNNNNNNNNNNNNNNNNNNNNNNNNNNNN
ZFIA2      1001  CCCAGCACAGGTGCTGCATGGCCGTCGTCAGCTCGTGTCGTGAGATGTTG
                 *               *


Vi7        1001  CCTAGCACAGGTGCTGCATGGCCGTCGTCAGCTCGTGTCGTGAGATGTTG
NMX2A1     1001  CCCAGCACAGGTGCTGCATGGCCGTCGTCAGCTCGTGTCGTGAGATGTTG
                 *


T.aq YT1   1001  CCTAGCACAGGTGCTGCATGGCCGTCGTCAGCTCGTGTCGTGAGATGTTG
YS025      1001  CCTAGCACAGGTGCTGCATGGCCGTCGTCAGCTCGTGTCGTGAGATGTTG
YS052      1001  CCTAGCACAGGTGCTGCATGGCCGTCGTCAGCTCGTGTCGTGAGATGTTG


HSA1       1001  CCTAGCACAGGTGCTGCATGGCCGTCGTCAGCTCGTGCCGTGAGGTGTTG
T.th HB8*  1001  CCTAGCACAGGTGCTGCATGGCCGTCGTCAGCTCGTGCCGTGAGGTGTTG
T.th HB8   1001  CCTAGCACAGGTGCTGCATGGCCGTCGTCAGCTCGTGCCGTGAGGTGTTG
T.th HB27  1001  CCTAGCACAGGTGCTGCATGGCCGTCGTCAGCTCGTGCCGTGAGGTGTTG
T.flavus   1001  CCTAGCACAGGTGCTGCATGGCCGTCGTCAGCTCGTGCCGTGAGGTGTTG


SPS14      1001  CCCAGCACAGGTGCTGCATGGCCGTCGTCAGCTCGTGTCGTGAGATGTTG


T.ruber    1001  CGTGAGACAGGTGCTGCATGGTCGTCGTCAGCTCGTGTCGTGAGATGTTG


D.rad      1001  -TAGACACAGGTGCTGCATGGCTGTCGTCAGCTCGTGTCGTGAGATGTTG
```

(continued)

most probably a strain of *T. brockianus*. The position of strain ZF1A1 from Iceland is more problematic. It is shown in Table II aligned with the *T. brockianus* cluster, although it shows 43 differences from the sequence of YS38. It also shows 56 differences from the sequence of Vi7 and 57 differences from NMX2A1, according to Saul *et al.*, 1993a. This intermediate state between ZHG1A1 and NMX2A1 is shown in the dendrogram of Saul *et al.* (1993b). As it is not clear what degree of diversity of 16SrDNA structure occurs between different strains of a species, ZF1A1 cannot be allocated at present, but it probably represents another species.

The sequence of the New Mexico, USA strain NMX2A1 shows only four differences (at 64, 67, 71, and 173) from that of strain Vi7 from

Table II. (*Continued*)

```
                                        REGION 41
                           1111111---22222222-------
                           ¦----V7-
         ────────────────────────────────────────────────────────────
T.fil      1051  GGTTAAGTCCCGCAACGAGCGCAACCCCTGCCCCTAGTTACCAGCGGGTG
Tok20A1    1051  GGTTAAGTCCCGCAACGAGCGCAACCCCTGCCCCTAGTTACCAGCGGGTG
OK6A1      1051  GGTTAAGTCCCGCAACGAGCGCAACCCCTGCCCCTAGTTACCAGCGGGTG
RT41A      1051  GGTTAAGTCCCGCAACGAGCGCAACCCCTGCCCCTAGTTACCAGCGGGTG
T351       1051  GGTTAAGTCCCGCAACGAGCGCAACCCCTGCCCCTAGTTACCAGCGGGTG
Tok8A1     1051  GGTTAAGTCCCGCAACGAGCGCAACCCCTGCCCCTAGTTACCAGCGGGTG
         ────────────────────────────────────────────────────────────
T.br YS38  1051  GGTTAAGTCCCGCAACGAGCGCAACCCCTGCCGTTAGTTGCCAGCGGGTT
ZHGIA1     1051  GGTTAAGTCCCGCAACGAGCGCAACCCCTGCCGTTAGTTGCCAGCGGGTT
OSRAM4     1051  NNNNNNNNNNNNNNNNNNNNNNNNNNNNNNNNNNNNNNNNNNNNNNNNNNN
ZFIA2      1051  GGTTAAGTCCCGCAACGAGCGCAACCCCTGCCGTTAGTTGCCAGCGGGTG
                                                                 *
         ────────────────────────────────────────────────────────────
Vi7        1051  GGTTAAGTCCCGCAACGAGCGCAACCCCTGCCCTTAGTTGCCAGCGGGAT
NMX2A1     1051  GGTTAAGTCCCGCAACGAGCGCAACCCCTGCCCTTAGTTGCCAGCGGGTT
                                                                 *
         ────────────────────────────────────────────────────────────
T.aq YT1   1051  GGTTAAGTCCCGCAACGAGCGCAACCCCTGCCGTTAGTTGCCAGCGGGTG
YS025      1051  GGTTAAGTCCCGCAACGAGCGCAACCCCTGCCGTTAGTTGCCAGCGGGTT
YS052      1051  GGTTAAGTCCCGCAACGAGCGCAACCCCTGCCGTTAGTTGCCAGCGGGTT
                                                                 *
         ────────────────────────────────────────────────────────────
HSA1       1051  GGTTAAGTCCCGCAACGAGCGCAACCCCCGCCGTTAGTTGCCAGC-GGTT
T.th HB8*  1051  GGTTAAGTCCCGCAACGAGCGCAACCCCCGCCGTTAGTTGCCAGC-GGTT
T.th HB8   1051  GGTTAAGTCCCGCAACGAGCGCAACCCCCGCCGTTAGTTGCCAGC-GGTT
T.th HB27  1051  GGTTAAGTCCCGCAACGAGCGCAACCCCCGCCGTTAGTTGCCAGC-GGTT
T.flavus   1051  GGTTAAGTCCCGCAACGAGCGCAACCCCCGCCGTTAGTTGCCAGC-GGTT
         ────────────────────────────────────────────────────────────
SPS14      1051  GGTTAAGTCCCGCAACGAGCGCAACCCCCGCCCTTAGTTGCCAGCGGGC-
         ────────────────────────────────────────────────────────────
T.ruber    1051  GGTTAAGTCCCGCAACGAGCGCAACCCCTGTTGCTAGTTGCCATC-AGTT
         ────────────────────────────────────────────────────────────
D.rad      1051  GGTTAAGTCCCGCAACGAGCGCAACCCTTGCTTCCAGTTGCCAGC--ATT
         ────────────────────────────────────────────────────────────
```

Portugal; therefore, this strain certainly belongs to *T. imahorii,* which has its known geographic range from Europe to North America.

The 16s RNA sequence of strain HSA1 (isolated in the South island of New Zealand, Hudson *et al.*, 1989) is closely similar to that of *T. thermophilus* HB8, there being 13 differences between the two sequences, indicating that this strain is a member of this widely distributed genospecies. The status of *T. flavus,* which was found to be a member of the *T. thermophilus* genospecies by DNA:DNA homology (Williams, 1989) and which shows six differences in the 16srDNA sequences from the sequence of *T. thermophilus*

```
                           REGION 41
           ----------------22222222--------------------1111111
                   -------¦
```

T.fil	1101	AAGCCGGGGACTCTAGGGGGACTGCCCGCGAAAGCGGGAGGAAGGCGGGG
Tok20A1	1101	AAGCCGGGGACTCTAGGGGGACTGCCCGCGAAAGCGGGAGGAAGGCGGGG
OK6A1	1101	AAGCCGGGGACTCTAGGGGGACTGCCCGCGAAAGCGGGAGGAAGGCGGGG
RT41A	1101	AAGCCGGGGACTCTAGGGGGACTGCCCGCGAAAGCGGGAGGAAGGCGGGG
T351	1101	AAGCCGGGGACTCTAGGGGGACTGCCCGCGAAAGCGGGAGGAAGGCGGGG
Tok8A1	1101	AAGCCGGGGACTCTAGGGGGACTGCCCGCGAAAGCGGGAGGAAGGCGGGG

T.br YS38	1101	AAGCCGGGCACTCTAACGGGACTGCCTGCGAAAGCAGGAGGAAGGCGGGG
ZHGIA1	1101	AAGCCGGGCACTCTAACGGGACTGCCTGCGAAAGCAGGAGGAAGGCGGGG
OSRAM4	1101	NNNNNNNNNNNNNNNNNNNNNNNNNNNNNNNNNNNNNNNNGGAGGAAGGHGGGG
ZFIA2	1101	AGGCCGGGCACTCTAACGGGACTGCCTGCGAAAGCAGGAGGAAGGCGGGG

Vi7	1101	AGGCCGGGCACTCTAAGGGGACTGCCTGCGAAAGCAGGAGGAAGGCGGGG
NMX2A1	1101	AGGCCGGGCACTCTAAGGGGACTGCCTGCGAAAGCAGGAGGAAGGCGGGG

T.aq YT1	1101	AAGCCGGGCACTCTAACGGGACTGCCTGCGAAAGCAGGAGGAAGGCGGGG
YS025	1101	GAGCCGGGCACTCTAACGGGACTGCCTGCGAAAGCAGGAGGAAGGCGGGG
YS052	1101	GAGCCGGGCACTCTAACGGGACTGCCTGCGAAAGCAGGAGGAAGGCGGGG
		*

HSA1	1101	CGGCCGGGCACTCTAACGGGACTGCCCGCGAAAGCGGGAGGAAGGAGGGG
T.th HB8*	1101	CGGCCGGGCACTCTAACGGGACTGCCCGCGAAAGCGGGAGGAAGGAGGGG
T.th HB8	1101	CGGCCGGGCACTCTAACGGGACTGCCCGCGAAAGCGGGAGGAAGGAGGGG
T.th HB27	1101	CGGCCGGGCACTCTAACGGGACTGCCCGCGAAAGCGGGAGGAAGGAGGGG
T.flavus	1101	CGGCCGGGCACTCTAACGGGACTGCCCGCGAAAGCGGGAGGAAGGAGGGG

SPS14	1101	AAGCCGGGCACTCTAAGGGGACTGCCGGCGAAAGCCGGAGGAAGGCGGGG

T.ruber	1101	CGGCTGGGCACTCTAGCGAGACTGCCTACGAAAGTAGGAGGAAGGCGGGG

D.rad	1101	CAGTTGGGCACTCTGGAGGGACTGCCTGTGAAAGCAGGAGGAAGGCGGGG

(continued)

HB8, determined by Hartmann *et al.* (1989) (at positions 436, 989, 997, 1420, 1422, and 1462) is confirmed. The taxonomic position of strain HB27, a strain used as a host for cloning in *Thermus*, within *T. thermophilus* is confirmed as its 16SrDNA sequence is identical to that of *T. flavus*.

Strains YS025 and YS052 (Munster *et al.*, 1986), from the same geographical site as *T. aquaticus* YT1 and identified as members of that species, show almost identical 16 srDNA sequences. The sequences for five strains analyzed by Saul *et al.* (1993a), Fiji3A1 (L10067), W28 (L10068), Tok3A1 (L10069), YSPIDA1 (L10070) and ZHGIBA4 (L10071), were not accessible from Genbank at the time of writing. However, the number of differences

Table II. (*Continued*)

SPACER LEADING TO HELIX 43

```
T.fil      1151 ATGACGTCTGGTCATCATGGCCCTTACGGCCTGGGCGACACACGTGCTAC
Tok20A1    1151 ATGACGTCTGGTCATCATGGCCCTTACGGCCTGGGCGACACACGTGCTAC
OK6A1      1151 ATGACGTCTGGTCATCATGGCCCTTACGGCCTGGGCGACACACGTGCTAC
RT41A      1151 ATGACGTCTGGTCATCATGGCCCTTACGGCCTGGGCGACACACGTGCTAC
T351       1151 ATGACGTCTGGTCATCATGGCCCTTACGGCCTGGGCGACACACGTGCTAC
Tok8A1     1151 ATGACGTCTGGTCATCATGGCCCTTACGGCCTGGGCGACACACGTGCTAC
```

```
T.br YS38  1151 ACGACGTCTGGTCATCATGGCCCTTACGGCCTGGGCGACACACGTGCTAC
ZHGIA1     1151 ACGACGTCTGGTCATCATGGCCCTTACGGCCTGGGCGACACACGTGCTAC
OSRAM4     1151 ACGACGTCTGGTCATCAT--NNCTTACGGCMTGGGCGACACACGTGCTAC
ZFIA2      1151 ACGACGTCTGGTCATCATGGCCCTTACGGCCTGGGCGACACACGTGCTAC
                                 **
```

```
Vi7        1151 ACGACGTCTGGTCATCATGGCCCTTACGGCCTGGGCGACACACGTGCTAC
NMX2A1     1151 ACGACGTCTGGTCATCATGGCCCTTACGGCCTGGGCGACACACGTGCTAC
```

```
T.aq YT1   1151 ACGACGTCTGGTCATCATGGCCCTTACGGCCTGGGCGACACACGTGCTAC
                                 **
YS025      1151 ACGACGTCTGGTCATCATGGCCCTTACGGCCTGGGCGACACACGTGCTAC
YS052      1151 ACGACGTCTGGTCATCATGGCCCTTACGGCCTGGGCGACACACGTGCTAC
```

```
HSA1       1151 ACGACGTCTGGTCAGCATGGCCCTTAAGGCCTGGGCGACACACGTGCTAC
T.th HB8*  1151 ACGACGTCTGGTCAGCATGGCCCTTACGGCCTGGGCGACACACGTGCTAC
T.th HB8   1151 ACGACGTCTGGTCAGCATGGCCCTTACGGCCTGGGCGACACACGTGCTAC
T.th HB27  1151 ACGACGTCTGGTCAGCATGGCCCTTACGGCCTGGGCGACACACGTGCTAC
T.flavus   1151 ACGACGTCTGGTCAGCATGGCCCTTACGGCCTGGGCGACACACGTGCTAC
                                 *
```

```
SPS14      1151 ACGACGTCTGGTCATCATGGCCCTTACGGCCTGGGCGACACACGTGCTAC
```

```
T.ruber    1151 ATGACGTCTGATCCGCATGGCCCTTACGACCTGGGCGACACACGTGCTAC
```

```
D.rad      1151 ATGACGTCTAGTCAGCATGGTCCTTACGTCCTGGGCTACACACGTGCTAC
```

between the nucleotide sequences indicate the probable species
identification. Thus, YSPIDA1 shows only 15 differences from the *T. aquaticus* YT1 sequence, Fiji3A1 shows 9–12 differences with four members of
the *T. thermophilus* cluster; ZHGIBA4 has only one difference from
ZHGIA1; and both W28 and Tok3A1 fall within the *T. filiformis* group.

6. THE SPECIES OF *THERMUS*

The taxonomic confusion surrounding *Thermus* because of the variation in phenotypic tests between isolates, and the removal of valid species

```
              ---STEM 43----111-------LOOP 44------------111---
T.fil       1201 AATGCCTACTACAGAGCGATGCGACCTGGTGACAGGGAGCGAATCGCAAA
Tok20Al     1201 AATGCCTACTACAGAGCGATGCGACCTGGTGACGGGGAGCGAATCGCAAA
OK6Al       1201 AATGCCTACTACAGAGCGATGCGACCTGGTGACAGGGAGCGAATCGCAAA
RT41A       1201 AATGCCTACTACAGAGCGATGCGACCTGGTGACAGGGAGCGAATCGCAAA
T351        1201 AATGCCTACTACAGAGCGATGCGACCTGGTGACAGGGAGCGAATCGCAAA
Tok8Al      1201 AATGCCTACTACAGAGCGATGCGACCTGGTGACGGGGAGCGAATCGCAAA


T.br YS38   1201 AATGCCCACTACAGAGCGAGGCGACCTGGCGACAGGGAGCGAATCGCGGA
ZHGIAl      1201 AATGCCCACTACAGAGCGAGGCGACCTGGCGACAGGGAGCGAATCGCGGA
OSRAM4      1201 AATGCCCACTACAGAGCGAGGCGACCTGGCRACAGGGAGCGAATCGCGGA
ZFIA2       1201 AATGCCCACTACAGAGCGCTGCGACCCGGTGACGGGGAGCGAATCGCGGA
                           **        *  *    *


Vi7         1201 AATGCCCACTACAGAGCGATGCGACCCAGTGATGGGGAGCGAATCGCAAA
NMX2Al      1201 AATGCCCACTACAGAGCGAGGCGACCCAGCGATGGGGAGCGAATCGCAAA
                          *            *


T.aq YTl    1201 AATGCCCACTACAGAGCGAGGCGACCTGGCAACAGGGAGCGAATCGCAAA
YS025       1201 AATGCCCACTACAAAGCGAGGCGACCTGGCAACAGGGAGCGAATCGCAAA
YSO52       1201 AATGCCCACTACAAAGCGAGGCGACCTGGCAACAGGGAGCGAATCGCAAA


HSA1        1201 AATGCCCACTACAAAGCGAAGCCACCCGGCAACGGGGAGCTAATCGCAAA
T.th HB8*   1201 AATGCCC--TACAAAGCGATGCCACCCGGCAACGGGGAGCTAATCGCAAA
T.th HB8    1201 AATGCCCACTACAAAGCGATGCCACCCGGCAACGGGGAGCTAATCGCAAA
T.th HB27   1201 AATGCCCACTACAAAGCGATGCCACCCGGCAACGGGGAGCTAATCGCAAA
T.flavus    1201 AATGCCCACTACAAAGCGATGCCACCCGGCAACGGGGAGCTAATCGCAAA
                          **       *


SPS14       1201 AATGCCCGCCACAAAGCGACGCAACCCGGCAACGGGAAGCCAATCGCAAA

T.ruber     1201 AATGCCTGCCACAAAGCGCTGCGACCCGGTAACGGGAAGCCAATCGCACAA

D.rad       1201 AATGGATAGGACAACGCGCAGCAAACATGTGAGTGTAAGCGAATCGCTGA
```

(continued)

status for *T. thermophilus,* has led to a tendency to name all isolates as strains of the genus *Thermus* sp. The alternative of attributing all strains to the type species *T. aquaticus* in the absence of any proof that they belong to that species, and sometimes even when there is clear evidence to the contrary, is to be discouraged. No taxonomic significance should be attached to experimental results obtained from such loosely, or improperly, named strains until the true taxonomic position of them can be clarified. *T. ruber* strains, which are easily distinguished by their red pigment and low optimum growth temperature from other *Thermus,* have a worldwide distribution. The DNA:DNA homology is high within the red strains, but low

Table II. (*Continued*)

```
T.fil      1251 AAGGTAGGCTCAGTTCGGATTGCAGTCTGCAACTCGACTGCATGAAGTCG
Tok20A1    1251 AAGGTAGGCTCAGTTCGGATTGCAGTCTGCAACTCGACTGCATGAAGTCG
OK6A1      1251 AAGGTAGGCTCAGTTCGGATTGCAGTCTGCAACTCGACTGCATGAAGTCG
RT41A      1251 AAGGTAGGCTCAGTTCGGATTGCAGTCTGCAACTCGACTGCATGAAGTCG
T351       1251 AAGGTAGGCTCAGTTCGGATTGCAGTCTGCAACTCGACTGCATGAAGTCG
Tok8A1     1251 AAGGTAGGCTCAGTTCGGATTGCAGTCTGCAACTCGACTGCATGAAGTCG


T.br YS38  1251 AAGGTGGGCGTAGTTCGGATTGGGGTCTGCAACCCGACCCCATGAAGCCG
ZHGIA1     1251 AAGGTGGGCGTAGTTCGGATTGGGGTCTGCAACCCGACCCCATGAAGCCG
OSRAM4     1251 AAGGTGGGNGTAGTTCGGATTGGGGTCTGCAACCCGACCCCATGAAGCCG
ZFIA2      1251 AAGGTGGGCGTAGTTCGGATTGGGGTCTGCAACCCGACCCCATGAAGCCG


Vi7        1251 AAGGTGGGCGTAGTTCGGATTGGGGTCTGCAACCCGACCCCATGAAGCCG
NMX2A1     1251 AAGGTGGGCGTAGTTCGGATTGGGGTCTGCAACCCGACCCCATGAAGCCG


T.aq YT1   1251 AAGGTGGGCGTAGTTCGGATTGGGGTCTGCAACCCGACCCCATGAAGCCG
YS025      1251 AAGGTGGGCGTAGTTCGGATTGGGGTCTGCAACCCGACCCCATGAAGCCG
YS052      1251 AAGGTGGGCGTAGTTCGGATTGGGGTCTGCAACCCGACCCCATGAAGCCG


HSA1       1251 AAGGTGGGCCCAGTTCGGATTGGGGTCTGCAACCCGACCCCATGAAGCCG
T.th HB8*  1251 AAAGTGGGCCCAGTTCGGATTGGGGTCTGCAACCCGACCCCATGAAGCCG
T.th HB8   1251 AAGGTGGGCCCAGTTCGGATTGGGGTCTGCAACCCGACCCCATGAAGCCG
T.th HB27  1251 AAGGTGGGCCCAGTTCGGATTGGGGTCTGCAACCCGACCCCATGAAGCCG
T.flavus   1251 AAGGTGGGCCCAGTTCGGATTGGGGTCTGCAACCCGACCCCATGAAGCCG
                     *


SPS14      1251 AAAGCGGGCCCAGTTCGGATTGGGGTCTGCAACCCGACCCCATGAAGCCG

T.ruber    1251 AAGC-AGGCTCAGTTCGGATTGGGGTCTGCAACTCGACCCCATGAAGCCG

D.rad      1251 AACCTATCCCCAGTTCAGATCGGAGTCTGCAACTCGACTCCGTGAAGTTG
```

between them and the yellow and colorless strains (Sharp and Williams, 1988; Ruffett, 1992).

6.1. The Yellow-Pigmented and Colorless Species

Only two yellow species of the genus *Thermus* are validly described at present, but there are several others that should be recognized.

1. *T. aquaticus,* the type species of the genus, has a number of representatives among the numerical taxonomy groups 1a and 1b (Munster *et al.*, 1986), which come from the same site as the type strain (Brock and Freeze, 1969).

2. *T. filiformis* was originally described as a single strain, which has a stable filamentous morphology and an extra outer cell wall layer. There

```
                                                       ggcc
T.fil       1301 GAATCGCTAGTAATCGCGGATCAGCCATGCCGCGGTGAATACGTTCCCGG
Tok20A1     1301 GAATCGCTAGTAATCGCGGATCAGCCATGCCGCGGTGAATACGTTCCCGG
OK6A1       1301 GAATCGCTAGTAATCGCGGATCAGCCATGCCGCGGTGAATACGTTCCCGG
RT41A       1301 GAATCGCTAGTAATCGCGGATCAGCCATGCCGCGGTGAATACGTTCCCGG
T351        1301 GAATCGCTAGTAATCGCGGATCAGCCATGCCGCGGTGAATACGTTCCCGG
Tok8A1      1301 GAATCGCTAGTAATCGCGGATCAGCCATGCCGCGGTGAATACGTTCCCGG


T.br YS38   1301 GAATCGCTAGTAATCGCGGATCAGCCATGCCGCGGTGAATACGTTCCCGG
ZHGIA1      1301 GAATCGCTAGTAATCGCGGATCAGCCATGCCGCGGTGAATACGTTCCCGG
OSRAM4      1301 GAATCGCTAGTAATCGCGGATCAGCCATGCCGCGGTGAATACGTTCCCGG
ZFIA2       1301 GAATCGCTAGTAATCGCGGATCAGCCATGCCGCGGTGAATACGTTCCCGG


Vi7         1301 GAATCGCTAGTAATCGCGGATCAGCCATGCCGCGGTGAATACGTTCCCGG
NMX2A1      1301 GAATCGCTAGTAATCGCGGATCAGCCATGCCGCGGTGAATACGTTCCCGG


T.aq YT1    1301 GAATCGCTAGTAATCGCGGATCAGCCATGCCGCGGTGAATACGTTCCCGG
YS025       1301 GAATCGCTAGTAATCGCGGATCAGCCATGCCGCGGTGAATACGTTCCCGG
YS052       1301 GAATCGCTAGTAATCGCGGATCAGCCATGCCGCGGTGAATACGTTCCCGG


HSA1        1301 GAATCGCTAGTAATCGCGGATCAGCCATGCCGCGGTGAATACGTTCCCGG
T.th HB8*   1301 GAATCGCTAGTAATCGCGGATCAGCCATGCCGCGGTGAATACGTTCCCGG
T.th HB8    1301 GAATCGCTAGTAATCGCGGATCAGCCATGCCGCGGTGAATACGTTCCCGG
T.th HB27   1301 GAATCGCTAGTAATCGCGGATCAGCCATGCCGCGGTGAATACGTTCCCGG
T.flavus    1301 GAATCGCTAGTAATCGCGGATCAGCCATGCCGCGGTGAATACGTTCCCGG


SPS14       1301 GAATCGCTAGTAATCGCGGATCAGCCACGCCGCGGTGAATACGTTCCCGG


T.ruber     1301 GAATCGCTAGTAATCGCGGATCAGCCATGCCGCGGTGAATACGTTCCCGG


D.rad       1301 GAATCGCTAGTAATCGCGGGTCAGC-ATACCGCGGTGAATACGTTCCCGG
```

(continued)

seem to be no other characteristics that distinguish it clearly from other yellow strains. DNA:DNA homology shows that a number of New Zealand isolates that are not filamentous should be considered as the same species despite the inappropriateness of the species name to the morphology of all of these isolates (Georganta *et al.*, 1993). This conclusion is supported by 16s RNA sequences, because a number of strains isolated in New Zealand have 16s RNA sequences (see Table II) almost identical to that of the *T. filiformis* type strain.

3. *T. thermophilus* was validly described (Oshima and Imahori, 1971), but subsequently not included in the Approved List of Bacterial Species (1980). The type strain HB8 has a high DNA:DNA homology with "*T. caldophilus*" GK24, *T. flavus* AT62, the Icelandic strain B (Williams, 1975),

Table II. (*Continued*)

```
cggaacatgtgtgg 1390rev
                                                          ¦ -
```

T.fil	1351	GCCTTGTACACACCGCCCGTCACGCCATGGGAGCGGGTTCTACCCGAAGT
Tok20A1	1351	GCCTTGTACACACCGCCCGTCACGCCATGGGAGCGGGTTCTACCCGAAGT
OK6A1	1351	GCCTTGTACACACCGCCCGTCACGCCATGGGAGCGGGTTCTACCCGAAGT
RT41A	1351	GCCTTGTACACACCGCCCGTCACGCCATGGGAGCGGGTTCTACCCGAAGT
T351	1351	GCCTTGTACACACCGCCCGTCACGCCATGGGAGCGGGTTCTACCCGAAGT
Tok8A1	1351	GCCTTGTACACACCGCCCGTCACGCCATGGGAGCGGGTTCTACCCGAAGT

T.brYS38	1351	GCCTTGTACACACCGCCCGTCACGCCATGGGAGCGGGTTCTNNCCGAAGT
ZHGIA1	1351	GCCTTGTACACACCGCCCGTCACGCCATGGGAGCGGGTTCTACCCGAAGT
OSRAM4	1351	GNNNNNNNNNNNNNNNNNNNNNNNNNNNNNNNNNNNNNNNNNNNNNNNNNN
ZFIA2	1351	GCCTTGTACACACCGCCCGTCACGCCATGGGAGCGGGTTCTACCCGAAGT

Vi7	1351	GCCTTGTACACACCGCCCGTCACNCCATGGGAGCGGGTTCTNNCCGAAGT
NMX2A1	1351	GCCTTGTACACACCGCCCGTCACGCCATGGGAGCGGGTTCTACCCGAAGT

T.aq YT1	1351	GCCTTGTACACACCGCCCGTCACGCCATGGGAGCGGGTTCTACCCGAAGT
YS025	1351	GCCTTGTACACACCGCCCGTCACGCCATGGGAGCGGGTTCTACCCGAAGT
YS052	1351	GCCTTGTACACACCGCCCGTCACGCCATGGGAGCGGGTTCTACCCGAAGT

HSA1	1351	GCCTTGTACACACCGCCCGTCACGCCATGGGAGCGGGCTCTACCCGAAGT
T.th HB8*	1351	GCCTTGTACACACCGCCCGTCACGCCATGGGAGCGGGCTCTACCCGAAGT
T.th HB8	1351	GCCTTGTACACACCGCCCGTCACGCCATGGGNNNNNNNNNNNNNNNNNNGT
T.th HB27	1351	GCCTTGTACACACCGCCCGTCACGCCATGGGAGCGGGCTCTACCCGAAGT
T.flavus	1351	GCCTTGTACACACCGCCCGTCACGCCATGGGAGCGGGCTCTACCCGAAGT

SPS14	1351	GTCTTGTACACACCGCCCGTCACGCCATGGGAGCGGGCTCTACCCGAAGT

T.ruber	1351	GCCTTGTACACACCGCCCGTCAAGCCATGGGAGTGGGTTTTGCCTGAAGT

D.rad	1351	GCCTTGTACACACCGCCCGTCACACCATGGGAGTAGATTGCAGTTGAAAC

the halotolerant Icelandic isolate IB21 (Kristjansson *et al.*, 1986), strains RQ1 and RQ3 from the Azores (Santos *et al.*, 1989), and strains ZK1, ZK2, and ZK3 from Japan (Takase and Horikoshi, 1988, 1989). The sequences of 16SrDNA indicate that the Japanese isolate HB27 and New Zealand strain HSA1 also show that they belong to this genospecies. There seems no longer any reason to doubt that *T. thermophilus* comprises a distinct genospecies, with characteristic phenotypic properties, and a wide geographical distribution of strains.

4. *T. brockianus* (in honor of T. D. Brock). A taxon was identified by DNA:DNA homology (Williams, 1989) among the numerical taxonomy cluster 2 of strains isolated from springs in Yellowstone National Park (Munster *et al.*, 1986), where it coexists with *T. aquaticus*. The name *T.*

```
          --------V9--------------¦

T.fil       1401  CGCCGGGAGCCTAAGGGCAGGCGCCGAGGGTAGGGCTCGTGACTGGGGCG
Tok20A1     1401  CGCCGGGAGCCTAAGGGCAGGCGCCGAGGGTAGGGCTCGTGACTGGGGCG
OK6A1       1401  CGCCGGGAGCCTAAGGGCAGGCGCCGAGGGTAGGGCTCGTGACTGGGGCG
RT41A       1401  CGCCGGGAGCCTAAGGGCAGGCGCCGAGGGTAGGGCTCGTGACTGGGGCG
T351        1401  CGCCGGGAGCCTAAGGGCAGGCGCCGAGGGTAGGGCTCGTGACTGGGGCG
Tok8A1      1401  CGCCGGGAGCCTAAGGGCAGGCGCCGAGGGTAGGGCTCGTGACTGGGGCG


T.br YS38   1401  CGCCGGGAGCCTTAGGGCAGGCGCCGAGGGTAGGGCCCGTGACTGGGGCG
ZHGIA1      1401  CGCCGGGAGCCTTAGGGCAGGCGCCGAGGGTAGGGCCCGTGACTGGGGCG
OSRAM4      1401  NNNNNNNNNNNNNNNNNNNNNNNNNNNNNNNNNNNNNNNNNNNNNNNNNNN
ZFIA2       1401  CGCCGGGAGCCTTAGGGCAGGCGCCGAGGGTAGGGCTCGTGACTGGGGCG


Vi7         1401  CGCCGGGANCCTTAGGGCAGGCGCCGAGGGTNNGGCTCGTGACTGGGGCG
NMX2A1      1401  CGCCGGGAGCCTTAGGGCAGGCGCCGAGGGTAGGGCTCGTGACTGGGGCG


T.aq YT1    1401  CGCCGGGAGCCTTAGGGCAGGCGCCGAGGGTAGGGCCCGTGACTGGGGCG
YS025       1401  CGCCGGGAGCCTTAGGGCAGGCGCCGAGGGTAGGGCCCGTGACTGGGGCG
YS052       1401  CGCCGGGAGCCTTAGGGCAGGCGCCGAGGGTAGGGCCCGTGACTGGGGCG


HSA1        1401  CGCCGGGAGCCTGCGGGCAGGCGCCGAGGGTAGGGCCCGTGACTGGGGCG
T.th HB8*   1401  CGCCGGGAGCCTACGGGCAGGCGCCGAGGGTAGGGCCCGTGACTGGGGCG
T.th HB8    1401  CGCCGGGAGCCTACGGGCA--CGCCGAGGGTAGGGCCCGTGACTGGGGCG
T.th HB27   1401  CGCCGGGAGCCTACGGGCAGGCGCCGAGGGTAGGGCCCGTGACTGGGGCG
T.flavus    1401  CGCCGGGAGCCTACGGGCAGGCGCCGAGGGTAGGGCCCGTGACTGGGGCG
                             *        **

SPS14       1401  CGCCCACCCCTACAGGG-AGGCGCCCAGGGTAGGGCTCGTGACTGGGGCG

T.ruber     1401  CGCCGGGAGCC-ACAGGCAGGCGCCTAGGGTAAGGCTCATGACTGGGGCT

D.rad       1401  CGCCGGGAGCCTCACGGCAGGCGTCTAGACTGTGGTTTATGACTGGGGTG
```

(continued)

brockianus seems preferable to "*T. brockii*" for this taxon, to avoid possible confusion of abbreviated names with *Thermoanaerobium brockii* (Zeikus *et al.*, 1979). Phenotypic characters distinguishing *T. brockianus* from *T. aquaticus* are: the former produces pale yellow spreading colonies on *Thermus* agar, and strains commonly grow on fructose and galactose, while the latter form deep yellow nonspreading colonies and generally degrade starch, gelatin, and casein. The Icelandic strains ZHG1A1 and ZF1A1 also belong to this genospecies, as indicated by 16SrDNA sequence.

5. *T. imahorii* (in honor of K. Imahori) is the name tentatively proposed for a genospecies detected by DNA:DNA homology (Williams, 1989) that corresponds to the numerical taxonomy clusters A, B, C, D of Santos *et al.* (1989). The close similarity of 16s rDNA sequences (Table II) of Vi7

Table II. *(Continued)*

```
T.fil       1451 AAGTCGTAACAAGGTAGCTGTACCG
Tok20Al     1451 AAGTCGTAACAAGGTAGCTGTACCG
OK6Al       1451 AAGTCGTAACAAGGTAGCTGTACCG
RT41A       1451 AAGTCGTAACAAGGTAGCTGTACCG
T351        1451 AAGTCGTAACAAGGTAGCTGTACCG
Tok8Al      1451 AAGTCGTAACAAGGTAGCTGTACCG
```

```
T.br YS38   1451 AAGTCGTAACAAGGTAGCTGTACCGGAAGGTGC
ZHGIAl      1451 AAGTCGTAACAAGGTAGCTGTACCG
OSRAM4      1451 NNNNNNNNNNNNNNNNNNNNNNNNN
ZFIA2       1451 AAGTCGTAACAAGGTAGCTGTACCG
```

```
Vi7         1451 AAGTCGTAANNNNNNNNNNNNNNNNN
NMX2Al      1451 AAGTCGTAACAAGGTAGCTGTACCG
```

```
T.aq YT1    1451 AAGTCGTAACAAGGTAGCTGTACCG
YS025       1451 AAGTCGTAACAAGGTAGCTGTACCGGAAGG
YS052       1451 AAGTCGTAACAAGGTAGCTGTACCGGAAGGT
```

```
HSA1        1451 AAGTCGTAACAAGGTAGCTGTACCG
T.th HB8*   1451 AAGTCGTAACAAGGTAGCTGTACCGGAAGGTGCGGCTGGATCACCTCCTTT
T.th HB8    1451 AAGTCGTAACA-GGTAGCTGTACCGGAAGGTGCGGCTGGATCACCTCCTTT
T.th HB27   1451 AAGTCGTAACAAGGTAGCTGTACCG
T.flavus    1451 AAGTCGTAACAAGGTAGCTGTACCG
                                  *
```

```
SPS14       1451 AAGTCGTAACAAGGTAGCTGTACCGGAAGGTG
```

```
T.ruber     1451 AAGTCGTAACAAGGTAGCTGTACCG
```

```
D.rad       1451 AAGTCGTAACAAGGTAACTGTACCGGAAGGTGCGGTTGGATCACCTCCTTT
```

[a]We are grateful to Dr. T. M. Embley, who did the alignment of the strains described in Embley *et al.*. (1993) on which this table is based. The numbering is based on the sequences deposited by Saul, but the numbering of the primers follows the *E. coli* sequence numbering. The N residues indicate where the sequence has not been determined, a dash indicates a gap introduced to improve the alignment. The * indicates where there are differences between the individual sequences within a block of strains that probably represent a species.

The two signature sequences are those described by Weisburg *et al.* (1989).

The 16sRNA sequences of *Thermus* strains aligned here are *T. aquaticus*, YT1, (L09659): *T. flavus*, AT62, TTHAT62, (L09660); *Thermus* sp., HSA.1, TTHHS, (L09670); *Thermus ruber*, TTHRUBER, (L09672); *Thermus* sp., NMX2A1, TTHNMX2, (L09661); *Thermus* sp., OK6A1, TTHOK6, (L09668); *Thermus* sp., Rt41A, TTHRT4, (L09669); *Thermus filiformis*, TTHFIL1, (L09667); *Thermus* sp., T351, TTHT351, (L09671); *Thermus* sp., Tok8A1, TTHTOK8, (L09666); *Thermus* sp., Tok20A1, TTHTOXK20, (L09665); *Thermus* sp., ZFIA2, TTHZFI, (L09662); *Thermus* sp., ZHGIA1, TTHZHGI, (L09664) all determined in the laboratory of Bergquist (Saul *et al.*, 1993). The sequences for five strains analyzed by Saul *et al.* (1993), Fiji3A1 (L10067), W28 (L10068), Tok3A1 (L10069), YSPIDA1 (L10070) and ZHGIBA4 (L10071), were not accessible from Genbank at the time of writing. The sequences of *T. ruber*, TR16SRRN, (Z15059), *T. oshimai*, SPS14, TS16SRRNA, (Z15060), *T. imahorii*, Vi7 (incorrectly described as Vi7), TS16SRRNB, (Z15061), *T. brockianus*, YS38, TS16SRRNC, (Z15062), are extensive sequences of which all but the *T. ruber* are used above (Embley *et al.*, 1993).

Two sequences are aligned for *T. thermophilus*: *T. thermophilus*, HB8, TTRN16S, (X07998) (Murzina *et al.*, 1988, listed as HB8*) and *T. thermophilus*, HB8, TT16SRAA, (M26923,M26924) (Hartmann *et al.*, 1989; Hartmann and Erdmann, 1989).

T. flavus, AT62, TFLA16S, (X58341), *Thermus* sp., X1, TAQU16S1, (X58340), *T. aquaticus*, YT1, TAQU16S2, (X58343), *Thermus* sp. OSRAM4, TSPE16S, (X58344) *T. filiformis*, Wai33A1, TFIL16S, (X58345), *T. ruber*, ATCC 35948, TRUB16S, (X58346) are all partial sequences deposited by Wolters, J., and used by Bateson *et al.* (1990). Of these only the sequence OSRAM4 is aligned.

Three sequences are aligned for *T. aquaticus*: that of Saul *et al.* (1993a) and sequences YS025 and YS052 were determined at the DSM for R. J. S. These strains from Yellowstone National Park were identified as *Thermus aquaticus* by Munster *et al.* (1988).

from Portugal and NMX2A1 from New Mexico reemphasizes the inappropriateness of speculation about local distribution of species before wide studies of taxonomy and ecology have been completed.

6. *T. oshimai* (in honor of T. Oshima) is also a new genospecies detected by DNA:DNA homology (Williams, 1989), which includes the isolates from Sao Pedro do Sul in Portugal and corresponds to the numerical taxonomy clusters E and F of Santos *et al.* (1989). The species also includes JK66 and JK91 isolates from Iceland.

6.2. The Taxonomic Position of *Thermus ruber*

The red-pigmented strains of the species *T. ruber* are rather homogeneous in their properties. *T. ruber* strains are easily distinguished from other *Thermus* by their red pigment and lower growth temperature optima. They all prefer to grow at temperatures lower than any of the yellow or colorless isolates, and have been isolated from worldwide sources. They also have a higher proportion of iso-C15 than iso-C17 fatty acids in the cellular lipids (Donato *et al.*, 1991). They are closely related genetically as indicated by DNA:DNA homology (Sharp and Williams, 1988; Ruffett, 1992), high within the red strains, but low with yellow and colorless strains (Sharp and Williams, 1988; see section 7.2.). The 16s RNA sequence is distinct from those of the other species of the genus, and the low thermophilic bias of this macromolecule is significant (Embley *et al.*, 1993) in rooting the tree that confirms the relationship between *Thermus* and *Dienococcus*. The phenotypic distinctness of low DNA:DNA homology of these species with the other *Thermus* species and together with comparative 16s RNA data indicate that these strains should be transferred to a new genus.

REFERENCES

Ado, Y., Kawamoto, T., Masunaga, I., Takayama, K., Takasawa, S., and Kimura, K., 1982, Production of 1-malic acid with immobilized thermophilic bacterium *Thermus rubens nov. sp, Enz. Eng.* **6**:303–304.

Alfredsson, G. A., Baldursson, S., and Kristjansson, J. K., 1985, Nutritional diversity among *Thermus spp.* isolated from Icelandic hot springs, *Systemat. Appl. Microbiol.* **6**:308–311.

Alfredsson, G. A., Kristjansson, J. K., Hjorleifsdottir, S., and Stetter, K. O., 1988, *Rhodothermus marinus, gen. nov., sp. nov.,* a thermophilic, halophilic bacterium from submarine hot springs in Iceland, *J. Gen. Microbiol.* **134**, 299–306.

Baldursson, S., and Kristjansson, K., 1990, Analysis of nitrate in food extracts using a thermostable formate-linked nitrate reductase enzyme system, *Biotechnol. Tech.* **4**: 211–214.

Becker, R. J., and Starzyk, J. J., 1984, Morphology and rotund body formation in *Thermus aquaticus, Microbios.* **41**:s–129.

Berenguer, J., Faraldo, M. L., and de Pedro, M. A., 1988, Ca2+ stabilized oligomeric protein complexes are major components of the cell envelope of *"Thermus thermophilus"* HB8, *J. Bacteriol.* **170:**2441–2447.

Breter, H.-J., and Zahn, R. K., 1973, A rapid separation of the four major deoxynucleosides and deoxyinosine by high-pressure cation-exchange liquid chromatography. *Analyt. Biochem.* **54:**346–352.

Brock, T. D., 1978, The Genus *Thermus*, in Microorganisms and Life at High Temperatures, Springer Verlag, New York, chap. 4.

Brock, T. D., and Edwards, M. R., 1970, Fine structure of *Thermus aquaticus*, an extreme thermophile, *J. Bacteriol.* **104:**509–517.

Brock, T. D., and Freeze, H., 1969, *Thermus aquaticus* gen. n. and sp. n., a nonsporulating extreme thermophile, *J. Bacteriol.* **98:**289–297.

Collins, M. D., and Jones, D., 1981, Distribution of isoprenoid quinone structural types in bacteria and their taxonomic implications. *Microbiol. Rev.* **45:**316–354.

Cometta, S., Sonnleitner, B., and Fiechter, A., 1982a, The growth behaviour of *Thermus aquaticus* in continuous culture. *Eur. J. Appl. Microbiol. Biotechnol.* **15:**69–74.

Cometta, S., Sonnleitner, B., Sidler, W., and Fiechter, A., 1982b, Population distribution of aerobic extremely thermophilic microorganisms in an Icelandic natural hot spring, *J. Appl. Microbiol. Biotechnol.* **16:**151–156.

Cossar, D., and Sharp, R. J., 1989, Loss of Pigmentation in *Thermus sp*, in *Microbiology of Extreme Environments and its Potential for Biotechnology*, da Costa, M. S., Duarte, J. C., and Williams, R. A. D., Eds, FEMS Symposium No. 49, Elsevier, London, p. 385.

Degryse, E., Glansdorff, N., and Pierard, A., 1978, A comparative analysis of extreme thermophilic bacteria belonging to the genus *Thermus*, *Arch. Microbiol.* **117:**189–196.

Denman, S., Hampson, K., Patel, B. K., 1991, Isolation of strains of *Thermus aquaticus* from the Australian Artesian Basin on a simple and rapid procedure for the preparation of their plasmids, *FEMS Microbiol. Lett.* **66:**73–78.

Donato, M. M., Seleiro, E. A., and da Costa, M. S., 1990, Polar lipid and fatty acid composition of strains of the genus *Thermus. System. Appl. Microbiol.* **13:**234–239.

Donato, M. M., Seleiro, E. A., and da Costa, M. S., 1991. Polar lipid and fatty acid composition of strains of *Thermus ruber. System. Appl. Microbiol.* **14:**235–239.

Embley, T. M., Thomas, R. H., and Williams, R. A. D., 1993, Reduced thermophilic bias in the 16s rDNA sequence from *Thermus ruber* provides further support for a relationship between *Thermus* and *Deinococcus Systemat Appl. Microbiol.* **16:**25–29.

Georganta, G., Smith, K. E., and Williams, R. A. D., 1993, DNA:DNA homology and cellular components of *Thermus filiformis* and other strains of *Thermus* from New Zealand hot springs. *FEMS Microbiol Letts.* **107:**145–150.

Hartmann, R. K., Wolters, J., Kroger, B., Schultze, S., Specht, T., and Erdmann, V. A., 1989, Does *Thermus* represent another deep eubacterial branching? *Systemat. Appl. Microbiol.* **11:**243–249.

Hensel, R., Demharter, W., Kandler, O., Kroppenstedt, R. M., and Stackebrandt, E., 1986, Chemotaxonomic and molecular-genetic studies of the genus *Thermus*: evidence for a phylogenetic relationships of *Thermus aquaticus* and *Thermus ruber* to the genus *Deinococcus, Internat. J. Systemat. Bacteriol.* **36:**444–453.

Hudson, J. A., Morgan, H. W., and Daniel, R. M., 1986, A numerical classification of some Thermus isolates, *J. Gen. Micro.* **132:**531–540.

Hudson, J. A., Morgan, H. W., and Daniel, R. M., 1987a, Numerical classification of some *Thermus* isolates from Icelandic hot springs, *Systemat. Appl. Microbiol.* **9:**218–223.

Hudson, J. A., Morgan, H. W., and Daniel, R. M., 1987b, *Thermus filiformis* sp. nov., a filamentous caldoactive bacterium, *Internat. J. Systemat. Bacteriol.* **37:**431–436.

Hudson, J. A., Morgan, H. W., and Daniel, R. M., 1989, Numerical classification of *Thermus* isolates from globally distributed hot springs, *Systemat. Appl. Microbiol.* **11:**250–256.

Ip, C. Y., Ha, D., Morris, P. W., Puttemans, M. L., and Venton, D. L., 1985, Separation of nucleosides and nucleotides by reversed-phase high-performance liquid chromatography with volatile buffers allowing sample recovery *Analyt. Biochem.* **147:**180–185.

Kraepelin, G., and Gravenstein, H. U., 1980, Experimental induction of rotund bodies in *Thermus aquaticus, Zeitschr. Allgem. Mikrobiol.* **20:**33–45.

Kristjansson, J. K., and Alfredsson, G. A., 1983, Distribution of *Thermus spp.* in Icelandic hot springs and a thermal gradient, *Appl. Environ. Microbiol.* **45:**1785–1789.

Kristjansson, J. K., Hreggvidsson, G. O., and Alfredsson, G. A., 1986, Isolation of halotolerant *Thermus spp.* from submarine hot springs in Iceland, *Appl. Environ. Microbiol.* **52:**1313–1316.

Loginova, L. G., and Egorova, L. A., 1976, An obligatory thermophilic bacterium *Thermus ruber* from hot springs in Kamchatka, *Mikrobiologiya* **44:**593–597.

Loginova, L. G., Egorova, L. A., Golovacheva, R. S., Seregina, L. M., 1984, *Thermus ruber sp. nov., nom. rev. Internat. J. Systemat. Bacteriol.* **34:**498–499.

Munster, M. J., Munster, A. P., Woodrow, J. R., and Sharp, R. J., 1986, Isolation and preliminary taxonomic studies of *Thermus* strains isolated from Yellowstone National Park, USA, *J. Gen. Microbiol.* **132:** 1677–1683.

Murzina, N. V., Vorozheykina, D. P., and Matvienko, N. I., 1988, Nucleotide sequence of *Thermus thermophilus* HB8 gene coding 16S rRNA, *Nucleic Acids Res.* **16:**8172.

Neefs, J.-M., Van de Peer, Y., De Rijk, P., Goris, A., and De Wachter, R., 1991, Compilation of small ribosomal subunit RNA sequences. *Nucleic Acids Res.* **19:**Suppl., 1987–2015.

Nordstrom, K. M., and Laakso, S. V., 1992, Effect of growth temperature on fatty acid composition of ten *Thermus* strains, *Appl. Environ. Microbiol.* **58:**1656–1660.

Olsen, G. J., Larsen, N., and Woese, C. R., 1991, The ribosomal RNA data base project. *Nucleic Acids Res.* **19:**Suppl., 2017–2021.

Oshima, M., and Yamakawa, T., 1974, Chemical structure of a novel glycolipid from an extreme thermophile, *Flavobacterium thermophilum, Biochemistry* **13:**1140–1146.

Oshima, T., and Imahori, K., 1971, Isolation of an extreme thermophile and thermostability of its transfer ribonucleic acid and ribosomes, *J. Gen. Appl. Microbiol.* **17:**513–517.

Oshima, T., and Imahori, K., 1974, Description of *Thermus thermophilus* (Yoshida and Oshima) comb.nov., a nonsporulating thermophilic bacterium from a Japanese thermal spa, *Internat. J. Systemat. Bacteriol.* **24:**102–112.

Oshima, M., 1978, Structure and function of membrane lipids in thermophilic bacteria, in: *Biochemistry of Thermophily* (S. M. Friedman ed.), Academic Press, New York, pp. 1–10.

Owen, R. J., and Hill, L. R., 1979, The estimation of base composition, base pairing and genome sizes of bacterial DNA. pp. 277–296 in S. A. B. Technical Series No. 14, *Identification methods for microbiologists.* F. A. Skinner, D. W. Lovelock (Eds) Academic, London.

Pask-Hughes, R. A., and Shaw, N., 1982, Glycolipids from some extreme thermophilic bacteria belonging to the genus *Thermus, J. Bacteriol.* **149:**54–58.

Pask-Hughes, R., and Williams, R. A. D., 1975, Extremely thermophilic gram-negative bacteria for hot tap water, *J. Gen. Microbiol.* **88:**321–328.

Pask-Hughes, R. A., and Williams, R. A. D., 1977, Yellow-pigmented strains of *Thermus* spp. from Icelandic hot springs, *J. Gen. Microbiol.* **102:**375–383.

Pask-Hughes, R. A., and Williams, R. A. D., 1978, Cell envelope components of strains belonging to the genus *Thermus, J. Gen. Microbiol.* **107:**65–72.

Peckova, M., 1990, Occurrence of hyperthermophile microorganisms belonging to the genus *Thermus* in Carlsbads vridlo, *Biologia,* **45:**219–228.

Prado, A., da Costa, M. S., and Madeira, V. M. C., 1988, Effect of growth temperature on the lipid composition of two strains of *Thermus sp.*, *J. Gen. Microbiol.* **134:**1653–1660.

Ray, P. H., White, D. C., and Brock, T. D., 1971a, Effect of temperature on the fatty acid composition of *Thermus aquaticus*, *J. Bacteriol.* **106:**25–30.

Ruffett, M., 1992, *Taxonomy of pink-pigmented strains of Thermus*, M. Phil thesis. CNAA, Nottingham Trent University.

Ruffett, M., Hammond, S., Williams, R. A. D., and Sharp, R. J., 1992, A taxonomic study of red pigmented gram negative thermophiles, Abstracts of Thermophiles:Science and Technology, Reykjavik, Iceland.

Saiki, T., Kimura, R., and Arima, K., 1972, Isolation and characterization of extremely thermophilic bacteria from hot springs, *Agric. Biol. Chem.* **36:**2357–2366.

Santos, M. A., Williams, R. A. D., and da Costa, M. S., 1989, Numerical taxonomy of *Thermus* isolates from hot springs in Portugal, *Systemat. Appl. Microbiol.* **12:**310–315.

Saul, D. J., Rodrigo, A. G., Reeves, R. A., Williams, L. C., Borges, K. M., Morgan, H. W., and Bergquist, P. L., 1993a, Phylogeny of twenty *Thermus* isolates constructed from 16S rRNA gene sequence data, *Int. J. Syst. Bacteriol.* **43(4):**754–760.

Saul, D. J., Reeves, R., Williams, L. C., Ridrigo, A., Morgan, H., and Bergquist, P. L., 1993b, GENBANK sequences, L09660–L09672.

Sharp, R. J., and Williams, R. A. D., 1988, Properties of *Thermus ruber* strains isolated from Icelandic hot springs and DNA:DNA Homology of *Thermus ruber* and *Thermus aquaticus*, *Appl. Environ. Microbiol.* **54:**2049–2053.

Sonnleitner, B., Cometta, S., and Feichter, A., 1982, Growth kinetics of *Thermus thermophilus*, *Eur. J. Appl. Microbiol. Biotechnol.* **54:**2049–2053.

Taguchi, H., Yamashita, M., Matsuzawa, H., and Ohta, T., 1982, Heat stable and fructose 1,6-biphosphate activated L-lactate dehydrogenase from an extremely thermophilic bacterium, *J. Biochem.* **91:**3–1348.

Takase, M., and Horikoshi, K., 1988, A thermostable β-glucosidase isolated from a bacterial species of the genus *Thermus*. *Appl. Microbiol. Technol.* **29:**55–60.

Takase, M., and Horikoshi, K., 1989, Purification and properties of a β-glucosidase from *Thermus sp.*, Z1. *Agric. Biol. Chem.* **53:**559–560.

Williams, R. A. D., 1975, Caldoactive and thermophilic bacteria and their thermostable proteins, *Sci. Prog.* **62:**373–393.

Williams, R. A. D., 1989, Biochemical taxonomy of the genus *Thermus*, in *Microbiology of Extreme Environments and its Potential for Biotechnology*, da Costa, M. S., Duarte, J. C., and Williams, R. A. D., Eds., FEMS Symposium No. 49, Troia, Portugal, September 18–23, 1988. Elsevier, London, p. 82.

Zeikus, J. G., Hegge, P. W., and Anderson, M. A., 1979, *Thermoanaerobium brockii* gen. nov. and sp. nov., a new chemoorganotrophic caldoactive, anaerobic bacterium, *Arch. Microbiol.* **122:**41–48.

Zillig, W., Stetter, K. O., Wunderl, S., Schultz, W., Preiss, H., and Scholz, I., 1980, The *Sulfolobus-Calderiella* group: taxonomy on the basis of the structure of DNA-dependent RNA polymerases. *Arch. Microbiol.* **125:**259–269.

Ecology, Distribution, and Isolation of *Thermus*

2

GUDNI A. ALFREDSSON
and JAKOB K. KRISTJANSSON

1. ECOLOGY AND DISTRIBUTION

1.1. Characteristics of the Thermal Environment

Temperature is one of the most important environmental factors determining the physiological activities of organisms and their evolution. High temperatures can be tolerated to varying degrees by different organisms. Many complex multicellular organisms are unable to withstand a temperature of 50 °C, even for very short periods, whereas many microorganisms survive, and even thrive, in much higher temperatures for extended periods.

Certain biological processes in prokaryotes seem to have distinct upper temperature limits. For photosynthesis, the limit seems to be 73–75 °C. Species diversity is very limited at higher temperatures and this reflects the limitation that high temperatures impose upon the adaptation and evolution of organisms (Brock, 1978).

Areas with geothermal springs are widely found on the earth's surface, but the "hot spots" where thermal environments are abundant are restricted to areas such as the USA (Yellowstone National Park), USSR, Iceland, Italy, New Zealand, and Japan. Hot springs and fumaroles are often associated with past or present volcanic activity. The temperature of hot springs varies greatly, as does the chemical composition of the water. The main chemical and physical characteristics can be determined using conventional measuring techniques but some adaptations may be neces-

GUDNI A. ALFREDSSON and JAKOB K. KRISTJANSSON • University of Iceland, Institute of Biology, Microbiology Laboratory, Ármúli 1A IS-108 Reykjavik, Iceland

Thermus Species, edited by Richard Sharp and Ralph Williams. Plenum Press, New York, 1995.

sary. Thus changes in the environment can be monitored, recorded, and possibly related to differences in the microbial populations observed in similar niches.

The pH values of hot springs fall into two major ranges: acid springs (pH 2–4) and alkaline springs (pH 8–10). Springs with water close to neutral pH seem to be less common. The acidic springs are characterized by high sulfuric acid content and the alkaline springs by high bicarbonate/carbonate and silica content. Many studies of hot spring chemistry have shown that the nature of each spring can differ widely from that of other springs close by. Although many springs have relatively constant water temperature and water level, in some places the ground water of the surrounding soil affects the springs. Earthquakes may change the water flow to the extent that hot springs may disappear and then reappear again. Variations in the volcanic activity and drilling in a geothermal area (for surveying or water utilization) can also greatly affect the water temperature of hot springs.

1.2. Different Thermal Systems

Some thermal environments of the earth are natural, others are man-made. Natural systems such as geothermal environments (e.g., hot springs, hot pools, steam holes, thermal gradients) are often fairly stable, but solar heated systems (e.g., soil or shallow water pools receiving direct solar heating) are generally unstable and often show great diurnal temperature variations. In some man-made thermal systems heating or combustion takes place (e.g., coal refuse piles, self-heated hay stacks, silage, and piled organic waste materials such as compost heaps). Hot effluent water from power stations or industrial plants and from geothermal heating systems are also man-made thermal environments.

The natural geothermal environment is unusual in many respects, showing great diversity as a habitat. In hot springs the water temperature and pH covers a wide range (30–100 °C and pH 1–11) (Waring, 1965; Brock, 1978; Kristjansson and Alfredsson, 1983). The chemical composition of hot spring water is also very variable, and quite different in sulfur-rich acidic mud pools compared with that of clear neutral or alkaline springs. Some clear-water springs have a very low silica content, whereas others have very high concentrations, evident by extensive and beautiful deposits of siliceous minerals where spring water flows over the edges of the hot spring basin.

The geothermal environment is often constant in temperature and flow at the source, but the outflow water may change drastically. However, no hot spring is completely constant and temperature variations within

5–10 °C may result in fluctuations in the solubility of organic and inorganic compounds, including oxygen. In hot springs where the water flows away from the source along a channel, a thermal gradient is established. The temperature of the water is reduced as the water flows away from the source and many factors affect the rate of cooling of the water. These are mainly the volume of water, the water temperature at the source, the weather (mainly air temperature and wind speed) and the depth and width of the outflow channel (Brock, 1978; Kristjansson and Alfredsson, 1983).

1.3. Habitats of *Thermus*

It is most likely that the primary habitat of *Thermus* is the hot spring ecosystem, since this is an ancient thermal environment widely distributed on the earth. More recently, various man-made thermal habitats have been created that also harbor these bacteria, often in considerable quantities (Brock, 1978). Brock and Boylen (1973) showed that hot water heaters could contain from less than 40 to above 2400 thermophilic bacteria per 100 ml water (MPN determination). *Thermus* strains isolated from such water heaters are often nonpigmented (see section 1.6.3.).

Detailed studies of the hot spring chemistry are lacking in most geothermal areas, particularly for components relevant to the microbiologist. Such data, if available, might shed some light on the effect of the chemistry on the distribution of *Thermus* in different hot springs. Extensive analytical data (usually obtained by geochemists interested in the utilization of water) is available for hot springs in some areas (see Table I and Brock, 1978; Höll, 1971), but data for the chemical elements of biological significance is often lacking. While these analytical data are of some interest, microbiologists need additional data on nitrogen, phosphorus, organic carbon, dissolved gases, sulfide, etc. (Brock, 1978). Furthermore, the water that has been analyzed sometimes comes from bore holes and may not be representative of the hot springs at the surface. Often the pH is lower and the sulfide content higher in the source than at the surface.

Geology may be important in determining the phenotypes of *Thermus* that predominate in different geographical areas (Hudson *et al.*, 1987a). A particular hot spring chemistry may be characteristic for a particular thermal region. The average pH of Icelandic hot pools yielding *Thermus* strains is higher (pH 8.54) than for the pools in New Zealand containing *Thermus*. There were also differences in the mean temperature of hot pools containing *Thermus* sp. in these two countries, although the differences were small and thought adequate to explain the difference in phenotypes that was observed (Hudson *et al.*, 1987a).

Table I. Composition of Icelandic Hot Springwater

	1963 Hveragerdi mg/L[a]	1968 Hveragerdi mg/L[a]	1970 Hveravellir mg/L[a]	1983 Hveragerdi mg/L[b]
Sodium	158.4	195.0	141.5	180.0
Lithium	0.42	—	0.5	—
Potassium	10.1	13.1	16.6	12.0
Calcium	1.4	0.3	0.88	2.8
Magnesium	1.8	0.65	8.51	<0.01
Ammonium	0.07	0.02	0.05	—
Iron	—	—	0.03	0.06
Fluoride	24	1.75	2.6	2.15
Chloride	172.9	189.5	70.9	166.0
Sulphate	2.5	31.0	139.9	78.7
Bicarbonate	146.4	173.7	143.4	55.8
Phosphate	0.003	—	0.1	—
Silicic acid	182.0	208.0	253.5	—
Boric acid	3.9	3.9	—	—

[a]Höll, 1971.
[b]Stainthorpe, 1986.

1.4. Sources of Organic Compounds for *Thermus*

As far as is known *Thermus* sp. are strictly heterotrophic and require sources of organic substrates for their growth. They are probably the most common heterotrophs in the temperature range 70–75 °C in natural thermal systems. Few data are available on the possible sources of organic nutrients for *Thermus* in nature. More studies are required on the sources of nutrients for *Thermus*, and their utilization in natural and man-made systems.

1.4.1. Hot Spring Ecosystems

1.4.1a. The Algal-Bacterial Mat

The decomposing algal-bacterial mat at the air interface of many hot springs is probably a major source of organic substrates for *Thermus*. The growth is thick and extensive, and the floating mat acts as an insulator, allowing the growth of organisms with lower temperature optimal at the surface. Important components of such mats are microorganisms such as *Phormidium*, *Mastigocladus*, *Chloroflexus*, and *Synechococcus*. These photosynthetic microorganisms are probably the most important primary producers of organic matter in the hot spring ecosystem up to the temperature limit of photosynthesis—73–75 °C (Brock, 1978). It is possible that the living algal-bacterial mat may release organic substances that are used for growth by heterotrophs like *Thermus* (Daft, 1988). It is not clear which organisms are the primary producers in the hot spring ecosystem at temperatures above the limit of photosynthesis (see below).

1.4.1b. Heterotrophs as Sources of Nutrients

In hot springs both attached and free-floating heterotrophs die and decompose in the water and may thus become organic substrates for *Thermus* (Stramer and Starzyk, 1978). Such organisms include the aerobes like *Bacillus* and *Thermus* and the anaerobes *Clostridium* and *Thermoanaerobium*.

1.4.1c. Chemoautotrophs as Sources of Nutrients

Many types of chemoautotrophic bacteria are known to exist in hot spring ecosystems. The species diversity and ecology in alkaline and neutral hot springs has received little study. These bacteria are probably the only primary producers growing above the limit of photosynthesis (73–75 °C). When the chemoautotrophs die and lyse, their cellular constituents provide organic substrates for use by heterotrophs (Stramer and Starzyk, 1978). Chemoautotrophic thermophilic bacteria such as the hydrogen oxidizers (Aragno, 1992; Kristjansson *et al.*, 1985; Kawasumi *et al.*, 1984), thermophilic methanogens and other thermophilic archaea (Stetter, 1986; Kristjansson and Stetter, 1992) provide a source of nutrients for heterotrophs. The main inorganic chemical energy sources that are available to chemoautotrophs in hot springs are hydrogen and hydrogen sulfide, which, together with carbon dioxide, are the main components in the volcanic gas released in geothermal areas.

1.4.1d. Influx of Nutrients from Surrounding Soil and Vegetation

Mesophiles, both animal and plant, and their products such as the leaves of trees, both decomposed and fresh, can be carried into the hot springs and their effluent channels by wind and water. Organic material may also wash into thermal springs from the adjacent soil. The importance of the soil and vegetation surrounding the hot springs depends on the type of soil and its texture and richness in organic materials. The rate of influx of organic compounds would probably be dependent on the amount of rainfall in the area. Hot springs often have clouds of insects and small snails associated with the algal-bacterial mats that also contribute their excreta and bodies to the organic content of the water.

1.4.2. Man-Made Systems

1.4.2a. Geothermal Power Plants and Heating Systems. *Thermus* has been isolated from industrial plants (Marteinsson and Kristjansson, 1991; Hjörleifsdóttir, 1984). It is unlikely that organic material in these systems originates from surrounding vegetation or soil, as the geothermal water usually comes from deep bore holes or large, clear, hot water springs. Photosynthetic primary producers cannot grow in these closed dark environments so the most likely primary producers of nutrients in this habitat are the thermophilic chemoautotrophs.

1.4.2b. Domestic and Industrial Water Heaters. *Thermus* has been found in both domestic and industrial heating systems (Brock and Boylen, 1973; Pask-Hughes and Williams, 1975; Stramer and Starzyk, 1981). Water heaters that take in cold water may receive considerable amounts of organic matter that can serve as nutrients for *Thermus* and other heterotrophs. Thermophilic chemoautotrophs that grow in these heaters can also provide organic substrates.

1.5. *Thermus* and Oligotrophy

No research seems to be available on the nutrition of *Thermus* to determine whether they are oligotrophic. Determinations of dissolved organic carbon (DOC) or total organic carbon (TOC) for hot spring ecosystems seem to be rare. For Firehole Pool, Yellowstone (Brock, 1981) the total organic carbon was estimated as only 2ppm or 2 µg/liter (Stahl *et al.*, 1985). This low concentration was thought sufficient to support a reasonable standing crop of bacteria, especially since no grazers consume the biomass at the prevalent high temperatures (72–73 °C).

Of the terms used to describe oligotrophic bacteria, those of Poindexter (1981) seem to be clear and logical. The copiotrophs are able to grow on high concentrations of organic carbon and are distinguished from the oligotrophic bacteria, which have two subgroups. Obligate oligotrophs are rare in nature and can only grow on low concentrations of carbon ($<$1–6 mg carbon l^{-1}). The facultative oligotrophs that can grow at both low and high concentrations of carbon are more common. However, definitions and groupings for oligotrophs are still difficult since little is known about their physiology, abundance and distribution in nature. Most research has been on oligotrophs from cold marine and fresh water environments. The open ocean water contains very little organic material (DOC 0.1–1 mg carbon l^{-1}), whereas coastal water and some fresh waters usually have much higher concentrations [DOC 1.5–15 mg carbon l^{-1} (Fry, 1990)].

Thermus strains should probably be regarded as oligotrophic because they occur in nutrient-limited environments and grow best on media with rather low nutrient concentrations. They should be classified as facultative oligotrophs because they are not restricted to low nutrient conditions. In particular, *T. thermophilus* strains can grow in media rich enough to be inhibitory to other *Thermus* strains.

To the best of our knowledge, no minimum value has been determined for the Ks concentration of a substrate required for growth of *Thermus* (Fry, 1990). We therefore have no clear indications of the oligotrophic potential of these bacteria. For oligotrophs the Ks values are always low, usually in the range of about 1–150 µg carbon l^{-1}. However, many *Thermus* strains grow on glutamate and the K_m for glutamate transport by *T. thermophilus* strain B is surprisingly high at 23mM, corresponding to 3.38 g glutamate l^{-1} or 1.38 g carbon l^{-1} (Holtom *et al.*, 1993). More research on these aspects of *Thermus* growth and nutrient uptake appears to be necessary.

Agar plates with low concentrations of organic material are regarded as the best procedure for the isolation of oligotrophs from marine and fresh water environments, and for distinguishing oligotrophs from copiotrophs (Fry, 1990).

The determination of the DOC in the hot spring water is important to the understanding of *Thermus* ecology and nutrition. Such measurements would provide important new, and microbiologically valuable data, that may have a bearing on the heterotrophic populations in the hot spring water. Reliable measurements for the assimilable organic carbon (AOC) for hot springs would be more valuable, but these estimations are fairly recent developments in water chemistry. To our knowledge such AOC assays have not been applied to hot spring water. Fry (1990) has recently reviewed oligotrophic microorganisms, with discussion of the significance of the TOC, DOC, and AOC. The different methods for estimating AOC concentrations in potable water are all bioassay methods that use either mixed natural populations, or pure cultures of facultative oligotrophs, as inocula. By employing appropriate testing techniques measurements of AOC are possible over a very wide range of concentrations (Fry, 1990). Presumably a new assay technique would have to be developed for the hot spring water and *Thermus* used as inoculum.

1.6. Effects of Some Environmental Factors

1.6.1. Temperature and pH

The temperature and pH values of the sources from which *Thermus* has been isolated are quite varied. Brock and Freeze (1969) originally isolated *Thermus* in Yellowstone National Park, USA, from hot springs with temperatures of 53–86 °C and pH values of 8–9. In Iceland the strains were isolated from sources from 64–84 °C and pH 7.2–8.9 (Williams, 1975; Pask-Hughes and Williams, 1977). More recently *Thermus* strains were isolated from sites in Yellowstone with temperatures of 30–90 °C and pH values of 6–10.5 (Munster *et al.*, 1985, 1986). In extensive studies, *Thermus* strains were found in Icelandic sources (Fig. 1) with temperatures of 55–85 °C and a pH range of 6.5–10 or more (Kristjansson and Alfredsson, 1983; Hudson *et al.*, 1987a). In Portugal *Thermus* occurs in hot springs and other hot sources in the temperature range of 50–85 °C, with pH values in the

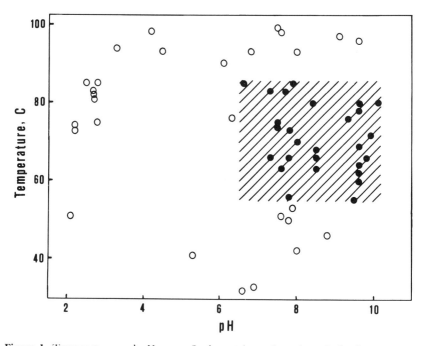

Figure 1. Temperature and pH range for hot springs of southern Iceland, Symbols: ●, *Thermus*-containing samples; ○, *Thermus*-free samples. The shaded area represents the pH and temperature distribution range of *Thermus* in natural hot springs. Reproduced with permission of the American Society for Microbiology, Journals Division.

range of 5.1–9.0 (Santos *et al.*, 1989). From these results it appears that *Thermus* sources have rather wide overall temperature and pH ranges from about 30 °C to around 90 °C and pH 5.1–10.5.

It is doubtful whether *Thermus* can exist for extended periods at or above 90 °C or pH 10.5, but it would survive transiently at these extremes. *Thermus* growing in milder conditions at the edges of pools or effluents or in the algal-bacterial mats, may be periodically washed into more extreme environments.

1.6.2. Oxygen and Nitrate

Thermus strains are thermophilic, obligately aerobic heterotrophs found in oxygenated hot waters containing some organic matter. At high temperatures, however, little oxygen is dissolved in the water and it is likely that the growth is often limited by oxygen rather than carbon. Fourteen *Thermus* strains from Yellowstone National Park (Munster *et al.*, 1986) were facultative anaerobes in the presence of nitrate and presumably nitrate reductase enabled them to use nitrate as a terminal electron acceptor. Nitrate reducing strains have also been isolated from Icelandic hot springs (Baldursson and Kristjansson, 1990; Kristjansson *et al.*, unpublished; Hardardóttir, 1987; Hollocher and Kristjansson, 1992). Hudson *et al.* (1986, 1987a, 1989) reported nitrate reduction and nitrite reduction by many strains, mainly from New Zealand, Iceland, and Yellowstone Park. Thus, the ability to reduce nitrate to nitrite seems to be widespread. Many strains may reduce the nitrite further, but more research is required for confirmation.

Some strains of *Thermus* form pellicles at the surface of static cultures. Oxygen is accessible at the surface of the culture while the deeper layers soon become anaerobic as growth proceeds. *Thermus* can be isolated from man-made systems such as geothermal municipal heating systems in which the dissolved oxygen is usually low. Special precautions are often taken to ensure that dissolved oxygen is minimized in piped water, in order to avoid corrosion of radiators. Usually, sulfite is added to make the water anaerobic, but *Thermus* can be isolated from such reduced water (Marteinsson and Kristjansson, 1991; Hjörleifsdóttir, 1984). The question of survival and possibly of growth in such "oxygen free" environments at 75–80 °C have not yet been addressed. It is possible that the strains living under these conditions can utilize nitrate for anaerobic respiration (Munster *et al.*, 1986). Whether nitrate reducing strains are more common in these closed anaerobic systems than in natural open hot springs has not, to our knowledge, been tested.

Munster *et al.* (1986) reported that some strains of *Thermus* grew anaerobically in *Thermus* broth medium and Santos *et al.* (1989) have isolated strains that could grow anaerobically in *Thermus* medium. One of nine strains in cluster E was reported to grow anaerobically without nitrate, as were strains RQ-1 and RQ-3. This anaerobic growth without nitrate is unusual and requires further study.

1.6.3. Sulfide and Light

Thermus strains have been isolated from thermal environments with high and low sulfide content, but little is known about the sulfide tolerance of *Thermus* or how sulfide affects the growth, survival, and distribution of *Thermus* in nature. Ramaley and Bitzinger (1975) did not find *Thermus* spp. in high-sulfide springs (40 mg of H_2S per liter) in Colorado. Kristjansson and Alfredsson (1983) measured the sulfide content and counted *Thermus* in water samples from hot springs in Iceland and from a thermal gradient. High concentrations of sulfide did not affect the number of *Thermus* very much. The sulfide content in the thermal gradient decreased rapidly with increased distance from the source. Sulfide undergoes rapid chemical aerobic oxidation in such gradients, and oxidation by some microorganisms may also reduce its concentration (Kristjansson and Alfredsson, 1983), (Fig. 2).

No direct quantitative measurements of the effect of light on *Thermus* seem to have been made. Brock (1984) pointed out that the carotenoid pigments of *Thermus* strains may be photoprotective. Ramaley and Hixon (1970) showed that nonpigmented strains are common and Brock and Boylen (1973) described isolates from hot water heaters that were nonpigmented. If the pigments of *Thermus* have a photoprotective role, then most strains isolated from natural sources that are exposed to sunlight should be pigmented. Those from closed and dark hot water heaters and heating systems might more often be nonpigmented. However, the source of the water for such heaters and heating systems may influence the frequency of pigmented *Thermus* strains there. If the heating systems receive geothermal water then pigmented isolates may be quite common (Hjörleifsdóttir, 1984; Marteinsson and Kristjansson, 1991). The actual colony color of *Thermus* isolates may also depend on the media used for their culture.

Thermus YS-45 growing in chemostat culture under certain nutrient limitations and in low light formed a reduced amount of pigment (Cossar and Sharp, 1989). They concluded that, at least under these nutrient limitations, pigment production was a response to light, although the nature of the response (i.e., a direct or a secondary response) was not clear.

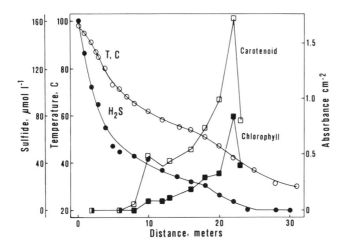

Figure 2. Properties of the thermal gradient in spring HV-11. ○, temperature; ●, sulfide; □, absorbance at 478 nm/cm^2; ■, absorbance at 662 nm/cm^2. The absorbances at 478 and 662 nm in the acetone extract represent chlorophyll and carotenoid, respectively. Reproduced with permission of the American Society for Microbiology, Journals Division.

Nutrient-limited cells may use all the available nutrients for growth, and therefore pigment production may decline. In complex nutrient-rich media no such limitation is imposed and pigment production may occur freely.

1.6.4. The Effect of Salinity

Most strains of *Thermus* have been isolated from terrestrial hot springs that are neutral or alkaline with low salinity. A few terrestrial hot springs contain some NaCl originating from seawater [Reykjanes in northwest Iceland (Kristjansson *et al.*, 1986)]. Hot spring water may go through salty layers on its way to the surface. Where hot springs emerge on the seashore or below sea-level at a shallow depth (Kristjansson *et al.*, 1986; Manaia and da Costa, 1991) the salinity may be quite low, perhaps 10% or less of that of seawater.

Thermus strains from such sources have been compared to strains isolated from terrestrial springs. The marine isolates were generally more halotolerant than most of the terrestrial *Thermus* strains, both in the proportion of strains showing halotolerance, and in the growth-limiting concentration of NaCl. The maximum salt concentration in which growth occurred was 6% (Kristjansson *et al.*, 1986). There are no reports of isolations of *Thermus* from deep-sea hydrothermal vents. Hot geothermal brines occur as runoffs from geothermal power plants but such systems can hardly be called "natural" and have only existed on earth for a few decades. They have very high salinity, and often have toxic concentrations of heavy metals (Ellis and Mahon, 1977), and are therefore very unlikely habitats for living organisms. None of the halotolerant *Thermus* strains are halophilic, because growth is not stimulated by salt. This is opposite to what was found for the slightly halophilic thermophile, *Rhodothermus marinus*, which has been isolated from similar shallow water marine hot springs in Iceland and in the Azores (Alfredsson *et al.*, 1988; Nunes *et al.*, 1992). The growth of typical terrestrial strains of *Thermus* such as *T. aquaticus* YT1 is slightly inhibited, even at 0.5% NaCl, and most strains cannot grow above 1% NaCl. The halotolerant strains, however, are unaffected by 1% NaCl, can all grow at 3% and many at 5% NaCl (Kristjansson *et al.*, 1986; Manaia and da Costa, 1991), and seem very similar to their terrestrial counterparts in all other respects.

1.7. Rotund Bodies and Their Possible Function

Rotund body formation in *Thermus* is an interesting phenomenon that has been observed by several authors (Brock and Freeze, 1969; Brock and Edwards, 1970; Brock, 1984; Pask-Hughes and Williams, 1978; Golovachevea, 1977; Kraepelin and Gravenstein, 1980; Becker and Starzyk, 1984). These unusual morphological structures have been observed using phase contrast microscopy and electron microscopic methods. One type of rotund body is described as an "aggregation and association" of a number of separate cells. These are held together by part of the outer wall layer of one cell, which joins up with the same wall layer of adjacent cells. The average number of cells per rotund body was around 14 and the diameter of each rotund body was up to 20 μm (Brock and Edwards, 1970; Brock, 1984; Becker and Starzyk, 1984). Much larger rotund bodies, consisting of a greater number of cells have also been described by Becker and Starzyk (1984). This second type of rotund body is sometimes described as "vesicular" and arises from a bleb on the outer cell layer of a single *T. aquaticus* cell. These seem to be observed less frequently than the "aggregate" type (Kraepelin and Gravenstein, 1980; Becker and Starzyk, 1984). Rotund bodies do not seem to be formed by *Thermus* cultures on solid media

(Becker and Starzyk, 1984; Golovachevea, 1977) and it is possible that these structures are only required and formed in liquid cultures. Rotund bodies have not been observed in *Thermus* in specimens taken directly from hot springs, although they have probably not been sought.

Whether rotund bodies are of ecological or survival significance for *Thermus* in nature, or whether some physiological significance be attached to these morphological peculiarities, is not known. Golovachevea (1977) suggested that the "complex spherical bodies" of *Thermus ruber* may act as floating bladders to retain the bacteria in the aerated surface layer of the water or culture. Becker and Starzyk (1984) have pointed out that since a single rotund body (often containing many cells) gives rise to a single colony on solid media, these bodies could affect the outcome of viable counts and may explain the apparently extended stationary phase some-times seen for *T. aquaticus* (Becker and Starzyk, 1984). Rotund bodies are formed in the logarithmic phase and the early stationary phase, and viable counts determined then would underestimate the number of individual viable cells. Later in the stationary phase, when the free viable cells de-crease (i.e., cells die), the reduction in the viable count may be counter-acted by the dissociation of rotund bodies to release individual cells. Thus the viable count could remain constant, or even rise slightly, so giving an apparently extended stationary phase of the culture. If rotund bodies are formed in the hot spring environment, then they may have some survival value for *Thermus*. It is hoped that these speculations will lead to further work in this field.

1.8. Distribution of *Thermus*

Thermus sp. are widely distributed in natural hot springs all over the world and are probably the most common heterotrophs in the tempera-ture range 70–75 °C in such natural systems. It has also been widely documented that *Thermus* can be isolated from all natural and man-made systems at this temperature if the pH is above 5–6.

Thermus can also be isolated from rivers and lakes which receive run-offs from hot springs or other thermal systems. To demonstrate the pre-sence of *Thermus* far away from thermal sources may be difficult as it is clear that the numbers will be very low (Hjörleifsdóttir *et al.*, 1989).

Kristjansson and Alfredsson (1983) showed the wide distribution of *Thermus* in Icelandic hot springs with broad temperature and pH ranges, but emphasized that the *Thermus* isolates were not necessarily growing in all these sites. Alfredsson *et al.* (1985) also showed a considerable nutri-tional diversity among the Icelandic *Thermus* strains. This has been supported by studies of others on Icelandic *Thermus* strains (Hudson *et al.*, 1987a, 1989).

Recently, *Thermus*-like bacteria have been isolated from some hot deep sea vents (Daniel Prieur, personal communication).

2. ISOLATION OF *THERMUS*

2.1. Introduction

The name *Thermus* was given to isolates from hot springs in Yellowstone National Park, USA, although strains from Japan had already been described as *Flavobacterium thermophilum*. Pools yielding *Thermus* isolates had temperatures between 53 and 86 °C and pH values of 8–9 (Brock and Freeze, 1969). Several different isolation and culture media that have been used are described in Chapter 3 (Table I).

2.2. Physiological and Nutritional Characteristics

Bacteria belonging to the genus *Thermus* are Gram negative, heterotrophic, nonmotile, and rodshaped and live aerobically (usually as obligate aerobes but see 1.6.2) at neutral pH and grow well at temperatures over 70 °C (Brock, 1981). *T. ruber* requires the lower temperature of 60 °C for optimal growth. The habitats where *Thermus* is usually found have low concentrations of organic matter. Thus, dilute nutrient media consisting of mineral salts together with low concentrations of organic nutrients such as tryptone and yeast extract (0.1–0.25%) seem to be best suited for growing *Thermus*.

T. aquaticus was found to have no special growth factor requirement and could grow on mineral salts media with NH_4^+ as a nitrogen source and with some single carbon sources like acetate, sucrose, citrate, succinate or glucose. However, the growth was slower in these media than in a complex medium. Nitrate was not used as nitrogen source by this strain and 0.1% tryptone and 0.1% yeast extract could not support growth unless the basal salts were present in the medium (Brock, 1978). For *T. aquaticus* the optimum pH for growth was 7.5–7.8. No growth occurred above pH 9.5 or below 6.0 (Brock, 1981).

2.3. Sampling and Enumeration

No special sampling techniques have been developed for *Thermus*. Normal aseptic techniques are necessary since thermophiles are widespread and sampling apparatus can easily become contaminated. A normal 15 min autoclaving at 121 °C is usually not sufficient, since spores of many thermophilic *Bacillus* can easily survive this procedure. Autoclaving may be extended to 30 min and 131 °C. The samples should be kept refrigerated if possible and processed as soon as possible. No quantitative study has been done concerning the survival of *Thermus* in samples, but storage time of more than a week is not advisable.

No special methods for the enumeration of *Thermus* have been developed or recommended. Marteinsson and Kristjansson (1991) used three different media and an MPN method for the enumeration of thermophilic bacteria. They used the filter technique for isolation. Their main conclusions were that the medium 162 of Degryse *et al.* (1978) seemed to be best for both *Thermus* and *Bacillus*, whereas medium R$_2$A seemed to be selective for *Thermus ruber*. The MPN method using liquid medium 162 always gave much higher numbers for viable thermophilic bacteria than did the filter-and-agar method.

2.4. Culture Media and Isolation

2.4.1. General Considerations

No single medium is suitable for isolation of all the heterotrophic thermophiles. The medium selected depends on the type of microorganisms to be studied. The media components used in cultivation of mesophiles may be suitable for thermophiles, especially if their physiological characters are similar. However, some special points must be kept in mind when media for thermophiles are being prepared. The first is that the bicarbonate/CO$_2$ ratio in the medium may change drastically at higher temperatures (above 50 °C) as CO$_2$ is released from the bicarbonate, causing the pH to increase. The second point concerns the choice of solidifying agent for plates (see below) and the need to seal plates to prevent evaporation.

2.4.2. Enrichment in Liquid Culture

No special enrichment procedure seems to be necessary for *Thermus*. However, liquid media with a suitable mineral salts mixture, low concentration of organic constituents, a pH between 7 and 8 and incubation at 70–75 °C are all important factors in ensuring good growth of *Thermus*, whether pigmented or colorless. Sometimes 80 °C has been used for incubation but strains isolated at that temperature usually grow better at 70–75 °C (Brock and Freeze, 1969). *T. ruber* requires a lower incubation temperature, usually 55–65 °C and optimum temperature for growth at 60 °C (Sharp and Williams, 1988; Loginova et al., 1984). T. thermophilus is reported to grow up to 85 °C and has optimum temperature between 65 and 72 °C (Oshima and Imahori, 1974), and is not inhibited by concentrations of organic matter that prevent the growth of other Thermus species. Thus the emphasis in this procedure is on two points: low nutrient concentration and high temperature of incubation. This higher temperature tends to suppress the growth of most Bacillus strains that often dominate in cultures at lower temperatures.

The liquid isolation media for T. aquaticus and most other Thermus strains usually contains 0.1–0.3% tryptone and 0.1% yeast extract. However, most of these bacteria will grow at concentration up to 1.0% tryptone and yeast extract. Brock showed that the growth of T. aquaticus was inhibited at 1% tryptone and yeast extract (Brock, 1981; Brock and Freeze, 1969). It is not uncommon finding that aquatic bacteria are inhibited by high concentration of organic matter as some of them may be oligotrophic. T. thermophilus grows well in higher concentrations of organic compounds such as 1% polypeptone plus 1% yeast extract. The standard medium used by the Japanese for T. thermophilus is 0.8% polypeptone, 0.4% yeast extract, 0.2% sodium chloride. This has been used by other workers with several strains of this genospecies. However, a strange contrast is provided by the statement that T. thermophilus was inhibited by meat extract as low as 0.07% at very low steady state biomass concentration, but not carbon-limited, in continuous culture (Sonnleitner et al., 1982). This unexplained result also conflicts with the finding (McKay et al., 1982) that the same strain of T. thermophilus achieved carbon-limited growth on 6mM (0.108%) glucose.

Degryse et al. (1978) reported that Thermus strains could be selectively isolated by aerobic incubation at 65 °C in their medium 162 supplemented with a single organic compound, such as malate or acetate. This medium had low nitrogen concentration in the form of NH_4Cl or glutamate.

2.4.3. Enrichment by "in Situ" Methods

Recently a new "in situ" direct enrichment method has been developed for isolation of thermophiles that can tolerate or utilize certain chemicals. This method is based on the notion that thermophiles with special characteristics can be enriched under natural conditions (e.g., in hot spring effluent) by introducing new factors into the existing "natural" environment. Slow and gradual (rather than sudden and drastic) changes allow the microorganisms to adapt to new environmental factors and grow in their presence.

Ordinary laboratory enrichment media presumably prove much

harsher for the microorganisms when they are placed in an "unnatural" medium where the conditions are often very different from the "natural" environment. It is therefore likely that the "*in situ*" method when used as a first stage or primary enrichment can be valuable in selecting new microorganisms, especially from environments where no natural or manmade selective factors exist. Such "*in situ*" enrichment methods have recently been used to isolate thermophilic acetate-utilizing bacteria from a hot spring. A cloth-swab was placed in a small hot spring pond at 70 °C and the concentrated enrichment solution allowed to flow very slowly into the swab through a narrow tubing at the rate of 0.7–1ml/min. This created high concentration in the swab, but in the pond the concentration was only about 2mM (Konradsdóttir *et al.*, 1991; Kristjansson, 1989). In this study mostly Gram-positive microorganisms were obtained, but two *Thermus*-like bacterial strains were isolated, so presumably similar methods could be used to isolate *Thermus* strains with other specific and useful characteristics (Baldursson and Kristjansson, 1990). *Thermus* strains resistant to high concentrations of metal salt have also been obtained by "*in situ*" enrichment. A bijou bottle containing a saturated solution of the salt was placed inside a larger bottle containing tryptone and yeast extract, both bottles with perforated screw caps. The whole was placed in a hot spring for several days until growth could be seen in the outer bottle. Plating on TYE agar with the appropriate metal salt gave colonies of resistant *Thermus* (personal communication, R. A. D. Williams).

2.4.4. Solid Media for Isolation

Agar has been the most commonly used solidifying agent for growth and isolation of *Thermus*. Normally agar can only be used up to 72 °C. After prolonged incubation at higher temperatures it cracks or melts. The evaporation from the agar medium can also become a severe problem. These problems are overcome by the use of 8–10 mm thick agar medium in the plates and agar concentration of 2.5–3%. After drying the plates well, they should be stacked between two uninoculated plates and placed in a double layer of plastic bags. In this way ordinary plastic petri dishes can be incubated for 5–7 days at 72 °C. With agarose plates incubation temperatures up to 89 °C can be used, but since agarose is more purified and more expensive than agar this is not practical on a routine basis. Solidifying agents such as silica gel can be used, but preparation is time-consuming and difficult and it seems to be unsuitable for certain extreme thermophiles. Starch has also been tried but it is susceptible to enzymic hydrolysis and its use is not widespread. A new solidifying agent called Gelrite has been used at higher temperatures with good results (Lin and Casida,

1984), but as it is organic compound it may be metabolized by some microorganisms. Gelrite is well suited for heterotrophic thermophiles but it is of no use for verification of autotrophic growth for which silica gel would have to be used (Wiegel, 1986). The membrane filter method (see section 2.5.3) has proved useful for the isolation and enumeration of *Thermus* and other aerobic thermophiles.

2.4.5. Minimal Agar Media

To determine the ability of *Thermus* strains to utilize different compounds as single carbon sources, various minimal media have been used. One such medium is based on the medium of Brock (1978) in which the mineral salts solution is that of Castenholz (1969), originally used for culturing algae from hot springs. This medium has been used by several workers (Pask-Hughes and Williams, 1975; Munster *et al.*, 1986; Sharp and Williams, 1988).

Another minimal medium is based on the salts solution of medium 162 of Degryse *et al.* (1978), which was used in nutritional studies of *Thermus* (Alfredsson *et al.*, 1985; Degryse *et al.*, 1978). Cometta *et al.* (1982a) used a synthetic medium of Kuhn *et al.* (1979) for testing utilization of single carbon sources. All these three minimal media have been used in numerical taxonomy studies of *Thermus* strains isolated from hot springs in various parts of the world (Cometta *et al.*, 1982b; Munster *et al.*, 1986; Hudson *et al.*, 1986, 1987a, 1989; Santos *et al.*, 1989).

Degryse *et al.* (1978) reported difficulties in using the salts medium D (Castenholz, 1969; Brock, 1981) as a base for minimal media plates intended for testing utilization of single carbon sources. This led Degryse *et al.* (1978) to design a new and improved minimal medium, named Medium 162, which had more buffering power, an important consideration for reproducible results for carbohydrate utilization. Under these growth conditions and at the high temperature of incubation, CO_2 was thought to be critical for growth initiation. In medium D with glucose the pH dropped considerably (chemical acidification), from the initial pH of 7.2 down to pH 5.5 at 65 °C after 40 hrs incubation. Both the high temperature and the drop in pH would make less CO_2 available for growth initiation on the minimal medium D (Brock, 1984). For these reasons it was considered essential to devise a new minimal medium that would give more reliable and reproducible results than minimal medium D (Degryse *et al.*, 1978).

It would certainly be useful if the research groups testing *Thermus* strains for numerical taxonomy, or for other purposes, would compare the merits of different minimal media and agree on one basic minimal medium in the future. This would make such research results more comparable, which is important for the future of the taxonomy of *Thermus*.

2.5. Growth Conditions

2.5.1. Some Basic Considerations

Isolation and cultivation of thermophiles requires that some basic considerations are taken into account. Melting or deformation of plastic petri dishes and instability of agar above 70 °C may cause problems. Gases such as oxygen have lower solubility at higher temperatures; this may be counteracted by increased aeration rate in liquid culture. Certain media components may change under prolonged high temperature incubation: caramelization of sugars, destruction of vitamins, inactivation of antibiotics, and others. Evaporation from plates and from culture vessels, especially if aeration is used, can cause problems due to the concentration of the medium. Plates have to be properly sealed in closed containers or double plastic bags. Compensation must be made for condensation lost from culture vessels. Fermenters have to be equipped with suitable condensers to reduce water loss.

All the above points can affect the outcome of the isolation and testing of *Thermus* strains and should therefore be kept in mind.

2.5.2. Incubation Conditions

For the type species *Thermus aquaticus* (strain YT-1) the optimum temperature for growth is 70 °C, the maximum is 79 °C, and the minimum is 40 °C (Brock, 1981). Many other *Thermus* strains have very similar optimum temperature and temperature ranges. The new species *Thermus filiformis*, for instance, has an optimum temperature of 73 °C, a minimum of 37 °C, and a maximum of 80 °C (Hudson *et al.*, 1987b). Two species are considerably different from those above. One is *Thermus thermophilus*, which has an optimum temperature between 65 and 72 °C, but is reported to grow up to 85 °C and has minimum growth temperature of 47 °C (Oshima and Imahori, 1974). The other is *Thermus ruber*, which grows optimally at 60 °C, has minimum growth temperature at 35–40 °C, and a maximum growth temperature at 70 °C (Loginova *et al.*, 1984; Sharp and Williams, 1988).

The pH of the two culture media most widely used for growing *Thermus* is pH 8.2 for medium D (Brock, 1978) and pH 7.2 for medium 162 (Degryse *et al.*, 1978). The medium of Oshima and Imahori (1974) has pH of 7.5, and the potato peptone medium of Loginova and Egorova (1975) has pH of 8. Thus, the pH of all these media is from about pH 7 to about pH 8, and *Thermus* normally grows best in this slightly alkaline pH range. A variety of incubation equipment can be used to cultivate thermophilic microorganisms like *Thermus*. Suitable hot air incubators are probably most commonly used both for plate cultures and liquid cultures. Such incubators are sometimes equipped with special devices like door switches to turn off the fans in order to reduce heat loss when the doors are open. For liquid cultures high temperature water baths can also be used up to 70–80 °C. These must have good lids and air-filled plastic balls, or other suitable floating material should be used to counteract the extensive evaporation. At higher temperatures (80–90 °C) it is necessary to replace the water in the bath with other liquids like glycerol. At still higher temperatures electric sandbaths with accurate temperature controls have been used. Liquid cultures in tubes can also be incubated in special electrically heated aluminum blocks (Wiegel, 1986).

Because *Thermus* is an obligate aerobe and the concentration of dissolved oxygen in liquid cultures is low in high temperature conditions, careful attention to aeration is important. In ordinary small flask cultures the best aeration is achieved by using thin liquid layer (i.e., high ratio of surface to liquid volume), baffled culture flasks, and orbital shaker for the flasks.

2.5.3. Membrane Filter Methods

The evaporation from agar plates at high temperatures and the resulting condensation of the water, which often spreads over the plate, makes direct isolation on plates difficult. It is also very common that some organisms, especially some thermophilic *Bacillus*, spread very quickly over the plate and overgrow the often slower-growing *Thermus* colonies. The method of filtering water samples containing thermophilic bacteria such as *Thermus* and *Bacillus*, and then placing the filter on the agar plates before incubation, inhibits the spreading and slows down the growth of thermophilic *Bacillus*, allowing the colonies of *Thermus* time to grow. The filter method gave higher total number of colonies than plating directly on agar and the colonies were smaller and more confined (Suzuki *et al.*, 1981). Higher numbers of more diverse strains can therefore be obtained in less time than with other isolation methods (Suzuki *et al.*, 1981; Kristjansson and Alfredsson, 1983; Kristjansson *et al.*, 1986; Hjörleifsdóttir *et al.*, 1989).

A new detection and enumeration method for *Thermus* using membrane filters and fluorescent antibody technique has recently been introduced (Cochran-Stafira and Starzyk, 1989).

ACKNOWLEDGMENTS. The research work of the authors presented in this chapter has been supported over a number of years by the Icelandic Science Foundation, the Research Fund of the University of Iceland, the Icelandic Research Council, and the Nordic Fund for Industrial Research and Development. All this support is gratefully acknowledged.

REFERENCES

Alfredsson, G. A., Baldursson, S., and Kristjansson, J. K., 1985, Nutritional diversity among *Thermus* spp. isolated from Icelandic hot springs, *Systemat. Appl. Microbiol.* **6:**308–311.

Alfredsson, G. A., Kristjansson, J. K., Hjörleifsdóttir, S., and Stetter, K. O., 1988, *Rhodothermus marinus*, gen.nov., sp.nov., a thermophilic, halophilic bacterium from submarine hot springs in Iceland, *J. Gen. Microbiol.* **134:**299–306.

Aragno, M., 1992, Aerobic, chemolithoautotrophic, thermophilic bacteria, in: *Thermophilic bacteria* (J. K. Kristjansson, ed.), CRC Press, Inc., Boca Raton, Florida, pp. 77–103.

Baldursson, S., and Kristjansson, J. K., 1990, Analysis of food extracts, using thermostable formate linked nitrate reductase system, *Biotechnol. Techniques* **4:**211–214.

Becker, R. J., and Starzyk, M. J., 1984, Morphology and rotund body formation in *Thermus aquaticus*, *Microbios.* **41:**115–129.

Brock, T. D., 1978, The genus *Thermus*, in: *Microorganisms and Life at High Temperatures* (T. D. Brock). Springer Verlag, New York, pp. 72–91.

Brock, T. D., 1981, Extreme thermophiles of the genus *Thermus* and *Sulfolobus*, in: *The Prokaryotes: A Handbook on Habitats, Isolation and Identification of Bacteria* (M. P. Starr, H. Stolp, H. G. Trüper, A. Balows, and H. G. Schlegel, eds.). Springer-Verlag, New York, pp. 978–984.

Brock, T. D., 1984, Genus *Thermus*, Brock and Freeze, 1969, 295[AL], in: *Bergey's Manual of Systematic Bacteriology*, Volume 1 (N. R. Krieg and J. G. Holt, eds.). Williams and Wilkins, Baltimore, pp. 333–337.

Brock, T. D., and Boylen, K. L., 1973, Presence of thermophilic bacteria in laundry and domestic hot-water heaters, *Appl. Microbiol.* **25:**72–76.

Brock, T. D., and Edwards, M. R., 1970, Fine structure of *Thermus aquaticus*, an extreme thermophile, *J. Bacteriol.* **104:**509–517.

Brock, T. D., and Freeze, H., 1969, *Thermus aquaticus* gen.n. and sp.n., a nonsporulating extreme thermophile, *J. Bacteriol.* **98:**289–297.

Castenholz, R. W., 1969, Thermophilic blue-green algae and the thermal environment, *Bacteriol. Reviews* **33:**476–504.

Cochran-Stafira, D. L., and Starzyk, M. J., 1989, Membrane filter-fluorescent antibody technique for the detection of the genus *Thermus* in water, *Microbios.* **60:**159–165.

Cometta, S., Sonnleitner, B., and Fiechter, A., 1982a, The growth behaviour of *Thermus aquaticus* in continuous cultivation, *Eur. J. Appl. Microbiol. Biotechnol.* **15:**69–74.

Cometta, S., Sonnleitner, B., Sidler, W., and Fiechter, A., 1982b, Population distribution of aerobic extremely thermophilic microorganisms in an Icelandic natural hot spring, *Eur. J. Appl. Microbiol. Biotechnol.* **16:**151–156.

Cossar, D., and Sharp, R. J., 1989, Loss of pigmentation in *Thermus* sp., in: *Microbiology of Extreme Environments and its Potential for Biotechnology* (M. S. da Costa, J. C. Duarte, and R. A. D. Williams, eds.), FEMS Symposium No. 49, Troia, Portugal, September 18–23, 1988. Elsevier, London, pp. 385.

Daft, M. J., 1988, Cyanobacteria: isolation, interactions and ecology, in: *Methods in Aquatic Bacteriology* (B. Austin, ed.), John Wiley & Sons, Chichester, England, pp. 241–268.

Degryse, E., Glansdorff, N., and Pierard, A., 1978, A comparative analysis of extreme thermophilic bacteria belonging to the genus *Thermus*, *Arch. Microbiol.* **117:**189–196.

Ellis, A. J., and Mahon, W. A. J., 1977, *Chemistry and Geothermal Systems*, Academic Press, New York.

Fry, J. C., 1990, Oligotrophs, in: *Microbiology of Extreme Environments* (C. Edwards, ed.), Open University Press, Milton Keynes, England, pp. 93–116.

Golovachevea, R. S., 1977, Complex spherical bodies of *Thermus ruber, Mikrobiologiya* (Trans) **46:**506–512.

Hardardóttir, F. Th., 1987, Investigation of nitrate respiration in strains of *Thermus*, B. S. research project at the Division of Biology, University of Iceland, Reykjavik.

Hjörleifsdóttir, S., 1984, Isolation of *Thermus* strains from municipal hot water in Reykjavik and growth on amino acids, B. S. research project at the Division of Biology, University of Iceland, Reykjavik.

Hjörleifsdóttir, S., Kristjansson, J. K., and Alfredsson, G. A., 1989, Thermophilic organisms in submarine freshwater hot springs in Iceland, in: *Microbiology of Extreme Environments and its Potential for Biotechnology* (M. S. da Costa, J. C. Duarte, and R. A. D. Williams, eds.), FEMS Symposium No. 49, Troia, Portugal, (September 18–23 1988), London, Elsevier, pp. 109–112.

Höll, K., 1971, Die heissen quellen und geysire Islands, ihre chemische beschaffenheit und verwendbarkeit. Research Institute Nedri As, Hveragerdi, Iceland, Bulletin No. 6.

Hollocher, T. C., and Kristjansson, J. K., 1992, Thermophilic dentrifying bacteria: A survey of hot springs in Southwestern Iceland, *FEMS Microbiol. Ecol.* **101:**113–119.

Holtom, G. J., Sharp, R. J., and Williams, R. A. D., 1993, Sodium-stimulated transport of glutamate by *Thermus thermophilus* strain B, *J. Gen. Microbiol.* **139:**2245–2250.

Hudson, J. A., Morgan, H. W., and Daniel, R. M., 1986, A numerical classification of some *Thermus* isolates, *J. Gen. Microbiol.* **132:**531–540.

Hudson, J. A., Morgan, H. W., and Daniel, R. M., 1987a, Numerical classification of some *Thermus* isolates from Icelandic hot springs, *Systemat. Appl. Microbiol.* **9:**218–223.

Hudson, J. A., Morgan, H. W., and Daniel, R. M., 1987b, *Thermus filiformis* sp.nov., a filamentous caldoactive bacterium, *Internat. J. Systemat. Bacteriol.* **37:**431–436.

Hudson, J. A., Morgan, H. W., and Daniel, R. M., 1989, Numerical classification of *Thermus* isolates from globally distributed hot springs, *Systemat. Appl. Microbiol.* **11:**250–256.

Kawasumi, T., Igarashi, Y., Kodama, T., and Minoda, Y., 1984, *Hydrogenobacter thermophilus* gen.nov., sp.nov., an extremely thermophilic, aerobic, hydrogen-oxidising bacterium, *Internat. J. Systemat. Bacteriol.* **34:**5–10.

Konradsdóttir, M., Perttula, M., Pere, J., Viikari, L., and Kristjansson, J. K., 1991, In situ enrichment of thermophilic acetate utilizing bacteria. *Systemat. Appl. Microbiol.* **14:**190–195.

Kraepelin, G., and Gravenstein, H. U., 1980, Experimental induction of rotund bodies in *Thermus aquaticus, Zeitschr. Allgem. Mikrobiol.* **20:**33–45.

Kristjansson, J. K., 1989, Thermophilic organisms as source of thermostable enzymes, *Trends Biotechnol.* **7:**349–353.

Kristjansson, J. K., and Alfredsson, G. A., 1983, Distribution of *Thermus* spp. in Icelandic hot springs and a thermal gradient, *Appl. Environ. Microbiol.* **45:**1785–1789.

Kristjansson, J. K., Ingason, A., and Alfredsson, G. A., 1985, Isolation of thermophilic obligately autotrophic hydrogen-oxidising bacteria, similar to *Hydrogenobacter thermophilus*, from Icelandic hot springs, *Arch. Microbiol.* **140:**321–325.

Kristjansson, J. K., Hreggvidsson, G. O., and Alfredsson, G. A., 1986, Isolation of halotolerant *Thermus* spp. from submarine hot springs in Iceland, *Appl. Environ. Microbiol.* **52:**1313–1316.

Kristjansson, J. K., and Stetter, K. O., 1992, Thermophilic bacteria, in: *Thermophilic bacteria* (J. K. Kristjansson, ed.). CRC Press, Inc., Boca Raton, Florida, pp. 1–18.

Kuhn, H. J., Friedrich, U., and Fiechter, A., 1979, Defined minimal medium for a thermophilic *Bacillus* sp. developed by a chemostat pulse and shift technique, *Eur. J. Appl. Microbiol. Biotechnol.* **6:**341–349.

Lin, C. C., and Casida, L. E., 1984, GELRITE as a gelling agent in media for the growth of thermophilic microorganisms, *Appl. Environ. Microbiol.* **47:**427–429.

Loginova, L. G., and Egorova, L. A., 1975, An obligately thermophilic bacterium *Thermus ruber* from hot springs in Kamchatka, *Mikrobiologiya* (Trans.) **44:**661–665.

Loginova, L. G., Egorova, L. A., Golovachevea, R. S., and Seregine, L. M., 1984, *Thermus ruber* sp.nov., nom.rev., *Internat. J. Systemat. Bacteriol.* **34:**498–499.

Manaia, C. M., and da Costa, M. S., 1991, Characterization of halotolerant *Thermus* isolates from shallow marine hot springs on S. Miguel, Azores, *J. Gen. Microbiol.* **137:**2643–2648.

Marteinsson, V. Th., and Kristjansson, J. K., 1991, Enumeration of thermophilic bacteria in district heating systems in Iceland, in: *Bakterier i varmtvannsystemer. Report no. 544*, Nordisk Ministerråd, Copenhagen, Appendix.

McKay, A., Quilter, J., and Jones, C. W., 1982, Energy conservation in the extreme thermophile *Thermus thermophilus* HB8, *Arch. Microbiol.* **131:**43–50.

Munster, M. J., Munster, A. P., and Sharp, R. J., 1985, Incidence of plasmids in *Thermus* spp. isolated in Yellowstone National Park, *Appl. Environ. Microbiol.* **50:**1325–1327.

Munster, M. J., Munster, A. P., Woodrow, J. R., and Sharp, R. J., 1986, Isolation and preliminary taxonomic studies of *Thermus* strains isolated from Yellowstone National Park, USA, *J. Gen. Microbiol.* **132:**1677–1683.

Nunes, O. C., Donato, M. M., and da Costa, M. S., 1992, Isolation and characterization of *Rhodothermus* strains from S. Miguel, Azores, *Systemat. Appl. Microbiol.* **15:**92–97.

Oshima, T., and Imahori, K., 1974, Description of *Thermus thermophilus* (Yoshida and Oshima) comb.nov., a nonsporulating thermophilic bacterium from a Japanese thermal spa, *Internat. J. Systemat. Bacteriol.* **24:**102–112.

Pask-Hughes, R., and Williams, R. A. D., 1975, Extremely thermophilic Gram-negative bacteria from hot tap water, *J. Gen. Microbiol.* **88:**321–328.

Pask-Hughes, R. A., and Williams, R. A. D., 1977, Yellow-pigmented strains of *Thermus* spp. from Icelandic hot springs, *J. Gen. Microbiol.* **102:**375–383.

Pask-Hughes, R. A., and Williams, R. A. D., 1978, Cell envelope components of strains belonging to the genus *Thermus*, *J. Gen. Microbiol.* **107:**65–72.

Poindexter, J. S., 1981, Oligotrophy: fast or famine, *Adv. Microbiol. Ecol.* **5:**63–89.

Ramaley, R. F., and Bitzinger, K., 1975, Types and distribution of obligate thermophilic bacteria in man-made and natural thermal gradients, *Appl. Microbiol.* **30:**152–155.

Ramaley, R. F., and Hixson, J., 1970, Isolation of a nonpigmented, thermophilic bacterium similar to *Thermus aquaticus*, *J. Bacteriol.* **103:**527–528.

Santos, M. A., Williams, R. A. D., and da Costa, M. S., 1989, Numerical taxonomy of *Thermus* isolates from hot springs in Portugal, *Systemat. Appl. Microbiol.* **12:**310–315.

Sharp, R. J., and Williams, R. A. D., 1988, Properties of *Thermus ruber* strains isolated from Icelandic hot springs and DNA:DNA homology of *Thermus ruber* and *Thermus aquaticus*, *Appl. Environ. Microbiol.* **54:**2049–2053.

Sonnleitner, B., Cometta, S., and Fiechter, A., 1982, Growth kinetics of *Thermus thermophilus*, *Eur. J. Appl. Microbiol. Biotechnol.* **15:** 75–82.

Stahl, D. A., Lane, D. J., Olsen, G. J., and Pace, N. R., 1985, Characterization of a Yellowstone hot spring microbial community by 5S rRNA sequences, *Appl. Environ. Microbiol.* **49**:1379–1384.

Stainthorpe, A. C., 1986, Thermophilic, anaerobic, cellulolytic bacteria from Icelandic hot springs, Ph.D. Thesis, University of London.

Stetter, K. O., 1986, Diversity of extremely thermophilic Archaebacteria, in: *Thermophiles—General, Molecular and Applied Microbiology* (T. D. Brock, ed.), Wiley–Interscience, New York, pp. 39–74.

Stramer, S. L., and Starzyk, M. J., 1978, Improved growth of *Thermus aquaticus* on cellular lysates, *Microbios* **23**:193–198.

Stramer, S. L., and Starzyk, M. J., 1981, The occurrence and survival of *Thermus aquaticus*, *Microbios* **32**:99–110.

Suzuki, T., Iijima, S., Saiki, T., and Beppu, T., 1981, Membrane filter method for isolation and characterization of thermophilic bacteria, *Agric. Biol. Chem.* **45**:2399–2400.

Waring, G. A., 1965, Thermal springs of the United States and other countries of the world. A summary. U.S. Geological Survey Professional Paper No. 492, 383 pp.

Wiegel, J., 1986, Methods for isolation and study of thermophiles, in: *Thermophiles—General, Molecular and Applied Microbiology* (T. D. Brock, ed.), Wiley–Interscience, New York, pp. 17–37.

Williams, R. A. D., 1975, Caldoactive and thermophilic bacteria and their thermostable proteins, *Sci. Prog. Oxf.* **62**:373–393.

Physiology and Metabolism of *Thermus*

3

RICHARD SHARP, DOUG COSSAR,
and RALPH WILLIAMS

1. ENZYMES OF THE MAIN METABOLIC PATHWAYS

The metabolism of thermophilic bacteria was extensively reviewed in 1979 by Ljungdahl. More recent books, monographs, and reviews that bear on the physiology of *Thermus* include those of Gould and Corry (1986), Brock (1986), Herbert and Codd (1986), Bergquist *et al.* (1987), Krulwich and Ivey (1990), Edwards (1990), Kristjansson (1992), and Herbert and Sharp (1992).

Most studies of enzymes from *Thermus* have concentrated on stability and related physical properties of the isolated proteins. It is rare that any attempt has been made to put these properties in either a physiological or an ecological context.

1.1. Glycolysis and Gluconeogenesis

The Embden-Meyerhof pathway appears functionally complete in *Thermus*, with the glyoxylate shunt and phosphoenolpyruvate carboxylase (PEPC), but not pyruvate carboxylase (PC), acting as anaplerotic pathways, and phosphoenolpyruvate carboxykinase (PEPCK) as a gluconeogenic enzyme. All enzymes of glycolysis have only been systematically measured in *T. thermophilus* strain HB8 (Yoshizaki and Imahori, 1979c), but a significant number have been purified and characterized from several strains. We

RICHARD SHARP • Centre for Applied Microbiology and Research, Porton Down, Salisbury, Wiltshire SP4 0JG, United Kingdom.   DOUG COSSAR • Cangene Corporation, 3403 American Drive, Mississauga, Ontario, Canada L4V 1T4.   RALPH WILLIAMS • Queen Mary & Westfield College, Biochemistry Department, Faculty of Basic Medical Sciences, Mile End Road, London E14 NS, United Kingdom.

Thermus Species, edited by Richard Sharp and Ralph Williams. Plenum Press, New York, 1995.

have found no reports of purification or kinetic studies on enzymes of glycolysis and gluconeogenesis, hexokinase, phosphoglucose isomerase, triose phosphate isomerase, and phosphoglycerate mutase; however, these enzymes have been assayed in extracts of *T. thermophilus* HB8.

The 6-phosphofructokinase (PFK) of *Thermus* (Yoshida *et al.*, 1971; Cass and Stellwagen, 1975; Hengartner *et al.*, 1976) exists in two forms. PFK-1 is constitutive and allosteric (Yoshida, 1972; Stellwagen and Thompson, 1979), being stimulated by monovalent cations (of which NH_4^+ and K^+ seem to be physiologically the most significant), as well as by fructose-6-phosphate and ADP. In these respects, the PFK-1 enzyme of *Thermus* resembles the mammalian PFK-1. The *Thermus* enzyme is strongly inhibited by phosphoenolpyruvate, and the PFK-1 tetramer is unusual in being reversibly dissociated into dimers by its allosteric effectors (Xu *et al.*, 1990). The PFK-1 gene has been cloned and sequenced for structural studies (Xu *et al.*, 1991b). PFK-2 is induced by glucose and is not allosteric (Xu *et al.*, 1991a). The *T. aquaticus* YT-1 aldolase is a class II enzyme, although of double the usual M_r (Freeze and Brock, 1970). Glyceraldehyde-3-phosphate dehydrogenase (Harris *et al.*, 1980; Hocking and Harris, 1973, 1980) is a tetrameric enzyme whose subunits are smaller than most enzymes. The kinetics were determined at low temperature because of the instability of the substrate. Compared with the rabbit muscle enzyme, the K_m for glyceraldehyde 3-phosphate is rather high at 0.3mM and is unaffected by salts. The K_m for NAD was 0.04mM, rising to 0.1mM NAD in 90 mM NH_4Cl; however, this would not occur under physiological conditions. This salt concentration activated the enzyme 25-fold, despite reducing the affinity for NAD (Fujita *et al.*, 1976). Phosphoglycerate kinase of strain HB8 is similar in molecular size and kinetic properties to the enzyme in eucaryotes and bacilli. The K_m values for ATP and 3-phosphoglycerate are 0.28 mM and 1.79 mM, respectively (Nojima *et al.*, 1979). The enzyme has been crystallized for structural investigation (Littlechild *et al.*, 1987; Bowen *et al.*, 1988). The enolases from strain X1 (Barnes and Stellwagen, 1973) and *T. aquaticus* YT1 are unremarkable in their properties. Pyruvate kinase (Yoshizaki and Imahori, 1979b) is allosterically activated by glucose-6-phosphate, fructose-6-phosphate, triose phosphates, and fructose 1,6-bisphosphate; but ATP is not an allosteric inhibitor. The pyruvate dehydrogenase complex (Maas *et al.*, 1992) is inducible by pyruvate, of the type II or gram-positive type, and is not controlled by reversible phosphorylation of E1 like the eucaryote enzyme, nor by allostery as in the gram-negative bacteria. There is no indication of substrate inhibition, but other 2-oxo acids, including glyoxylate, are competitive with pyruvate. The enzyme complex is inactivated *in vitro* at 70 °C, a surprising finding in a bacterium that grows at 80 °C. Lactate dehydrogenase (Taguchi *et al.*,

1982; Taguchi *et al.*, 1984; Kunai *et al.*, 1986) of strains YT1 and GK24 is allosterically regulated by fructose-1,6-bisphosphate (FBP), which reduces the K_m for both pyruvate and lactate. This allosteric effector does not change the K_m of LDH for NADH (65μM) or NAD^+ (15μM) (Machida *et al.*, 1985a, 1985b). The FBP-binding site contains two positive charges and replacement of his-188 by mutation abolishes the stimulation of LDH by FBP (Schroeder *et al.*, 1988). Despite the presence of LDH, *Thermus* does not appear to ferment glucose to lactate; this would be explained if LHD were inhibited by pyruvate, like the heart muscle form of the enzyme.

Phosphoglucomutase has been detected in strain HB8 (Yoshizaki *et al.*, 1971). We have not found any reports of other enzymes related to glycogen metabolism nor the storage of energy and carbon as reserve materials such as poly-β-hydroxybutyrate or glycogen by any *Thermus* strain.

The gluconeogenic enzyme fructose 1:6-bisphosphatase is inhibited by AMP as in mammalian muscle, and is stimulated by phosphoenolpyruvate (Yoshida and Oshima, 1971; Yoshida *et al.*, 1973). Together with phosphoenolpyruvate carboxykinase, this enzyme will allow for gluconeogenesis from oxaloacetate or any substrate that gives rise to it. PEP carboxylase requires acetyl CoA and is inhibited by citrate and ATP (Bridger and Sundaram, 1976; Sundaram and Bridger, 1979; Degryse and Glansdorff, 1981); its role is in oxaloacetate synthesis from glycolytic intermediates. The absence of pyruvate carboxylase appears to preclude gluconeogenesis from pyruvate, lactate, and alanine by the usual route (i.e., PC and PEPCK). However, glyoxylate shunt as a bypass to the CO_2-generating reactions of the tricarboxylic acid cycle may provide an alternative pathway for gluconeogenesis from 2 of the three-carbon atoms of these substrates. One mole of pyruvate gives rise to one mole of phosphoenolpyruvate with a substantial production of energy from reduced coenzymes. PEP carboxykinase favors the synthesis of phosphoenolpyruvate from oxaloacetate for gluconeogenesis. The K_m for bicarbonate is 100 mM—too high for the enzyme to make oxaloacetate physiologically (Yoshizaki and Imahori, 1979a). PEP carboxykinase is not inducible and its activity is not affected by any of the metabolites or coenzymes tested.

Enzymes of the pentose phosphate pathway in *Thermus*, and glucose-6-phosphate dehydrogenase have not been studied (Yoshizaki and Imahori, 1979c).

1.2. Tricarboxylic Acid and Glyoxylate Cycles

Metabolic studies of *T. aquaticus* YT1 and four Icelandic strains (Pask-Hughes and Williams, 1977) included the assay for isocitrate lyase and malate synthase of the glyoxylate bypass and all the enzymes of the TCA cycle except succinate thiokinase. These studies agreed with preliminary results (Degryse and Glansdorff, 1976) for the Belgian isolate Z05, leading to a comprehensive study (Degryse and Glansdorff, 1981) in which the glyoxylate shunt was shown to be essential for growth on pyruvate and acetate. The succinate thiokinase of *Thermus* was subsequently shown to kinetically resemble that of pseudomonads (Weitzman and Jaskowska-Hodges, 1982). Extensive structural studies have been carried out on some of the enzymes of the TCA pathway, especially malate and isocitrate de-hydrogenases (see Chap. 4), but these studies did not elucidate the phys-iology of *Thermus*. Citrate synthase, aconitase, fumarase, and the dehydro-genases were assayed in *T. thermophilus* HB8 by Yoshizaki and Imahori (1979c). *Thermus rubens* has an unusually high activity of fumarase, and immobilized cells of this patent strain have been used to manufacture malate (Ado, 1982).

1.3. Amino Acid Metabolism

An amidase from strain YT1, specifically hydrolyzes asparagine, but it has a high K_m (8.6mM) increased to 20mM during inhibition of the enzyme by 20mM L-aspartate (Curran *et al.*, 1985). In the media commonly used to grow *Thermus*, the asparagine concentration would be below the K_m, and aspartate would also be present so that hydrolysis of asparagine would be limited.

Thermostable amino acid dehydrogenases have been purified from several thermophiles for use in biotechnological applications (Ohshima and Soda, 1989). An alanine dehydrogenase depending on NAD was found to be highly active in *T. thermophilus* HB8 (Vali *et al.*, 1980). Homo-serine dehydrogenase converts aspartate semialdehyde to homoserine in the biosynthetic pathway of methionine, threonine, and isoleucine. The enzyme from *T. flavus* AT62 is not inhibited by any of these amino acids, nor by lysine, alone or in combination, but it is inhibited by cysteine (Saiki *et al.*, 1973). All *Thermus* strains tested (Holtom *et al.*, 1989; Holtom, 1991) contained glutamate dehydrogenase (NAD), although the activity was low in a strain that did not grow on glutamate alone. Glutamate-oxaloacetate, glutamate-pyruvate, and glutamine-α-oxoglutarate (NAD) aminotrans-ferases are present in all strains. Alanine aminotransferase (Walker and Wang, 1993) of *T. aquaticus* is unusually thermostable ($t_{0.5}$ of inactivation about 6 h at 100 °C), and has a temperature optimum above 95 °C.

Threonine deaminase from strain X1 (Higa and Ramaley, 1973) also catalyses the deamination of serine, and is not inhibited by isoleucine. The

Thermus enzyme differs from *B. stearothermophilus*, but resembles that of *Rhodopseudomonas* and *Rhodospirillum*.

Acetohydroxy acid synthetase catalyses the synthesis of both α-acetolactate for valine synthesis and α-acetohydroxybutyrate for isoleucine synthesis. The enzyme is noncompetitively inhibited by valine, but is represented only when leucine, isoleucine, and valine are all present in the growth medium (Chin and Trela, 1973).

Two operons for the biosynthesis of amino acids have been investigated in relation to gene control and the use of these genes as markers in cloning vectors. Of the leu operon genes: *leuA*, isopropylmalate synthetase; *leuB*, isopropylmalate dehydrogenase; *leuC* and *leuD* (both required for isopropylmalate isomerase activity), all but *leuA* were cloned together on one plasmid (Tanaka *et al.*, 1981; Croft *et al.*, 1987; Kirino and Oshima, 1991; see Chapter 7). The isopropylmalate dehydrogenase gene has been cloned separately and its protein purified (Nagahari *et al.*, 1980; Tanaka *et al.*, 1981). The *trpB* gene of *T. thermophilus* overlaps the *trpA* gene (Sato *et al.*, 1988; Koyama and Furukawa, 1990). The pathways of biosynthesis of these amino acids seem not to be unusual.

The pathway for arginine synthesis from glutamate (Cunin *et al.*, 1986) involves recycling of acetyl groups from N-acetyl-ornithine to glutamate by ornithine acetyl transferase and also a glutamine-dependent carbamoyl phosphate transferase (Degryse *et al.*, 1976) and is otherwise unremarkable.

2. UTILIZATION OF SUBSTRATES FOR GROWTH AND ENERGY

2.1. Amino Acids

"*T. thermophilus*" HB8 was reported to require lysine for growth on agar plates (Degryse *et al.*, 1978), and yet was apparently prototrophic for lysine in continuous culture (Sonnleitner *et al.*, 1982), "*T. flavus*" AT62 [actually a strain of "*T. thermophilus*" see below] is auxotrophic for glutamate (Saiki *et al.*, 1972) and many *Thermus* strains utilize glutamate as a sole source of carbon.

T. thermophilus strain B, growing aerobically in continuous culture at 60 °C in a defined medium on glutamate (Holtom, 1991), was dependent on sodium for growth, and no glutamate uptake took place without sodium. The apparent K_m for sodium was approximately 23mM. In hot springs in Yellowstone National Park (Brock, 1978) and Iceland (Stainthorpe and Williams, unpublished) concentrations of sodium are 2–20mM, so the low affinity is surprising, and in the natural environment such a system may operate suboptimally. The uptake of glutamate is by sodium-symport, dependent on the proton motive force (Holtom *et al.*, 1993). Glutamate uptake is sensitive to collapse of both the electrochemical gradient ($\Delta\Psi$) and the chemical gradient of sodium ions (ΔpNa). Transport follows the artificial imposition of ΔpNa and is inhibited by compounds which cause its collapse. The apparent K_m for transport was 0.23μM glutamate. At pH 7.6 uptake increases up to 75 °C, but is negligible at 83 °C, which correlates with the maximum growth temperature of the strain ($T_{max} = 82$ °C), indicating that destabilization of the membrane may determine the upper limit for growth. At high pH the growth of *Thermus* B requires higher sodium concentrations and also yields more CO_2 per unit biomass compared with growth at or close to neutrality. More studies are required to clarify the sources and transport mechanisms for nutrients in *Thermus*.

2.2. Carbohydrates

Several common monosaccharides are used as single carbon sources by many strains of *Thermus* in the presence of yeast extract, but pentoses are not usually metabolized at all (Munster *et al.*, 1986; Hudson *et al.*, 1987; Alfredsson *et al.*, 1985). Although lactate dehydrogenase is active, and many strains of *Thermus* can utilize monosaccharides aerobically, fermentation of carbohydrates, such as glucose, does not seem to occur. Where disaccharides are utilized, the constituent monosaccharides may not necessarily be good substrates for growth. Thus, most strains from Iceland use sucrose and maltose, while only two use are reported to glucose, and none grow on fructose (Alfredsson *et al.*, 1985).

2.3. Carboxylic Acids

Most strains grow on acetate and pyruvate (Alfredsson *et al.*, 1985), but red-pigmented strains are generally unable to grow on these two substrates (Sharp and Williams, 1988). In this respect it is important to note that the glyoxylate bypass has not been demonstrated in *Thermus ruber*. On the other hand, most *Thermus* strains cannot grow on the carboxylic acids of the TCA cycle. Growth of strain ZO5 on pyruvate is stimulated by glucose and vice versa (Degryse and Glansdorf, 1981). Pyruvate induces pyruvate dehydrogenase (Maas *et al.*, 1992) in various media, but the effect was greatest in the semisynthetic M162 medium (Degryse *et al.*, 1978).

3. ENERGETICS

3.1. Electron Acceptors

The bacteria of the genus *Thermus* are aerobic obligate heterotrophes, using oxygen as the terminal electron acceptor. Nitrate is not reduced by *T. aquaticus, T. filiformis,* nor most *T. ruber* strains examined, but serves as a terminal electron acceptor for many strains other than these three species, several of which also reduce nitrite (Munster *et al.*, 1986; Hudson *et al.*, 1987; Brock, 1978). Moreover some yellow-pigmented strains grow anaerobically in the presence, but not in the absence, of nitrate. No *Thermus* strain has yet been shown to be capable of fermentation.

3.2. The Electron Transport Chain

The electron transport chain in *Thermus* strain T351 comprised membrane-bound NADH and succinate oxidases (Hickey and Daniel, 1979), with cytochromes a, b, o, and c in the membranes and two c-type cytochromes in the cytosol. Electron transport was sensitive to inhibitors including cyanide, azide, amytal, rotenone, and antimycin A. The respiratory chain of "*T. thermophilus*" was described as comprising NADH dehydrogenase, the menaquinone MK-8 and cytochromes b, c, aa_3, and o (McKay *et al.*, 1982). These authors found that the molar growth yield for glucose to be low compared with metabolically similar mesophiles, and suggested that this is due to the high permeability of the membrane to protons. The respiratory proteins of *T. thermophilus* have been reviewed by Fee *et al.* (1986). A soluble NADH dehydrogenase, $M_r = 50$ kDa from strain T351, is inhibited by physiological concentrations of ATP. The ATP is competitive with NADH (Walsh *et al.*, 1983). A soluble NADH oxidase from YT1 reduces oxygen in the purified form, without the need of cytochromes (Cocco *et al.*, 1988). Two forms of this enzyme occur in strain HB8 (Yagi *et al.*, 1988), but the physiological significance of this has not been elucidated. The Reiske Fe-S protein is reduced by the quinone-requiring NADH dehydrogenase. Three cytochromes *c* (Fee *et al.*, 1986) comprise soluble c-552 in the periplasmic space, c-555,549 in the membrane and c-549 or c_1 associated with aa_3. The single unit complex of aa_3 is four component cytochrome *c* oxidase containing Cu_A and Cu_B. Cytochrome *o*, a b-type CO-binding pigment, predominates over aa_3 at low oxygen concentrations (Fee *et al.*, 1986), contains two copper ions (Zimmerman *et al.*, 1988) and functions as an alternative oxidase.

Nitrate can act as a terminal electron acceptor in a significant number of strains. Strain M10 (Baldursson and Kristjansson, 1990) has been used for the production of formate-dependent nitrate reductase, which is formed in static but not aerobic shake-flask cultures, and best of all, in the presence of azide (Baldursson and Kristjansson, personal communication).

3.3. ATP Synthetase

A membrane ATPase of *T. thermophilus* HB8 seemed to be of the V-type, like the archaebacterial and eucaryotic vacuolar ATPases (Yokoyama *et al.*, 1990), but unlike the ATPases of other bacteria, mitochondria and chloroplasts. Features of the *Thermus* enzyme that characterized it as V-type were the large α-subunit and insensitivity to azide but sensitivity to nitrate inhibition. It has not been proved that this ATPase is responsible for ATP synthesis *in vivo* but it seems likely. Zakharov and Kuz'mina (1992) disagreed with these findings. The ATP-synthase of *T. thermophilus* was found to catalyze P_i-ATP exchange in liposome membranes, and the catalytic part (F1) consisted of four subunits with M_r of 69, 56, 30, and 20 kDa. The membrane part (F0) contained 13, 12, and 9 kDa polypeptides, of which the last was a DCCD-binding protein that forms a stable oligomeric complex (Mr = 68 kDa) resistant to SDS. The Mg^{2+}-dependent ATPase activity was rapidly lost during the reaction but was restored by high concentrations of sulfite. In the presence of sulfite, ATP hydrolysis was suppressed by nitrate but remained sensitive to azide. Without sulfite, nitrate had no inhibiting effect, whereas azide accelerated the enzyme inactivation. The analysis of this inhibition data did not support the relatedness of the *T. thermophilus* ATP-synthase to that of the archea.

3.4. Defenses against Oxygen Toxicity

Oxygen concentration in water at 70 °C is low, so oxygen toxicity might not be expected to be a risk for *Thermus*, but reactions that generate the superoxide radical anion, hydroxyl radical, and hydrogen peroxide are favored at high temperatures. Superoxide dismutase, catalase, and peroxidase were all active in *Thermus aquaticus* YT1, but at high oxygen tensions, peroxidase was reported (Allgood and Perry, 1986) induced by less than twofold and catalase by threefold. Paraquat induced catalase tenfold and was considered the major protective enzyme. In agreement, MacMichael (1988) found that changing from static to aerated culture resulted in induction of catalase by threefold, peroxidase twofold, and SOD by 70% by the time growth began again, after which the specific activity of catalase alone continued to increase.

4. MICROBIAL GROWTH

4.1. Medium Development and Growth Requirements

Thermus isolations have generally been carried out using complex media with protein hydrolysates and yeast extract. Growth has been examined on a wide range of substrates supplemented with low concentration of ammonium salts or yeast extract. Many carbohydrates, amino acids, some organic acids, carboxylic acids, peptides, and proteins have been utilized for growth (Table I). Aromatic substrates, alkanes, and C1 compounds are not metabolized (Degryse *et al.*, 1978).

Many strains are able to grow on a basal salts medium with ammonium as nitrogen source and an appropriate carbon source, although many require a vitamin supplement (Sharp and Williams, 1988; Alfredsson *et al.*, 1985).

Nutritional diversity has been observed with many substrates used by many strains (Pask-Hughes and Williams, 1977; Santos *et al.*, 1989; Alfredsson *et al.*, 1985; Williams and Sharp, 1988; Ruffett *et al.*, 1992). Hudson *et al.* (1989) reported substrate utilization by strains correlated with their geographical distribution, but this may only be apparent. Some strains are inhibited by high concentrations of organic substrates. It is possible that certain strains are selected by the chemical composition of the water and the availability of nutrients in particular hot springs.

Brock (1981) considered that the onset of inhibition by organic material occurred at 10 gl^{-1}. However, there is considerable variation between different species. A linear increase in biomass yield was observed for *Thermus* by Cometta with increasing concentrations of meat extract and peptone up to 42 gl^{-1}. Decreasing biomass production occurred above this concentration with no growth observed above 100 gl^{-1} (Cometta *et al.*, 1982).

T. thermophilus inhibition was reported to occur in chemostats at a level of 0.7 gl^{-1} of organic carbon, a release from substrate inhibition was observed after diluting medium with cell free filtrates from stationary cultures of *T. thermophilus* (Sonnleitner *et al.*, 1982). No explanation is available for this finding.

Thermus flavus strains AT62 and AT61 were reported auxotrophic for glutamic and aspartic acid, or aspartic acid, isoleucine, proline, biotin, folic acid, and p-amino benzoic acid (Saiki *et al.*, 1972). Brock and Freeze (1969) reported *T. aquaticus* to be prototrophic, but defined medium usually includes a number of amino acids and vitamins. Kenkel and Trela (1979) used mineral salts medium supplemented with biotin, nicotinic acid, and thiamine in which glutamic acid served as both a carbon and nitrogen source. *Thermus ruber* requires tyrosine, tryptophan, histidine, and aspartic

Table I. Media for the Growth of *Thermus* Species

Medium for T. aquaticus, originally proposed by Brock and Freeze (1969) (Ramaley and Hixson, 1970)

Tryptone	1 gm/l^{-1}
Yeast extract	1 gm/l^{-1}
Castenholz salts	100 mls/l^{-1}
Castenholz basal salts (Castenholz, 1969)	10× concentrate
Nitrilo triacetic acid	1.0 g
CaSO$_4$ (2H$_2$O)	0.6 g
MgSO$_4$ (7H$_2$O)	1.0 g
NaCl	0.08 g
KNO$_3$	1.03 g
NaNO$_3$	6.89 g
Na$_2$HPO$_4$	1.11 g
FeCl$_3$ soln (0.28 gl^{-1})	10.0 ml
Nitsch's trace elements soln	10.0 ml

Nitsch's trace elements soln is composed of

H$_2$SO$_4$	0.5 ml/l^{-1}
MnSO$_4$ H$_2$O	2.2 g/l^{-1}
ZnSO$_4$ 7H$_2$O	0.5 g/l^{-1}
H$_3$BO$_3$	0.5 g/l^{-1}
CuSO$_4$	0.016 g/l^{-1}
Na$_2$MoO$_4$ 2H$_2$O	0.025 g/l^{-1}
CoCl2 6H$_2$O	0.046 g/l^{-1}

The 10× concentrate of salts was adjusted to pH 8.2 with 1 N NaOH. Stock solutions of 1% tryptone and 3% yeast extract were autoclaved separately and added to give a final concentration of 0.1%. Final pH was adjusted to 7.6 (Ramaley and Hixson, 1972).

Medium for T. aquaticus YT-1 (Kaledin et al., 1990)

Tryptone	3.0 g/l^{-1}
Yeast extract	1.0 g/l^{-1}
Glutamic acid	3.0 g/l^{-1}
Castenholz basal salts	1000 ml

Medium for T. caldophilus GK24 (Taguchi et al., 1983)

Casamino acid	4.0 g/l^{-1}
Yeast extract	4.0 g/l^{-1}
Trytophan	0.2 g/l^{-1}
(NH$_4$)$_2$ SO$_4$	1.0 g/l^{-1}
Glucose	2.0 g/l^{-1}
K$_2$HPO$_4$	6.5 g/l^{-1}
K H$_2$PO$_4$	1.0 g/l^{-1}
Castenholz basal salts	1000 ml

Medium for Thermus strain B (Koh, 1985)

Difco Bacto tryptone	2 g/l^{-1}
Difco Bacto yeast extract	2 g/l^{-1}
Castenholz basal salts	1000 ml

pH adjusted to 7.6

Minimal medium for T. aquaticus YT-1 (Verhoeven et al., 1986)

Glutamic acid	3.0 g/l^{-1}
Biotin	0.1 mg/l^{-1}
Thiamine	0.1 mg/l^{-1}
Nicotinic acid	0.05 mg/l^{-1}
Castenholz basal salts	1000 ml/s

Medium 162 for growth of Thermus sp. (Degryse et al., 1978)

10× concentrate

Nitrilotriacetic acid	1.0 g/l^{-1}
Nitsch's trace elements	0.5 ml/l^{-1}
Fe citrate 0.01 M soln	5.0 ml/l^{-1}
CaSO$_4$ 2H$_2$O	0.4 g/l^{-1}
MgCl$_2$ 6H$_2$O	2.0 g/l^{-1}

Before use the following additions are made to the ten fold diluted medium.

NaHPO$_4$ 12H$_2$O (0.2 M)	60 ml
KH$_2$PO$_4$ (0.2 M)	20 ml

The pH is 7.2

NH$_4$Cl at 0.01 M final concentration used as nitrogen source.
Carbon sources used include glucose at 0.4%.

Medium for growth of T. rubens, T. lacteus, T. aquaticus and T. thermophilus (Ado et al., 1988)

Polypeptone	8.0 g/l^{-1}
Yeast extract	4.0 g/l^{-1}

pH adjusted to 7.2

Medium for T. flavus (Kaledin et al., 1981)

Peptone	20.0 g/l^{-1}
Yeast extract	2.0 g/l^{-1}
Beef extract	1.0 g/l^{-1}
Na lactate	2.5 g/l^{-1}
K succinate	0.5 g/l^{-1}
Glutamic acid	1.0 g/l^{-1}
K$_2$HPO$_4$	0.522 g/l^{-1}
MgSO$_4$	0.12 g/l^{-1}
CaCl$_2$	0.056 g/l^{-}
NaCl	0.5 g/l^{-1}
Tris	50 mM

pH adjusted to 8.0

Medium for T. flavus (Aleksandrushking and Egorova, 1978)

Potato decoction	20.0 %
Peptone	0.5 %
Yeast extract	0.1 %

pH adjusted to 8.2

(continued)

Table I. (*Continued*)

Medium for T. thermophilus (Hishinuma et al., 1977)

Polypeptone	5.0 g/l^{-1}
Yeast extract	4.0 g/l^{-1}
NaCl	2.0 g/l^{-1}
Glucose	1.0 g/l^{-1}

pH adjusted to 7.0

Medium for T. thermophilus (Vali et al., 1980)

Peptone	8.0 g/l^{-1}
Yeast extract	4.0 g/l^{-1}
NaCl	3.0 g/l^{-1}

pH adjusted to 7.5

Defined medium for Thermus aquaticus YT-1 (Kenkel and Trela, 1979)

Glutamic acid	3.0 g/l^{-1}
Biotin	0.1 mg/l^{-1}
Thiamine	0.1 mg/l^{-1}
Nicotinic acid	0.05 mg/l^{-1}
Castenholz basal salts	1000 ml

Complete medium M1 for growth of Thermus aquaticus YT-1 and Thermus thermophilus (Sonnleitner et al., 1982)

Peptone	3.0 g/l^{-1}
Meat extract	5.0 g/l^{-1}

Defined medium for growth of T. thermophilus (Sonnleitner et al., 1982)

$(NH_4)_2 SO_4$	1.0 g/l^{-1}
$NH_4 Cl$	1.0 g/l^{-1}
KH_2PO_4	0.3 g/l^{-1}
$MgSO_4 (7H_2O)$	0.024 g/l^{-1}
$CaCl_2 (2H_2O)$	0.024 g/l^{-1}

Trace elements of Kuhn *et al.* (1979) 0.15 ml^{-1}
Carbon source at 0.8 to 1 g l^{-1}

Defined medium for Thermus thermophilus

DSM 86 Castenholz medium (DSM medium 86)	
Nitrilotriacetic acid (Titriplex I)	100.0 mg
$CaSO_4 2H_2O$	60.0 mg
$MgSO_4 7H_2O$	100.0 mg
NaCl	8.0 mg
KNO_3	103.0 mg
$NaNO_3$	689.0 mg
$Na_2HPO_4 2H_2O$	140.0 mg
$FeCl_3 6H_2O$	0.47 mg
$MnSO_4 H_2O$	2.2 mg

$ZnSO_4\ 7H_2O$	0.5 mg
H_3BO_3	0.5 mg
$CuSO_4\ 5H_2O$	0.025 mg
$Na_2MoO_4\ 2H_2O$	0.025 mg
$CoCl_2\ 6H_2O$	0.046 mg
Tryptone	1.0 mg
Yeast extract	1.0 mg
Distilled water	1000.0 ml

pH adjusted to 8.2 with NaOH.

DSM 256 T. ruber medium (DMS medium 256)

Universal peptone (Merck)	5.0 g
Yeast extract	1.0 g
Starch, soluble	1.0 g
Agar	12.0 g
Distilled water	1000.0 ml

pH adjusted to 8.0

DSM 74 T. thermophilus medium (DSM medium 74)

Yeast extract	4.0 g
Polypeptone	8.0 g
NaCl	2.0 g
Distilled water	1000.0 ml

pH adjusted to 7.0

acid (Bogdanova and Loginova, 1986). *Thermus aquaticus* strain YS045 requires leucine, isoleucine, and valine when grown on lactose in continuous culture (Sharp and Cossar, personal observations).

The original isolation of *Thermus aquaticus* was made using Castenholz (1969) salts as basal medium. This medium and various modifications have since been widely used. It contains a high nitrate concentration and many groups have replaced it with ammonium sulphate as an improved nitrogen source (Pask-Hughes and Williams, 1977; Munster *et al.*, 1986; Santos *et al.*, 1989). Development of minimal salts medium 162 by Degryse *et al.* (1978) as an alternative to Medium D of Castenholz (1969) replaces nitrate with ammonium, removes the NaCl, and uses $NaHPO_4/KH_2PO_4$ as a buffer. *Thermus aquaticus* YT1 and Thermus strains NH and DI grow in the absence of added vitamins and amino acids with glucose, acetate or malate as the carbon source on solid media. *Thermus thermophilus* HB8 was found to have a requirement for lysine, however, subsequent studies in chemostat culture using medium 162 or Castenholz medium indicated no requirement for lysine (Sonnleitner *et al.*, 1982 and McKay *et al.*, 1982). Other medium formulations utilize only NaCl or phosphate buffer with $MgCl_2$ or KCl with complex substrates such as tryptone and yeast extract.

Thermus sp. YSO45 (Munster *et al.*, 1986) did not grow on basal *Thermus* medium supplemented with 0.1% (w/v) lactose and 2 M $(NH_4)_2SO_4$, however, addition of yeast extract and tryptone (0.1% w/v) each satisfied nutritional requirements. The auxotrophic requirements for this strain were resolved using a pulse–shift chemostat technique (Kuhn *et al.*, 1979) using a mineral salts basal medium with lactose (0.1%) supplemented with 20 amino acids (25 μM final concentration) and eight vitamins. When the culture had attained a "steady state," individual amino acids were aseptically added by filtration directly to the vessel at eight × their normal level. Initially, individual amino acids appeared to have no discernible effect, and so were added in groups based on their common biosynthetic pathway. A depression of dissolved oxygen tension was taken to indicate growth limitation was relieved by the added component. Growth limiting amino acids were identified were incorporated into the medium until lactose became the limiting nutrient. In this way, isoleucine, valine, lysine, and biotin were found essential for growth and were incorporated into the basal medium at 0.05 g l⁻¹ for the amino acids and 100 μg l⁻¹ for the vitamin. All other amino acids were not required and were omitted; however, the vitamin supplement which enhanced growth was retained. The defined vitamin requirements remain to be resolved (Cossar and Sharp, unpublished observation).

Growth of *T. thermophilus* HB8 on a defined medium with glucose as the sole carbon source (2 gl⁻¹) yielded 20 mg l⁻¹ cell dry weight at a dilution rate of 0.5 h⁻¹. It is not clear how much of the glucose was used. The cell yield was increased by the addition of meat extract, yeast extract, or peptones to the medium but the authors report "no increase" in the utilization of glucose. These complex substrates were not completely utilized and those medium components that stimulate growth rate or cell yield were investigated (Sonnleitner *et al.*, 1982). Amino acids, ten vitamins, organic acids, fatty acids, amines, polyamines, and various sugars were unable to replace the complex medium components. Incomplete substrate utilization was followed by regrowth of a fresh inoculum of the same strain on the cell-free filtrate of a stationary phase culture (Sonnleitner, 1983). This phenomenon remains to be clarified.

Thermus aquaticus YT1 was inhibited by growth in 2% NaCl (Brock and Freeze, 1969), and "*Thermus flavus*" strain AT62, "*T. lacteus*," *T. filiformis*, *T. ruber* BKMB and *Thermus aquaticus* YT-1 were all unable to grow in 1% NaCl (da Costa, Sharp, Williams, personal communications). *T. thermophilus* HB8 was unable to grow in the presence of 3% NaCl.

4.2. Growth Kinetics

The growth kinetics of *Thermus* species have not been widely studied and few growth parameters have been determined (Table II). Investigations have generally involved complex media. Biomass yields, where reported, are generally low and growth rates usually recorded as doubling time in complex medium. Oshima and Imahori (1974) reported a doubling time of 18–20 minutes for *T. thermophilus*, but the range for this parameter is more often 40–80 minutes at the optimum growth temperature. Sonnleitner *et al.* (1982) reported a μmax of 0.2 h^{-1} and 0.5 h^{-1} in shake flasks and controlled batch reactors respectively. Low cell yield and incomplete substrate utilization were both observed independent of the carbon source. Using chemostats, Sonnleitner determined the maximum specific growth rate (μmax) of strain HB8 as 2.7 h^{-1}, corresponding to a doubling time of 15.4 min. Growth of *T. aquaticus* YT1 in chemostats was examined in a synthetic defined medium and a complex medium. In the former, nonpigmented 'white' cells had a μmax of 1.62 h^{-1} (td of 27.7 min) with a Y (glucose) of 0.44 g g^{-1} at 68 °C. Such colorless "mutants" have not been reported by others. On complex medium μmax was 3.5 h^{-1} (td of 11.9 min) with a low Y of 0.022 g g^{-1} due to incomplete substrate utilization (Cometta *et al.*, 1982).

Growth of *Thermus aquaticus* strain YS045 (Munster *et al.*, 1986) on lactose limited media indicated that D^{crit} was approached at approximately 0.35 h^{-1}. Steady states could not be realized at dilution rates higher than 0.33 h^{-1}. The decline in culture optical density at D >0.22 h^{-1} was considered due to a change in mature cell shape rather than a true decline in population (Luscombe and Greig, 1971). The β-galactosidase activity increased slightly up to a D of 0.22 h^{-1} after which it showed a marked decline, concomitant with the increase in residual lactose in the culture supernatant. This was consistent with lactose metabolism providing increased intracellular glucose that repressed synthesis of β-galactosidase. In contrast with lactose-limited cultures of *E. coli* (Horiuchi *et al.*, 1962), long-term cultivation of *Thermus* sp. YS045 did not result in an increase in enzyme specific activity, nor a change in cell density that could be attributable to a change in enzyme synthesis (Cossar and Sharp, unpublished observation). This is consistent with the observation that lactose is, itself, a relatively poor inducer of β-galactosidase in other yellow-pigmented strains of *Thermus* sp. (Ulrich *et al.*, 1972).

Growth rate studies of various strains of *Thermus* are summarized in Table III.

4.3. Growth and Pigmentation

Strains of *Thermus* isolated from natural environments produce carotenoids exhibited by yellow or red pigmentation (Brock and Freeze, 1969; Loginova and Egorova, 1975). The absorption spectrum of the

Table II. Growth Kinetics of *Thermus* Species

Organism	Medium	μ_{max}	t_d (min)	T_{opt} (°C)	T_{max} (°C)	Yield (g g^{-1})	Reference
T. aquaticus YT-1	Complex	0.83	50		70	—	Brock and Freeze, 1969
T. aquaticus YT-1	Complex	0.35	—	69–71	—	—	Ramaley and Hixson, 1970
T. aquaticus	Complex	0.53	78	70–71	80.8	—	Pask-Hughes and Williams, 1977
T. aquaticus YT-1							
(shake flasks)	Complex	0.2	40	78	80	0.17[c]	Cometta *et al.*, 1982
(batch reactor)	Complex	0.5	—	—	—	0.12[c]	Cometta *et al.*, 1982
(chemostat culture)	Complex	3.5	—	—	—	0.05[c]	Cometta *et al.*, 1982
(chemostat culture)	Glucose	1.6	41	70	75	0.4[a]	Cometta *et al.*, 1982
T. aquaticus D1	Complex	0.63	66	68–72	77.8	—	Pask-Hughes and Williams, 1977
T. flavus st 71	Glucose/YE	—		70	—	0.14[a]	Posmogova, 1975
Thermus sp. strain K2	Complex	0.83	50	60	80	—	Ramaley *et al.*, 1975
T. thermophilus HB8	Complex	2.1	20	65–72	85	—	Oshima and Imahori, 1974
(shake flash)	Complex	0.2	—	—	—	0.06[c]	Sonnleitner *et al.*, 1982
(batch reactor)	Complex	0.5	—	75	—	0.06[c]	Sonnleitner *et al.*, 1982
(continuous culture)	Complex	2.7	—	75	85	0.12[c]	Sonnleitner *et al.*, 1982
(continuous culture)	Glucose	0.15	—	—	—	0.31[a]	McKay *et al.*, 1982
Thermus XI	Complex	0.72	58	70–71	77.6	—	Pask-Hughes and Williams, 1977
Thermus ZO5	Glucose	0.14	300	—	85		Degryse *et al.*, 1978
Thermus ZO5	Glutamate	0.69	60				Degryse *et al.*, 1978
Thermus ZO5	Glutamate	0.35	120				Degryse *et al.*, 1978

Thermus ZO5	Acetate	0.35	120			Degryse *et al.*, 1978
T. ruber **BKMB B-1258**	TYE medium	—	96		3.3[b]	Holtom, 1991
T. brockianus YS38	TYE medium	—	118	74	2.8[b]	Holtom, 1991
T. aquaticus YS9	TYE medium	—	114	70	3.36[b]	Holtom, 1991
T. thermophilus strain B	TYE medium	—	96	80	3.3[b]	Holtom, 1991
T. aquaticus YT1	TYE medium	—	89		3.2[b]	Holtom, 1991
T. ruber **BKM B-1258**	Casamino acids + vitamins	—	210	66	0.6[b]	Holtom, 1991
T. brockianus YS38	Casamino acids + vitamins	—	240	74	0.7[b]	Holtom, 1991
T. thermophilus strain B	Casamino acids + vitamins	—	210		1.1[b]	Holtom, 1991
T. aquaticus YT1	Casamino acids + vitamins	—	210		0.7[b]	Holtom, 1991

[a]Yield based on g biomass (g/substrate utilized)$^{-1}$.
[b]Yield based on optical density at 600 nm.
[c]Yield based on g biomass (g/organic carbon utilized)$^{-1}$.

pigments extracted in 90% v/v acetone, and water has a maxima of 453 and 480 nm with a shoulder at 425 nm.

Those strains from manmade and predominantly dark environments often lack pigment (Ramaley and Hixson, 1970; Pask-Hughes and Williams, 1975) (see also Chapter 2).

Cometta *et al.* (1982) reported the loss of pigmentation in *Thermus aquaticus* (YT1) in chemostat culture with glucose as growth limiting carbon source in a noncomplex medium. After 3–5 volume changes of medium the culture was observed to become predominantly white, addition of complex substrates did not result in recovery of pigmentation. Nonpigmented variants were supposedly confirmed to be strain YT-1 following studies of freeze fracture micrographs. Other physiological differences observed include the complete utilization of glucose by nonpigmented variants but not by yellow pigmented cells.

Studies by Cossar and Sharp (1989) of strain YS045 in lactose limited chemostats with defined medium, indicated a loss in intensity of pigmentation in the vessel. By contrast to the observations of Cometta *et al.* (1982), pigment-free variants were not found following subculture to agar plates. The chemostat was situated in continuous illumination, although partially shaded by an insulating jacket. In a subsequent study with lactose limitation, the vessel insulating jacket was omitted and the pigmentation was observed to be more intense. The extracted pigment showed the characteristic spectra of a carotenoid. There were two main peaks at 480 and 450 nm with a shoulder at 426 nm. When the vessel was covered with aluminum foil, the levels of pigment were shown to decline. On returning the culture to the light the pigmentation returned, although not to its previous level. Subsequent light–dark transition demonstrated the same effect. In an induced gene system, the removal of pigment (light–dark) was a slower process than its *de novo* production (dark–light). The dissolved oxygen concentration was observed to transiently increase each time the culture was returned to the light and to decline on covering of the vessel with foil. Increasing the illumination incident on the surface of the vessel using a 100 watt tungsten filament bulb resulted in wash out of the culture, which failed to recover even when returned to the dark. Comparative batch cultures grown in the light and dark were indistinguishable in growth rate and pigment content. Similarly, repeated subculture and incubation on solid media under continuous light or dark conditions failed to produce strains with reduced pigment content. Pigment loss (or failure to produce pigment) appears to be a feature of the more selective environment characteristic of the chemostat. However, growth in a carbon-limited chemostate for over 150 generations did not result in the isolation of strains unable to produce pigment. Growth of strain YS045 in an anaerobic chemostat using nitrate as electron acceptor was unstable and proved to be nitrate limited on the basis of total depletion of nitrate from the medium (Cossar and Sharp, 1986). Addition of a pulse of nitrate produced a transiently higher level of nitrite and a subsequent reduction in cell density due to its toxicity.

When grown in a chemostat under anaerobic conditions, methionine proved to be a growth requirement, and a pigment-free strain was isolated which grew as pale/transparent colonies at 60 °C. The strain was unstable and was lost during subculture. In the chemostat energy 'wasted' in producing pigment in the dark probably represents a significant drain on the cell. Under conditions of nutrient excess in batch culture, such a drain is not normally significant and hence pigment is produced under all conditions. Production of pigment appears to be under stringent control by a number of factors, including light. No direct effect of light on extracted pigment could be detected by exposure to a range of wavelengths in a spectrophotometer (250–700 nm) over a period of 60 min. However, if cells were extracted in 90% (v/v) acetone for more than 15 min, which resulted in extraction of cellular components rather than predominantly pigments, it was possible to observe bleaching of the pigment within 10 mm (at 250 nm) or 25 min (at 550 nm). This may provide some evidence for a role of pigment in protecting cellular components from damage by reactive species produced as a result of exposure to light. Our results indicated that this strain does not contain a significant concentration of glutathione, which is implicated as a protective agent against free radicals (Meister and Anderson, 1983). Other strains of *Thermus sp.* have been shown to contain catalase, peroxidase, and superoxide dismutase (Allgood and Perry, 1985). However, these may be primarily concerned with the cytoplasmic protective system. The carotenoid pigment may be more specifically concerned with countering the radiation-mediated damage to membranes (Suzuki *et al.*, 1982).

REFERENCES

Ado, Y., Kawamoto, T., Masunaga, I., Takayama, K., Takasawa, S., and Kimura, K., 1982, Production of 1-malic acid with immobilized thermophilic bacterium *Thermus rubens nov. sp*, *Enz. Eng.* **6**:303–304.

Aleksandrushkina, N. I., and Egorova, L. A., 1978, Nucleotide make-up of the DNA of the thermophilic bacteria of the genus *Thermus*, *Mikrobiol.* **47**:250–252.

Alfredsson, G. A., Baldursson, S., and Kristjansson, J. K., 1985, Nutritional diversity among *Thermus spp*, isolated from Icelandic hot springs, *Systemat. Appl. Microbiol.* **6**:308–311.

Allgood, G. S., and Perry, J. J., 1985, Paraquat toxicity and effect of hydrogen peroxide on thermophilic bacteria, *J. Free Radicals Biol. Med.* **1**:233–237.

Allgood, G. S., and Perry, J. J., 1986, Effect of methyl viologen and oxygen concentration on thermophilic bacteria, *J. Basic Microbiol.* **26:**379–382.

Baldursson, S., and Kristjansson, K., 1990, Analysis of nitrate in food extracts using a thermostable formate-linked nitrate reductase enzyme system, *Biotechnol. Tech.* **4:** 211–214.

Barnes, L. D., and Stellwagen, E., 1973, Enolase from the thermophile *Thermus* X1, *Biochemistry* **12:**1559–1565.

Bergquist, P. L., Love, D. R., Croft, J. E., Streiff, M. B., Daniel, R. M., and Morgan, W. H., 1987, Genetics and potential biotechnological applications of thermophilic and extremely thermophilic microorganisms, *Biotechnol. Genet. Eng. Rev.* **5:**199–244.

Bogdanova, T. I., and Loginova, L. G., 1986, Influence of amino acids and vitamins on the development of the thermophilic bacteria *Bacillus stearothermophilus* and *Thermus ruber* isolated on a medium containing paraffin, *Mikrobiologia.* **55:**570–574.

Bowen, D., Littlechild, J. A., Fothergill, J. E., Watson, H. C., and Hall, L., 1988, Nucleotide sequence of the phosphoglycerate kinase gene from the extreme thermophile *Thermus thermophilus.* Comparison of the deduced amino acid sequence with that of the mesophilic yeast phosphoglycerate kinase. *Biochem. J.* **254:**509–517.

Bridger, G. P., and Sundaram, T. K., 1976, Occurrence of phosphoenolpyruvate carboxylase in the extremely thermophilic bacterium *Thermus aquaticus, J. Bacteriol.* **125:**1211–1213.

Brock, T. D., and Freeze, H., 1969, *Thermus aquaticus* gen. n. and sp. n., a non-sporulating extreme thermophile, *J. Bacteriol.* **98:**289–297.

Brock, T. D., 1978, The Genus *Thermus*, in: *Microorganisms and Life at High Temperatures*, Springer Verlag, New York, chap. 4.

Brock, T. D., 1986, *Thermophiles: General Molecular, and Applied Microbiology*, New York: John Wiley.

Brock, T. D., and Brock, M. L., 1971, Temperature optimum of nonsulphur bacteria from a spring at 90 C, *Nature* **233:**494–495.

Brock, T. D., 1981, Extreme thermophiles of the genera *Thermus* and *Sulfolobus* in: *The Prokaryotes* (M. P. Starr, H. Stalp, H. G. Truper, A. G. Balows, H. G. Schlegel, eds.), Berlin, Heidelberg, New York: Springer, p. 978.

Cass, K. H., and Stellwagen, E., 1975, A thermostable phosphofructokinase from the extreme thermophile *Thermus* X1, *Arch. Biochem. Biophys.* **171:**682–694.

Castenholz, R. W., 1969, Thermophilic blue-green algae and the thermal environment, *Bacteriol. Rev.* **33:**476–504.

Chin, N. W., and Trela, J. M., 1973, Comparison of acetohydroxy-acid synthetases from two extreme thermophilic bacteria, *J. Bacteriol.* **114:**674–678.

Cocco, D., Rinaldi, A., Savini, I., Cooper, J. M., and Bannister, J. V., 1988, NADH oxidase from the extreme thermophile *Thermus aquaticus* YT1: purification and characterization, *Eur. J. Biochem.* **174:**267–271.

Cometta, S., Sonnleitner, B., Sidler, W., and Fiechter, A., 1982, Population distribution of aerobic extremely thermophilic microorganisms in an Icelandic natural hot spring, *Eur. J. Appl. Microbiol. Biotechnol.* **16:**151–156.

Cossar, D., and Sharp, R. J., 1986, Preliminary physiological studies on a new denitrifying strain of *Thermus.* Abstracts of XIV International Congress of Microbiology, **PII-9:**196.

Cossar, D., and Sharp, R. J., 1989, Loss of pigmentation in: *Thermus spp.*, in: *Microbiology of Extreme Environments and its Potential for Biotechnology* (M. S. da Costa, J. C. Duarte, and R. A. D. Williams, eds.), **385** London, Elsevier.

Croft, J. E., Love, D. R., and Bergquist, P. L., 1987, Expression of leucine genes from an extremely thermophilic bacterium in *E. coli, Mol. Gen. Genet.* **210:**490–497.

Cunin, R., Glansdorff, N., Pierard, A., and Stalon, V., 1986, Biosynthesis and metabolism of arginine in bacteria. *Microbiol. Rev.* **50:**314–352.

Curran, M. P., Daniel, R. M., Guy, G. R., and Morgan, H. W., 1985, A specific L-asparaginase from *Thermus aquaticus, Arch. Biochem. Biophys.* **241:**571–576.

Degryse, E., and Glansdorff, N., 1976, Metabolic function of the glyoxylic shunt in an extreme thermophilic strain of the genus *Thermus, Arch. Int. Physiol. Biochim.* **84:**598–599.

Degryse, E., and Glansdorff, N., 1981, Studies on the central metabolism of *Thermus aquaticus* an extreme thermophilic bacterium: Anaplerotic reactions and their regulation, *Arch. Microbiol.* **129:**173–177.

Degryse, E., Glansdorff, N., and Pierard, A., 1976, Arginine biosynthesis and degradation in an extreme thermophile strain ZO5, *Arch. Int. Physiol. Biochim.* **84:**599–601.

Degryse, E., Glansdorff, N., and Pierard, A., 1978, A comparative study of extreme thermophilic bacteria belonging to the genus *Thermus, Arch. Microbiol.* **117:**189–196.

Edwards, C., 1990, *Microbiology of Extreme Environments*, Open University Press, Milton Keynes, England.

Fee, J. A., Kuila, D., Mather, M. W., Yoshida, T., 1986, Respiratory proteins from extremely thermophilic aerobic bacteria, *Biochim. Biophys. Acta.* **853:**153–185.

Freeze, H., and Brock, T. D., 1970, Thermostable aldolase from *Thermus aquaticus, J. Bacteriol.* **101:**541–550.

Fujita, S. C., Oshima, T., and Imahori, K., 1976, Purification and properties of D-glyceraldehyde-3-phosphate dehydrogenase from an extreme thermophile, *Thermus thermophilus* strain HB8, *Eur. J. Biochem.* **64:**57–68.

Gould, G. W., and Corry, J. C. L., 1986, *Microbial Growth and Survival in Extremes of Environment,* Academic Press, London.

Harris, J. I., Hocking, J. D., Runswick, M. J., Suzuki, K., and Walker, J. E., 1980, D-glyceraldehyde-3-phosphate dehydrogenase. The purification and characterisation of the enzyme from the thermophiles *Bacillus stearothermophilus* and *Thermus aquaticus, Eur. J. Biochem.* **108:**535–547.

Hengartner, H., Kolb, E., and Harris, J. I., 1976, Phosphofructokinase from thermophilic microorganisms, in: *Enzymes and Proteins from Thermophilic Microorganisms* (H. Zuber, ed.), Basel, Birkhauser, pp. 199–206.

Herbert, R. A., and Codd, G. A., 1986, *Microbes in Extreme Environments.* Academic Press, London.

Herbert, R. A., and Sharp, R. J., 1992, *Molecular Biology and Biotechnology of Extremophiles,* Blackie, Glasgow.

Hickey, C. W., and Daniel, R. M., 1979, The electron transport system of an extremely thermophilic bacterium, *J. Gen. Microbiol.* **114:**195–200.

Higa, E. H., and Ramaley, R. F., 1973, Purification and properties of threonine deaminase from the X1 isolate of the genus *Thermus, J. Bacteriol.* **114:**556–562.

Hishinuma, F., Hiria, K., and Sakaguchi, K., 1977, Thermophilic polynucleotide phosphorylase from *Thermus thermophilus.* Purification and properties of an altered form of enzyme which lacks phosphorolytic activity to polynucleotide, *Eur. J. Biochem.* **77:**575–583.

Hocking, J. D., and Harris, J. I., 1973, Purification by affinity chromatography of thermostable glyceraldehyde-3-phosphate dehydrogenase from *Thermus aquaticus, FEBS Lett.* **34:**280–284.

Hocking, J. D., and Harris, J. I., 1980, D-glyceraldehyde-3-phosphate dehydrogenase. Amino acid sequence of the enzyme from the extreme thermophile *Thermus aquaticus. Eur. J. Biochem.* **108:**567–579.

Holtom, G. J., 1991, The physiology and biochemistry of glutamate transport and utilisation in the genus *Thermus*. Ph.D. Thesis, University of London.

Holtom, G. J., Cossar, D., Sharp, R., and Williams, R. A. D., 1989, Amino acid utilization by strains of the genus *Thermus*, in: *Microbiology of Extreme Environments and Its Potential for Biotechnology* (M. S. da Costa, J. C. Duarte, and R. A. D. Williams, eds.), Elsevier, London.

Holtom, G. J., Sharp, R. J., and Williams, R. A. D., 1993, Sodium-stimulated transport of glutamate by *Thermus thermophilus* strain B, *J. Gen. Microbiol.* **139:**2245–2250.

Horikoshi, K., and Grant, W. D., 1991, *Superbugs: Microorganisms in Extreme Environments.* Japan Scientific Societies: Tokyo, and Springer Verlag, Berlin.

Horiuchi, T., Tomizawa, J., and Novick, A., 1962, Isolation and properties of bacteria capable of high rates of β-galactosidase synthesis, *Biochem. Biophys. Acta.* **55:**152–163.

Hudson, J. A., Morgan, H. W., and Daniel, R. M., 1987, Numerical classification of some *Thermus* isolates from Icelandic hot springs, *Syste. Appl. Microbiol.* **9:**218–223.

Hudson, J. A., Morgan, H. W., and Daniel, R. M., 1989, Numerical classification of *Thermus* isolates from globally distributed hot springs, *Syst. Appl. Microbiol.* **11:**250–256.

Kaledin, A. S., Slyusarenko, A. G., and Gorodetskii, S. I., 1981, Isolation and properties of DNA polymerase from the extreme thermophilic bacterium *Thermus flavus*, *Biokhimiia*, **46:**1576–1584.

Kenkel, T., and Trela, J. M., 1979, Protein turnover in the extreme thermophile *Thermus aquaticus*, *J. Bacteriol.* **140:**543–546.

Kirino, H., and Oshima, T., 1991, Molecular cloning and nucleotide sequence of 3-iso-propylmalate dehydrogenase gene (leuB) from an extreme thermophile *Thermus aquaticus* YT1. *J. Biochem.* **109:**852–857.

Koh, C. L., 1985, Detection and purification of plasmids present in *Thermus* strains from Icelandic hot springs, *Mircen. J.* **1:**77–81.

Koyama, Y., and Furukawa, K., 1990, Cloning and sequence analysis of tryptophan synthetase genes of an extreme thermophile *Thermus thermophilus* HB27: Plasmid transfer from replica-plated *Escherichia coli* recombinant colonies to competent *T. thermophilus* cells, *J. Bacteriol.* **172:**3490–3495.

Kristjansson, J. K., 1992, *Thermophilic Bacteria,* Boca Raton: CRC Press.

Kristjansson, J. K., Hregvidsson, G. O., and Alfredsson, G. A., 1986, Isolation of halotolerant *Thermus spp.* from submarine hot springs in Iceland, *Appl. Environ. Microbiol.* **52:**1313–1316.

Krulwich, T. A., and Ivey, D. A., 1990, Bioenergetics in extreme environments, in: *The Bacteria, Vol XII, Bacterial Energetics* (T. A. Krulwich, ed.), Academic Press, New York, pp. 417–447.

Kuhn, H., Friedrich, U., and Fiechter, A., 1979, Defined minimal media for a thermophilic Bacillus sp. developed by a chemostat pulse and shift technique, *Eur. J. Appl. Microbiol. Biotechnol.* **6:**341–349.

Kunai, K., Machida, M., Matsuzawa, H., and Ohta, T., 1986, Nucleotide sequence and characteristics of the gene for L-lactate dehydrogenase of *Thermus caldophilus* GK24 and the deduced amino acid sequence of the enzyme, *Eur. J. Biochem.* **160:**433–440.

Littlechild, J. A., Davies, G. J., Gamblin, S. J., and Watson, H. C., 1987, Phosphoglycerate kinase from the extreme thermophile *Thermus thermophilus:*crystallization and preliminary X-ray data, *FEBS Lett.* **225:**123–126.

Loginova, L. G., and Egorova, L. A., 1975, An obligately thermophilic bacterium, *Thermus ruber* from hot springs in Kamchatka, *Microbiologiya* **46:**661–665.

Ljungdahl, L. G., 1979, Physiology of thermophilic bacteria, *Adv. Microb. Physiol.* **19:**149–243.

Luscombe, B. M., and Greig, T. R. G., 1971, Effect of varying growth rate on the morphology of *Arthrobacter*, *J. Gen. Microbiol.* **69:**433–434.

Maas, E., Popischal, H., Koplin, R., and Biswanger, H., 1992, Multi-enzyme complexes in thermophilic organisms: isolation and characterisation of the pyruvate dehydrogenase complex from *Thermus aquaticus* AT62, *J. Gen. Microbiol.* **138**:795–802.

Machida, M., Matsuzawa, H., and Ohta, T., 1985a, Fructose 1,6-bisphosphate dependent L-lactate dehydrogenase from *Thermus aquaticus* YT-1, an extreme thermophile: activation by citrate and modification reagents and comparison with *Thermus caldophilus* GK24 L-lactate dehydrogenase, *J. Biochem.* **97**:899–909.

Machida, M., Yokoyama, S., Matsuzawa, H., Miyazawa, T., and Ohta, T., 1985b, Allosteric effect of fructose 1,6-bisphosphate on the conformation of NAD$^+$ as bound to L-lactate dehydrogenase from *Thermus caldophilus* GK24, *J. Biol. Chem.* **260**:16143–16147.

MacMichael, G. J., 1988, Effects of oxygen and methyl viologen on *Thermus aquaticus*, *J. Bacteriol.* **170**:4995–4998.

McKay, A., Quilter, J., and Jones, C. W., 1982, Energy conservation in the extreme thermophile *Thermus thermophilus* HB8, *Arch. Microbiol.* **131**:43–50.

Meister, A., and Anderson, M. E., 1983, Glutathione, *Ann. Rev. Biochem.* **52**:711–760.

Munster, M. J., Munster, A. P., Woodrow, J. R., and Sharp, R. J., 1986, Isolation and preliminary taxonomic studies of *Thermus* strains isolated from Yellowstone National Park, USA, *J. Gen. Microbiol.* **132**:1677–1683.

Nagahari, K., Koshikawa, T., and Sakaguchi, K., 1980, Cloning and expression of the leucine gene from *Thermus thermophilus* in *Escherichia coli*, *Gene* **10**:137–145.

Nojima, H., Oshima, T., and Noda, H., 1979, Purification and properties of phosphoglycerate kinase from *Thermus thermophilus* HB8, *J. Biochem.* **85**:1509–1517.

Oshima, T., and Soda, K., 1989, Thermostable amino acid dehydrogenases: applications and gene cloning, *Trends Biotechnol.* **7**:210–214.

Oshima, T., and Imahori, K., 1974, Description of *Thermus thermophilus*, a non-sporulating thermophilic bacterium from a Japanese Thermal Spa, *Int. J. System. Bacteriol.* **24**:102–112.

Posmogora, I. N., 1975, *Mikrobiologiya* **44**:492.

Pask-Hughes, R., and Williams, R. A. D., 1975, Extremely thermophilic gram negative bacteria from hot tap water, *J. Gen. Microbiol.* **88**:321–328.

Pask-Hughes, R. A., and Williams, R. A. D., 1977, Yellow pigmented strains of *Thermus spp.* from Icelandic hot springs, *J. Gen. Microbiol.* **102**:375–383.

Ramaley, R. F., and Hixson, J., 1970, Isolation of a non-pigmented thermophilic bacterium similar to *Thermus aquaticus*, *J. Bacteriol.* **103**:527–528.

Ramaley, R. F., Bitzinger, K., Carroll, R. M., and Wilson, R. B., Isolation of a new, pink, obligately thermophilic gram-negative bacterium (K-2 Isolate), *Int. J. Syst. Bacteriol.* **25**:357.

Ruffett, M., Hammond, S., Williams, R. A. D., and Sharp, R. J., 1992, A taxonomic study of red pigmented gram negative thermophiles. Abstracts of Thermophiles: Science and Technology, Reykjavik, Iceland p74.

Saiki, T., Kimura, R., and Arima, K., 1972, Isolation and characterization of extremely thermophilic bacteria from hot springs, *Agric. Biol. Chem.* **36**:2357–2366.

Saiki, T., Shinshi, H., and Arima, K., 1973, Studies on homoserine dehydrogenase from an extreme thermophile, *Thermus flavus* AT62: Partial purification and properties, *J. Biochem.* **74**:1239–1248.

Santos, M. N. A., Williams, R. A. D., and da Costa, M. S., 1989, Numerical taxonomy of *Thermus* isolates from hot springs in Portugal, *Systemat. Appl. Microbiol.* **12**:310–315.

Sato, S., Nakada, Y., Kanaya, S., and Tanaka, T., 1988, Molecular cloning and nucleotide sequence of *Thermus thermophilus* HB8 trpE and trpG, *Biochim. Biophys. Acta.* **950**:303–312.

Schroeder, G., Matsuzawa, H., and Ohta, T., 1988, Involvement of the conserved histidine-188 residue in the L-lactate dehydrogenase from *Thermus caldophilus* GK24 in allosteric regulation by fructose 1,6-bisphosphate, *Biochem. Biophys. Res. Comm.* **152:**1236–1241.

Sonnleitner, B., 1983, Biotechnology of thermophilic bacteria—growth, products and applications, *Adv. Biochem. Eng. Biotechnol.* **28:**69–138.

Sonnleitner, B., Cometta, S., and Feichter, A., 1982, Growth kinetics of *Thermus thermophilus*, *Eur. J. Appl. Microbiol. Biotechnol.* **15:**75–82.

Sharp, R. J., and Williams, R. A. D., 1988, Properties of *Thermus ruber* strains isolated from Icelandic hot springs and DNA:DNA homology of *Thermus ruber* and *Thermus aquaticus*, *Appl. Environ. Microbiol.* **54:**2049–2053.

Stellwagen, E., and Thompson, S. T., 1979, Activation of *Thermus* phosphofructokinase by monovalent cations, *Biochim. Biophys. Acta.* **569:**6–12.

Sundaram, T. K., and Bridger, G. P., 1979, Regulatory characteristics of phosphoenolpyruvate carboxylase from the extreme thermophile, *Thermus aquaticus*, *Biochim. Biophys. Acta.* **570:**406–410.

Suzuki, S., Oshima, M., and Akamatsu, Y., 1982, Radiation damage to membranes of the thermophilic bacterium *Thermus thermophilus* HB-8: Membrane damage without concomitant lipid peroxidation, *Radiat. Res.* **91:**564–572.

Taguchi, H., Yamashita, M., Matsuzawa, H., and Ohta, T., 1982, Heat stable and fructose 1,6-bisphosphate activated L-lactate dehydrogenase from an extremely thermophilic bacterium, *J. Biochem.* **91:**1343–1348.

Taguchi, H., Hamaoki, M., Matsuzawa, H., Ohta, T., 1983, Heat stable extracellular proteolytic enzyme produced by *Thermus caldophilus* strain GK24 an extremely thermophilic bacterium, *J. Biochem.* **93:**7–13.

Taguchi, H., Matsuzawa, H., and Ohta, T., 1984, Lactate dehydrogenase from *Thermus caldophilus* GK24, an extremely thermophilic bacterium, *Eur. J. Biochem.* **145:**283–290.

Tanaka, T., Kawano, N., and Oshima, T., 1981, Cloning of 3-isopropylmalate dehydrogenase gene of an extreme thermophile and partial purification of the gene product, *J. Biochem.* **89:**677–682.

Ulrich, J. T., McFeters, G. A., and Temple, K. L., 1972, Induction and characterisation of β-galactosidase in an extreme thermophile, *J. Bacteriol.* **110:**691–698.

Vali, Z., Kilar, F., Lakatas, S., Venyaminov, S. A., and Zavodsky, P., 1980, L-Alanine dehydrogenase from *Thermus thermophilus*, *Biochim. Biophys. Acta.* **615:**34–47.

Venegas, A., Vicuna, R., Alonso, A., Valdes, F., and Yudelevich, A., 1980, A rapid procedure for purifying a restriction endonuclease from *Thermus thermophilus* (Tth 1), *FEBS Lett.* **109:**156–158.

Verhoeven, J. A., Schenk, K., Meyer, R. R., and Trela, J. M., 1986, Purification and characterisation of an inorganic pyrophosphatase from the extreme thermophilic *T. aquaticus*, *J. Bact.* **168:**318–321.

Walsh, K. A. J., Daniel, R. M., and Morgan, H. W., 1983, A soluble NADH dehydrogenase (NADH: ferricyanide oxidoreductase) from *Thermus aquaticus* strain T351, *Biochem. J.* **209:**427–433.

Walker, J. M., and Wang, Y. X., 1993, Purification of aspartate aminotransferase from *Thermus aquaticus*, *Biochem. Molec. Biol. Internat.* **29:**867–873.

Ward, O. P., and Moo-Young, M., 1988, Thermostable enzymes, *Biotechnol. Adv.* **6:**39–69.

Weitzman, P. J. D., and Jaskowska-Hodges, H., 1982, Patterns of nucleotide utilisation in bacterial succinate thiokinases, *FEBS Lett.* **143:**237–240.

Xu, J., Oshima, T., and Yoshida, M., 1990, Tetramer-dimer conversion of phosphofructokinase from *Thermus thermophilus* induced by its allosteric effects, *J. Mol. Biol.* **215:**597–606.

Xu, J., Oshima, T., and Yoshida, M., 1991a, Phosphoenol pyruvate insensitive phospho-fructokinase isozyme from *Thermus thermophilus HB8, J. Biochem.* **109:**199–203.

Xu, J., Seki, M., Denda, K., and Yoshida, M., 1991b, Molecular cloning of phosphofructo-kinase 1 gene from a thermophilic bacterium, *Thermus thermophilus, Biochim. Biophys. Acta.* **176:**1313–1318.

Yagi, T., Hon-Nami, K., and Ohnishi, T., 1988, Purification and characterization of two types of NADH-quinone reductase from *Thermus thermophilus* HB8, *Biochemistry* **27:**2008–2013.

Yokoyama, K., Oshima, T., and Yoshida, M., 1990, *Thermus thermophilus* membrane-asso-ciated ATPase. Indication of a eubacterial V-type ATPase, *J. Biol. Chem.* **265:**21946–21950.

Yoshida, M., Oshima, T., and Imahori, K., 1971, The thermostable allosteric enzyme: phos-phofructokinase from an extreme thermophile, *Biochem. Biophys. Res. Commun.* **43:**36–39.

Yoshida, M., and Oshima, T., 1971, The thermostable allosteric nature of fructose 1,6-bisphosphatase from an extreme thermophile, *Biochem. Biophys. Res. Comms.* **45:**495–500.

Yoshida, M., Oshima, T., and Imahori, K., 1973, Fructose-1,6-bisphosphatase of an extreme thermophile, *J. Biochem.* **74:**1183–1191.

Yoshida, M., 1972, Allosteric nature of thermostable phosphofructokinase from an extreme thermophilic bacterium, *Biochemistry* **11:**1087–1093.

Yoshizaki, F., Oshima, T., and Imahori, K., 1971, Studies on phosphoglucomutase from an extreme thermophile *Flavobacterium thermophilum* HB8, *J. Biochem.* **69:**1083–1089.

Yoshizaki, F., and Imahori, K., 1979a, Properties of phosphoenolpyruvate carboxykinase from an extreme thermophile, *Thermus thermophilus* HB8, *Agric. Biol. Chem.* **43:**397–399.

Yoshizaki, F., and Imahori, K., 1979b, Regulatory properties of pyruvate kinase from an extreme thermophile *Thermus thermophilus* HB8, *Agric. Biol. Chem.* **43:**527–536.

Yoshizaki, F., and Imahori, K., 1979c, Key role of phosphoenolpyruvate in the regulation of glycolysis-gluconeogenesis in *Thermus thermophilus* HB8, *Agric. Biol. Chem.* **43:**537–545.

Zakharov, S. D., and Kuz'mina, V. P., 1992, Subunit composition of ATP-synthase from the thermophilic bacterium *Thermus thermophilus, Biokhimia* **57:**777–786.

Zimmerman, B. H., Nitsche, C. I., Fee, J. A., Rusnak, F., and Munck, E., 1988, Properties of a copper-containing cytochrome ba$_3$: a second terminal oxidase from the extreme-thermophile *Thermus thermophilus, Proc. Natl. Acad. Sci. USA* **85:**5779–5783.

Enzymes of *Thermus* and Their Properties

4

MELANIE L. DUFFIELD and DOUG COSSAR

1. INTRODUCTION

Microorganisms have evolved and adapted to survive a great variety of conditions. The *Thermus* species are extreme thermophiles, which can be isolated from hot springs where they grow at temperatures up to 82 °C. Proteins, enzymes, organelles, and DNA purified from strains of *Thermus* are generally more resistant to heat than their mesophilic equivalents (Oshima *et al.*, 1976). This suggests that enhanced thermal stability is an intrinsic property of the molecules. Since mesophilic enzymes do not become thermostable when mixed with extracts from thermophiles, it is unlikely that thermal stability is achieved by transferable, intracellular, nonprotein-stabilizing factors. Genetic experiments support this. Genes from thermophiles still produce thermostable enzymes when cloned in mesophilic hosts (Jaenicke, 1981).

Thermus enzymes have a high resistance to destabilizing factors other than heat, such as extremes of pH, urea, and guanidine hydrochloride (GdmCl) (Table I). Daniel *et al.* (1982) reported that enzymes isolated from thermophilic bacteria were unusually resistant to proteolysis. Since protein unfolding increases the susceptibility to proteolysis, it was proposed that a rigid conformation would contribute to both high thermal stability and low susceptibility to proteolysis. It appears that the mechanisms of conferring protein stability are similar for both chemical and physical challenges.

Industrial and biotechnological processes use a number of enzymes under conditions where extreme temperatures and high concentrations of

MELANIE L. DUFFIELD • Centre for Applied Microbiology and Research, Porton Down, Salisbury, Wiltshire SP4 0JG, United Kingdom. DOUG COSSAR • Cangene Corporation, 3403 American Drive, Mississauga, Ontario, Canada, L4V 1T4.

Thermus Species, edited by Richard Sharp and Ralph Williams. Plenum Press, New York, 1995.

Table I. Various *Thermus* Enzymes and Their Stabilities[a]

Protein	Source	A	B	Reference
L-Alanine dehydrogenase	HB8	+	−	Vali *et al.*, 1980
Alkaline phosphatase	YT1	+	−	Yeh and Trela, 1976
Asparaginase	T351	+	−	Curran *et al.*, 1985
Aqualysin	YT1	+	−	Matsuzawa *et al.*, 1988
Caldolysin	T351	+	+	Cowan and Daniel, 1982
Cytochrome C552	HB8	+	+	Nojima *et al.*, 1978a
Cytochrome oxidase	HB8	+	−	Hon-nami and Oshima, 1980
DNA polymerase (Taq)	YT1	+	−	Sato *et al.*, 1977
Enolase	YT1	+	−	Stellwagen *et al.*, 1973
	X1	+	−	Stellwagen and Barnes, 1978
EFTu	HB8	+	+	Nakamura *et al.*, 1978
Ferredoxin	HB8	+	+	Sato *et al.*, 1981
Fumerase	X1	+	−	Cook and Ramaley, 1976
α-Glucosidase	HB8	+	−	Yang and Zhang, 1988
Glyceraldehyde-3 phosphate dehydrogenase	YT1	+	+	Hocking and Harris, 1976
	YT1	+	+	Oshima *et al.*, 1976
Homoserine dehydrogenase	AT62	+	−	Saiki *et al.*, 1978
Isocitrate dehydrogenase	AT62	+	+	Saiki *et al.*, 1978
	HB8	+	+	Eguchi *et al.*, 1989
3-Isopropylmalate dehydrogenase	HB8	+	−	Tanaka *et al.*, 1981
Lactate dehydrogenase	GK24	+	−	Taguchi *et al.*, 1984
	YT1	+	−	Machida *et al.*, 1985a
Malate dehydrogenase	YT1	+	+	Biffen and Williams, 1976
	AT62	+	+	Iijima *et al.*, 1984
NADH oxidase	YT1	+	−	Cocco *et al.*, 1988
Phosphofructose kinase	X1	+	+	Cass & Stellwagen, 1975
Phosphoglycerate kinase	HB8	+	−	Nojima *et al.*, 1978b
Pullulanese	YT1	+	−	Plant *et al.*, 1986
Superoxide dismutase	YT1	+	+	Sato and Harris, 1977
Phe-tRNA synthetase	HB8	+	−	Ankilova *et al.*, 1988
Trp-tRNA synthetase	HB27	+	−	Koyama and Furukawa, 1990
Val-tRNA synthetase ⎫ Met-tRNA synthetase ⎬ Ile-tRNA synthetase ⎭	HB8	+	−	Kohda *et al.*, 1984

[a]A = Reported heat stability; B = Reported resistance to other denaturing agents.

organic solvents are encountered. Mesophilic enzymes are unable to withstand many of the conditions and so are of limited use. In this context, the proteins and enzymes of thermophiles are of particular interest to the biotechnology industry. A more detailed understanding of the mechanisms of thermal stability and the use of genetic engineering to enhance this property would be of great value (Nosoh and Sekiguchi, 1990).

The DNA of *Thermus* has a high guanine and cytosine content, which may stabilize the dynamic structure formed during transcription and translation at high temperatures (Kushiro *et al.*, 1987). The total G+C content of the chromosomal DNA, their usage in coding regions and, in particular, for the third letter of a codon, are found to be higher in *Thermus* than in *E. coli* (Table II). The high G+C content of *Thermus* DNA also supports some observations on the amino acid content in *Thermus* proteins. The amino acids, such as alanine, arginine, and glutamic acid, that occur most frequently are those which correlate with a high G+C content of the DNA. It is uncertain, however, whether the increased usage of these amino acids is to conserve the %G+C of the DNA or whether they are there to

Table II. *Thermus* Proteins Showing a High G+C Content in Their Coding Region

		Thermophile		Mesophile equivalent
Source		%G+C total	%G+C 3rd letter	%G+C total
Overall	*T. aquaticus*[a]	67.4	—	50.0[l]
	T. thermophilus[b]	69.0	—	
	T. flavus[c]	68.0	—	34.4[m]
Protein				
Aqualysin	*T. aquaticus*[d]	64.6	75.8	—
EFTu	*T. thermophilus*[e]	62.9	84.5	53.3[l]
IPMDH	*T. thermophilus*[f]	70.1	84.9	56.6[n]
LDH	*T. aquaticus*[g]	70.9	91.0	51.9[o]
LDH	*T. caldophilus*[h]	74.1	95.5	51.9[o]
MDH	*T. flavus*[i]	68.0	54.7	50.7[l]
MDH	*T. aquaticus*[j]	68.5	95.7	50.7[l]
W-tRNA Synthetase	*T. thermophilus*[k]	69.3	92.3	54.0[l]

[a]Kunai *et al.*, 1986
[b]Brock and Freeze, 1969
[c]Saiki *et al.*, 1972
[d]Kwon *et al.*, 1988b
[e]Kushiro *et al.*, 1987; Seideler *et al.*, 1987
[f]Kagawa *et al.*, 1984
[g]Ono *et al.*, 1990
[h]Kunai *et al.*, 1986
[i]Nishiyama *et al.*, 1986
[j]Nicholls *et al.*, 1988
[k]Koyama and Furukawa, 1990
[l]*Escherichia coli*
[m]*Caldocellum saccharolyticum*
[n]*Bacillus caldotenax*
[o]*Bacillus stearothermophilus*

increase the thermal stability of the protein. Not all thermophiles show this trend of high %G+C in the DNA. The extreme thermophile *Caldocellum saccharolyticum* has an overall %G+C in DNA of 34%, while that of the β-glucosidase gene was 38% (Love *et al.*, 1988).

2. THERMAL STABILITY

Various proposals have been made to explain the high thermal stability of *Thermus* proteins. Oshima *et al.* (1976) suggested that thermal stability could be conferred on protein molecules by:

1. intrinsic stability, an inherent property of the primary structure of the molecule.
2. ligand induced stability, where the protein is stabilized by the presence of a specific protector molecule.

These mechanisms by which stability is conferred can be further subdivided to cover a wide diversity of thermostabilizing mechanisms operating on the primary, secondary, tertiary, and even the quaternary structure of the protein. Where thermal stability has been carefully studied, it is evident that different mechanisms appear to operate in different cases. Sometimes a cumulative effect of several different factors, including amino acid content, secondary structure, and ligand stability, causes the overall stabilization.

Several proteins isolated from various *Thermus* species were found to be more thermostable than those of the mesophiles. In some cases *Thermus* proteins showed greater stability to denaturing agents such as extremes of pH, organic solvents, urea, GdmCl, and sodium dodecyl sulfate (SDS). However, features of the proteins such as M_r, substrate specificity, and number of subunits were found to be remarkably similar in thermophiles and mesophiles. Since the three-dimensional structures of proteins with homologous functions are more conserved than their primary sequences (reviewed by Blundell *et al.*, 1987), it is likely that the three-dimensional structures of thermophilic enzymes will resemble their mesophilic counterparts. This implies that the high stability of thermophilic proteins must be due to minor differences in the amino acid sequence. No single mechanism or structural change confers thermal stability.

Stability is measured by effects on the catalytic activity of an enzyme. Thermophilic enzymes remain active at much higher temperatures than those of mesophiles. Loss of activity is usually brought about by the unfolding of the protein. In the case of oligomeric proteins, chain unfolding may occur either before or after subunit dissociation. Unfolding occurs by rapid, reversible conformational transitions with equilibria between the states (Figure 1). In thermostable proteins such equilibria are shifted to the unfolded state.

The fully unfolded, inactive state may be obtained by high temperature, the addition of solvents or proteolysis, and proceeds from intermediate stages of unfolding. Therefore, the more the equilibrium favors the fully folded functional form, the larger the input of energy that is required to produce the unfolded form. Once the inactive unfolded protein has been obtained the process cannot be reversed.

The activity of thermostable enzymes is usually quite low at mesophilic temperatures, but increases as the temperature rises. A thermostable protein may have the same degree of flexibility at high temperatures as a mesophile has at lower temperatures. Thermostability and activity can be regarded as a conflict between rigidity and flexibility. To withstand high temperatures, a thermophilic protein needs a rigid structure to compensate for thermal fluctuations. Nevertheless, a degree of flexibility must be maintained, especially in proteins that require conformational changes for their biological activity (Vihinen, 1987).

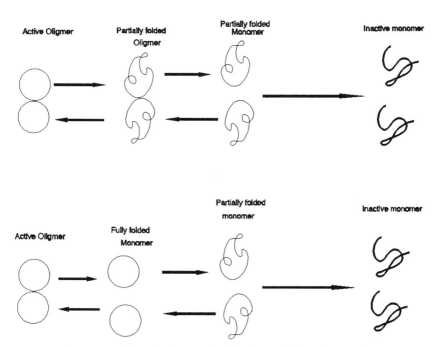

Figure 1. A schematic diagram showing the unfolding of a protein.

On the absolute temperature scale (K) the difference in optimum growth temperature between mesophiles and thermophiles is relatively small. Consequently, the additional free energy of stabilization needed to confer thermal stability on an enzyme is only a relatively small increment in the total interaction energy of a protein (Jaenicke, 1981). Only a small increase in activation energy is required to stabilize a thermophilic protein at elevated temperatures (Atkinson, 1976). Estimates of the differences between thermophiles and mesophiles in the change in free energy of activation for the denaturation, which may be in the order of 5 kcal/mole, may be obtained from the half-life for loss of biological activity. Such small differences in activation energy can be achieved by only one or two additional interactions in the protein structure (Perutz and Raidt, 1975). Therefore, three-dimensional structure determination is of fundamental importance to the understanding of protein thermal stability.

For the determination of three-dimensional crystal structures, purified protein must be crystalized. However, crystals of a purity and size sufficient for x-ray crystallography are often extremely difficult to produce. Hence only around a hundred crystal structures have been determined (not counting multiple versions of the same protein structure with different degrees of refinement or different ligands).

Several *Thermus* proteins have been crystallized, but very few three-dimensional structures have been determined. One is the Mn^{2+} superoxide dismutase from *T. thermophilus* (Stallings *et al.*, 1985), although a complete amino acid sequence was not available for this protein. The crystal structure was built into an electron density map at 2.4 Å resolution, using the amino acid sequences of Mn dismutases from *T. aquaticus* and *B. stearothermophilus*. The tetramer had monomer subunits folded in a manner homologous to the folds of the dimeric Fe superoxide dismutases.

2.1. Proposed Mechanisms of Thermal Stability

The various methods proposed by which thermal stability can be increased will now be considered in more detail.

2.1.1. Amino Acid Content

Attempts have been made to correlate amino acid content and thermal stability of proteins. Argos *et al.* (1979) compared the amino acid sequences of mesophilic and thermophilic molecules of ferredoxin, glyceraldehyde-3-phosphate dehydrogenase, and lactate dehydrogenase and concluded that Gly, Ser, Ser, Lys, and Asp in mesophiles are often replaced in thermophiles by Ala, Ala, Thr, Arg, and Glu, respectively. These replacements have the effect of increasing internal and decreasing external hydrophobicity and favor helix-stabilizing residues in helices. Such substitutions also require an increase in the G+C content of the DNA. Glycine lacks a β-carbon atom and has more backbone conformational flexibility than other amino acids, so replacement of glycine with another amino acid reduces the free energy change during folding. Similarly, proline residues decrease the free energy change during folding, since the ring of proline restricts it to fewer conformations than other amino acid residues (Matthews *et al.*, 1987). Such substitutions cannot be made randomly; for example, the introduction of prolines into helices and sheets may destabilize these secondary structures and so the protein.

Comparative studies on an enzyme from both mesophilic and thermophilic sources have shown a positive correlation between thermal stability and high arginine content (Merkler *et al.*, 1981). An Arg/Lys ratio greater than 1 is often seen in thermophiles (Eguchi *et al.*, 1989). Arginyl groups on the surface of a protein would attract water molecules creating a stable hydration shell, which may support the conformation in the solvent and help protect against denaturation (Rupley *et al.*, 1983). Chemical conversion of lysyl residues to homoargininyl residues in bovine serum albumin, yeast enolase, and alcohol dehydrogenase, increased the thermal stability at lower degrees of modification, but caused a decrease in thermal stability at greater degrees of modification (Qaw and Brewer, 1986). Arginine on the surface could also destabilize proteins, because arginine residues are more bulky than lysine side chains and may disrupt protein folding more. Thermophilic proteins often contain more cysteine than mesophilic proteins, but cysteines involved in essential S-S bridges, or otherwise in binding, are highly conserved.

When proteins fold, hydrophobic side chains are often buried in the core (Argos *et al.*, 1979). By optimizing the internal packing of hydrophobic amino acid residues, a protein will become more tightly folded and exclude water from the interior. Chemical modifications that increased the surface hydrophobicity of proteins have been shown to increase the thermostability (Kellis *et al.*, 1988; Pakula and Sauer, 1990). However, comparisons of hydrophobicity between mesophilic and thermophilic proteins have shown that although most thermophilic proteins have a greater hydrophobicity this is not the sole or major reason for the thermostability.

2.1.2. Secondary Structure

Interactions between amino acids that help to maintain secondary structure and stabilize proteins include hydrogen bonds, salt bridges and ring-ring interactions (for review see Burley and Petsko, 1988). Additional interactions can also confer thermal stability, and it has been predicted that the number of hydrogen bonds and salt bridges would increase with thermal stability (Stellwagen and Wilgus, 1978).

2.1.3. Tertiary Structure

Although few *Thermus* enzymes have had their crystal structures determined, *Thermus* amino acid sequences have been compared with the equivalent mesophilic enzyme for which a structure is known. When functionally similar proteins from two different sources have greater than 50% sequence identity, then the secondary and tertiary structures are also very similar (Stewart and Weiner, 1987). Using such amino acid alignments of mesophiles and thermophiles, it appears that no particular major secondary structure or feature confers thermal stability in all cases.

Salt bridges help to maintain the three-dimensional protein structure and also mediate the binding of charged ligands, substrates and, substrate analogues in proteins (Burley and Petsko, 1988). One extra internal salt bridge may contribute up to 3 kcal/mole (Perutz and Raidt, 1975) and could significantly stabilize a structure.

Hydrogen bonds are electrostatic dipole–dipole interactions between a donor and acceptor atom less than 3 Å apart. Such interactions can contribute -0.5 to -1.5 kcal/mole to the free energy of stabilization of a protein, if the donor and acceptor groups are only partially charged, and -3 to -5 kcal/mole if one of the pairs bears a full electronic charge (Burley and Petsko, 1989). Increasing the number of hydrogen bonds in a molecule will therefore increase the free energy of stability and so promote thermal stability.

The aromatic amino acids, phenylalanine, tryptophan, and tyrosine, may form bonds due to the partial electronic charge created about the aromatic ring, giving a δ^- π-electron cloud covering the face of the aromatic ring, and a δ^+ hydrogen atoms bound to each of the ring carbons. Therefore, the positively polarized hydrogen could interact with the π-electron cloud of a second ring, thus giving a ring/ring interaction (Burley and Petsko, 1989). Interactions between positively charged histidine, lysine or arginine with aromatic rings has also been considered important in the stability of proteins (Loewenthal *et al.*, 1992).

Disulfide bonds make large contributions to tertiary structure and can stabilize folded conformations by between 2 and 5 kcal/mole for each disulfide (for a review, see Creighton, 1988). Bovine pancreatic trypsin inhibitor is a very stable protein with a melting temperature of around 95 °C. However, it unfolds completely when its three S–S bonds are broken and the melting temperature can be lowered to 50 °C when two of the three bonds are broken (Johnson *et al.*, 1987). The introduction of cysteines and disulfide bridges does not necessarily increase thermal stability. Since such bonds have stringent structural requirements, satisfying these geometrical constraints for the introduction of new cysteines can be difficult. Under some conditions disulfides can be very labile and the thiol groups are susceptible to oxidation. Disulfide bonds can be potential sites for inactivation of an enzyme required to function aerobically at temperatures above 70 °C (Creighton, 1988). Therefore, attempts to introduce disulfide bridges into proteins to increase thermal stability by site-directed mutagenesis have not always been successful. It was proposed that the disulfide link should be introduced so that the loop formed is as large as possible. Any other interactions that stabilize the protein should be maintained (Matsumura *et al.*, 1988).

2.1.4. Ligand-Induced Stability

Stability can also be conferred on a protein by ligands and constituents other than amino acids. For instance, metal ions are known to affect the thermal stability of some proteins. Stability of the protease caldolysin depends on the presence of Ca^{2+} ions ($t\frac{1}{2}$ at 75 °C with 10mM calcium >193hrs, $t\frac{1}{2}$ at 75 °C with no calcium = 4.8 min). However, the activity of the enzyme remains unaffected by the concentration of calcium ions (Cowan and Daniel, 1982). The thermal stability of pullulanase is also dependent on the presence of Ca^{2+} ions (Plant *et al.*, 1986).

2.2. Examples of Thermal Stability

The thermal stability of some *Thermus* proteins has been investigated in detail. For examples of this, see section 3.1, 3.2 and 3.6.

3. OXIDOREDUCTASES

3.1. Lactate Dehydrogenase (EC 1.1.1.27)

L-lactate + NAD$^+$ + H$^+$ → pyruvate + NADH

Tetrameric lactate dehydrogenase (LDH) purified from *T. caldophilus* GK24 is allosteric, with fructose 1,6-bisphosphate (FBP) as the effector. The pH optimum for pyruvate reduction is 4.5, and the optimum temperature is at 80 °C while the optimum temperature for lactate oxidation is 95 °C. An arginine residue was involved in the binding of FBP and in the allosteric regulation of the enzyme activity (Taguchi *et al.*, 1984). The change of His188, the protonization of which is important for the interaction of the enzyme with the allosteric effect of FBP (Schroeder *et al.*, 1988), to Phe188 abolished the stimulatory effect of FBP. The binding of FBP to the *T. caldophilus* enzyme converted the ribose ring of the adenosine moiety of NAD^+ from the C2'-endo form to the C3'-endo form (Machida *et al.*, 1985b). The nucleotide sequence of the *T. caldophilus* enzyme translates to 310 amino acid residues with a calculated M_r of 32,808. The amino acid sequence shows a 40% identity in alignment with lactate dehydrogenase from *B. stearothermophilus, Lactobacillus,* and dogfish. The %G+C usage overall was 74.1% while at the third nucleotide it was 95.5% (Kunai *et al.*, 1986). Preliminary crystallographic data for this LDH has been reported (Koide *et al.*, 1991). The *T. aquaticus* YT1 enzyme is also FBP dependent, and its N-terminal sequence of 34 amino acids has a high homology with other bacterial lactate dehydrogenases (Machida *et al.*, 1985a). The LDH gene from *T. aquaticus* has a 70.9% usage of G+C overall, rising to 91.0% at the third base. It translated to an amino acid sequence 310 residues long with a calculated M_r of 33,210 and showed 87.1% sequence identity with lactate dehydrogenae from *T. caldophilus*. LDH from *T. aquaticus* showed a 4-to-16 fold greater affinity for the substrate (pyruvate), coenzyme (NADH), and FBP than did the *T. caldophilus* enzyme (Ono *et al.*, 1990). LDH from *T. caldophilus* shows considerable heat stability without FBP (t½ at 100 °C ≈ 5 min without FBP and ≈ 20 min with 0.5 mM FBP) (Taguchi *et al.*, 1982). The presence of FBP increased the thermal stability of this enzyme. The same is found with LDH from *T. aquaticus* (without FBP there is a 30% loss in activity at 97 °C, while with 100 µM FBP, 15% activity is lost after 20 min) (Machida *et al.*, 1985b). These enzymes are both tetramers with FBP bridging across the two dimeric structures in the native molecule, therefore both the activation and thermal stability of lactate dehydrogenase from *T. caldophilus* could reflect the stabilization of the tetrameric form.

Comparisons between the LDH enzymes from *T. aquaticus* and *T. caldophilus* show that citrate activates both enzymes in acidic conditions while FBP activates at both acidic and neutral pHs. High concentrations of citrate did not inhibit either enzyme, but FBP did cause inhibition because of the change in pH optimum (Table III) (Machida *et al.*, 1985a).

A thermophilic isolate from hot springs in China, thought to belong to the *Thermus* genus, had LDH with the optimum temperature for pyruvate reduction of 60 °C at pH 8.0. The t½ for thermal inactivation at 85 °C was 10 min (Yang *et al.*, 1991).

Table III. The Properties of Lactate Dehydrogenase from *T. aquaticus* and *T. caldophilus*

	T. aquaticus	*T. caldophilus*
M_r subunit	33,000	31,000
M_r native	110,000	120,000
pI	5.3	5.5
Vmax (units/mg)	77	7.4

3.2. Malate Dehydrogenase (EC 1.1.1.37)

$$\text{L-malate} + NAD^+ \rightarrow \text{oxaloacetate} + NADH + H^+$$

Malate dehydrogenase (MDH) isolated from *T. flavus* AT62 is a dimer of identical subunits, each of 327 amino acids, with a native M_r of 67,000. It is inhibited by oxaloacetate (but this decreases with increasing temperature) and is heat-stable, showing no loss of activity after 60 min at 90 °C (Iijima *et al.*, 1980). *T. flavus* MDH denatures at pH 2.0, losing its activity without dissociating, which suggests strong subunit–subunit interactions. At 5 mM GdmCl it is both denatured and dissociated, but both effects are reversible. Kinetic analysis of the reactivity of the enzyme after denaturation by GdmCl suggests that a change from inactive dimer to active dimer is the rate-limiting step (Iijima *et al.*, 1984). The nucleotide sequence of *T. flavus* AT62 MDH gene shows a usage of G+C of 68.6% and translation gives a subunit M_r of 35,000. Asp159 and His187 were identified as being essential residues to catalysis (Nishiyama *et al.*, 1986).

MDH from *T. aquaticus* YT1 has an M_r of 73,500 and a pI estimated at 4.4 (Biffen and Williams, 1976). The gene encoding MDH differed only by 20 nucleotides from the MDH sequence of *T. flavus* and gave rise to an identical amino acid sequence. The calculated M_r of the monomer was determined as 35,397. The %G+C content over the gene was 68.5%, rising to 95.7% at the third position in the codon (Nicholls *et al.*, 1989).

The x-ray crystal structure of *Thermus flavus* AT62 MDH complexed with NADH has been determined to a 1.9 Å resolution. Four ion pairs were identified in this structure, which may play a significant role in thermostability (Birkoft *et al.*, 1989). One of these is located at the interface between domains whilst the others are found at the subunit interface, which supports the observations made by Iijima *et al.* (1984). Only small changes were seen in the increased hydrophobicity and decreased flexibility in α-helical and interdomain regions, so their net contribution to thermostabilization was considered small (Kelly *et al.*, 1993).

3.3. NADP-Dependent Isocitrate Dehydrogenase (EC 1.1.1.42)

threo-Ds-isocitrate + NADP$^+$ → 2-oxoglutarate + CO_2 + NADPH + H$^+$

Isocitrate dehydrogenases (ICDH) exhibit significant diversity in physical and catalytic properties. A comparison of M_r of bacterial ICDH shows quite a range of size, although they are all dimers (Table IV). The *T. aquaticus* ICDH was found to be the most thermostable, but addition of 0.2 M KCl enhanced the stability, the half life at 66 °C going from 305 min to over 500 min. The *Thermus* enzyme had the highest arginine content of all the isocitrate dehydrogenases examined (Edlin and Sundaram, 1989). ICDH from *T. flavus* was highly stable, with no loss of activity after 60 min at 70 °C, although 20% and 80% of the activity were lost after 60 min at 80 °C and 90 °C, respectively (Saiki *et al.*, 1978). It has an M_r of 60–70,000 and an optimum pH of 8.0 (Ramaley and Hudock, 1973). NADP-dependent ICDH from *T. thermophilus* HB8 was a dimeric protein of M_r 115,000. Divalent cations such as Mn^{2+} and Mg^{2+} were essential to activity, the optimal pH was 7.8 at 55 °C, and the K_m for NADP and D-isocitrate were 6.3 and 8.8 μM respectively, with a V_{max} of 77.6 μmol/min/mg at 55 °C (Eguchi *et al.*, 1989).

The gene for this enzyme consisted of a single open reading frame of 1,485 base pairs. There was 65% usage of G+C overall, rising to 90.3% for the third letter of the codon. The deduced amino acid sequence gave an M_r of 54,189. The amino acid sequence compared with ICDH from *E. coli* showed a 37% overall similarity. However, there are an extra 141 residues at the C-terminal tail of the *Thermus* enzyme. The serine residue known to be phosphorylated in the *E. coli* enzyme was conserved (Miyazaki *et al.*, 1992).

Table IV. The Properties of Isocitrate Dehydrogenase from Various Sources

	M_r of monomer	M_r of dimer	pI
T. aquaticus	33,000	66,000	—
T. thermophilus	57,500	115,000	5.5
T. flavus	46,500	80–90,000	5.4
B. stearothermophilus	46,000	92,000	5.0
E. coli	43,000	80,000	4.4

3.4. 3-Isopropylmalate Dehydrogenase (EC 1.1.1.85)

3-isopropylmalate + NAD^+ → 2-ketoisocaproate + NADH + H^+ + CO_2

The gene coding for 3-isopropylmalate dehydrogenase (IPMDH), namely the *leuB* gene from *T. thermophilus* HB8 has been cloned and expressed in *E. coli*. The enzyme was a dimer of identical subunits of M_r 40,000 ± 500. The K_m for threo-Ds-3-isopropylmalate and NAD were estimated at 8.0×10^{-5} and 6.3×10^{-5} M respectively. The optimum pH at 75 °C in the presence of 1M KCl was around 7.2. The presence of Mg^{2+} or Mn^{2+} was essential for activity. The enzyme was activated 30-fold by 1M KCl or RbCl. The high salt concentration decelerated the thermal unfolding of the enzyme, and accelerated the aggregation of the unfolded protein (Yamada *et al.*, 1990). This enzyme has been crystallized (Katsube *et al.*, 1988).

The *leuB* gene consists of 1017 base pairs, which translates into a protein of 339 amino acids with a calculated M_r of 35,968. The M_r of the active dimeric protein has been determined at 73,000. The G+C usage of the coding region was 70.1%, and that of the third letter was 89.4% (Kagawa *et al.*, 1984).

The *leuB* gene from *T. aquaticus* YT1 has an open-reading frame of 1,035 base pairs and the G+C content of the coding region was found to be 69.9%, rising to 93.6% for the third letter. This encoded a subunit of 344 amino acids for which 87% homology was calculated with the amino acid sequence of ICDH from *T. thermophilus* (Kirino and Oshima, 1991).

The crystal structure of IPMDH from *T. thermophilus* has been determined at a 2.2 Å resolution. Comparison of this with the three-dimensional structure of mesophilic enzymes shows that the dimeric form is essential for enzymatic activity. The close subunit contact at the hydrophobic core is important for the thermal stability of the *Thermus* enzyme (Imada *et al.*, 1991), which can be denatured by GdmCl. At above 60 °C there is little or no refolding after dialysis, but the introduction of a chaperonin isolated from *Thermus* induces refolding in the presence of ATP, even above 70 °C (Taguchi *et al.*, 1991).

3.5. Ferredoxin (EC 1.6.7.1)

Reduced ferredoxin + NAD^+ → oxidized ferredoxin + NADH + H^+

Ferredoxins are low M_r electron transfer proteins which contain non-heme iron and acid-labile sulfur atoms. Ferredoxin from *T. thermophilus* has been compared with that from other mesophilic sources (Sato *et al.*,

1981). The sequences of ferredoxins from *T. thermophilus* and *Acetobacter vinelandii* show quite a high sequence identity, particularly at the cluster sites. The crystal structure of ferredoxin from *A. vinelandii* was determined at 2 Å resolution and the amino terminal residues (1–50) form a core containing the Fe-S cluster sites. Residues 51–107 wrap around this core (Ghosh *et al.*, 1982; Howard *et al.*, 1976). From the sequence identity it is likely that ferredoxin from *T. thermophilus* has a similar structure.

The optical absorption spectrum of *Thermus* ferredoxin has a peak at 280 nm and a broad maximum around 400 nm, similar to that of *A. vinelandii* and various *Clostridia* species, which have peaks at 280 and 390 nm. *Thermus* ferredoxin is extremely thermostable, with no decrease in the absorbance at 400 nm after a 45 min incubation at 65 °C (Sato *et al.*, 1981). Comparison of *C. pasteurianum*, *C. acidiurici*, *C. tartarivorum*, and *C. thermosaccharolyticum* shows that they are similar in size and absorption spectrum but different in resistance to heat. After incubation at 70 °C for one hour, the protein of *C. thermosaccharolyticum* loses only 5% activity, while those of *C. tartarivorum*, *C. acidiurici* and *C. pasteurianum* lose 15, 50, and 70% respectively under these conditions (Devanathan *et al.*, 1969). *T. thermophilus* and *Azotobacter vinelandii* ferredoxins have one trinuclear and one tetranuclear Fe-S cluster detected by magnetic circular dichroism (Johnson *et al.*, 1987). Ferredoxins from most other bacteria have one or two 4Fe-4S clusters, and plant ferredoxins have only one 2Fe-2S cluster (Table V), and there are differences in M_r. The amino acid sequences of ferredoxins from clostridia were compared (Perutz and Raidt, 1975) using the three-dimensional structure of ferredoxin from the mesophilic *Micrococcus aerogenes* for comparison. The higher thermal stability of the two thermophilic clostridia species was proposed to be due to more hydrogen bonds and salt bridges. In particular, the additional salt bridges were noted between His2 and Gln44 (or Glu44), and between Glu7 and Lys53. The amino acid sequence of ferredoxin from *T. thermophilus* has 78 amino acids rather than the 55 of *Clostridium* species, which makes the sequences and tertiary structures more difficult to compare. However, the distribution of the cysteine res-

**Table V. The Properties of Ferredoxin
from Various Sources**

Source of ferredoxin	M_r	Clusters
T. thermophilus	10,500	3Fe-3S, 4Fe-4S
A. vinlandii	14,000	3Fe-3S, 4Fe-4S
C. thermosaccharolyticum	≈6,000	4Fe-4S, 4Fe-4S
C. pasteurianum	≈6,000	4Fe-4S, 4Fe-4S

idues involved in the iron complexes is similar and the amino acid sequence around them can be aligned. Sato *et al.* (1981) suggested that His2′ in *T. thermophilus* ferredoxin, which is replaced by tyrosine or phenylalanine in homologous mesophilic ferredoxins, could also contribute to the heat stability of the enzyme. However, of the four residues thought to be involved in the thermal stability in *C. thermosaccharolyticum* ferredoxin, only His2 is conserved in the *Thermus* enzyme. Therefore, the thermal stability of *T. thermophilus* may be due to other interactions. These will only be evident once a three-dimensional crystal structure is determined.

3.6. Glyceraldehyde-3-Phosphate Dehydrogenase (EC 1.2.1.12)

D-glyceraldehyde-3-phosphate + orthophosphate + NAD^+ →
1:3-diphospho-glycerate + NADH $+H^+$

Glyceraldehyde-3-phosphate dehydrogenase (GAPDH) from *T. aquaticus* is a tetrameric molecule, like mesophilic enzymes, but it is extremely thermostable, with $t\frac{1}{2}$ < 90 min at 90 °C. It is also stable in denaturing solvents such as urea, GdmCl, and SDS (Hocking and Harris, 1976, 1980). A comparison of the properties of GAPDHs from *T. aquaticus* and *B. stearothermophilus* can be seen in Table VI. The nucleotide sequence of the GAPDH gene from *T. aquaticus* YT1 has a G+C content of 96% (Hecht *et al.*, 1989).

The three-dimensional structure of GAPDH from lobster and the moderate thermophile, *Bacillus stearothermophilus*, have been compared in an attempt to explain the extra thermal stability of the *B. stearothermophilus* enzyme (Biesecker *et al.*, 1977). There is a loop of polypeptide (called the S-loop) which is in contact with the co-enzyme NAD^+. In the tetramer, four S-loops form the core of the molecule, with their residues internal and making important interactions with the other subunits. In *B. stearothermo-*

Table VI. The Properties of Glyceraldehyde Dehydrogenase from Two Thermophiles

	T. aquaticus[a]	*B. stearothermophilus*[b]
M_r (subunit)	37,000	40,000
M_r (native)	150,000	163,500 ± 4,000
pI	3.5–4.0	4.3
K_m (M)	7.5×10^{-4}	

[a]Hocking and Harris, 1980
[b]Singleton *et al.*, 1969

philus GAPDH, changes in the residues of the S-loop cause structural changes that could contribute to the thermal stability of this enzyme. There is an extra ionic bond between Arg194 and Asp293. Also a double ionic interaction occurs between the Arg281 from the Q-axis related subunits, and the Glu201 from the R-axis and P-axis related subunits, that are not seen in lobster GAPDH. These interactions are shielded from the solvent and bring the S-loop from two subunits into contact with the other two subunits, making a major contribution to the stability of the *B. stearothermophilus* tetramer (Biesecker *et al.*, 1977). GAPDH from *Thermus* is even more thermostable than that from *B. stearothermophilus*. By fitting the sequence from *Thermus* to the three-dimensional structure of *B. stearothermophilus*, the possible source of this thermal stability was sought. The interaction between Arg194 and Asp293 described above occurs, but the other double ionic bonds do not. Predicted changes between the α^β and α^3 helices introduced more bulky and hydrophobic residues that would block the surface crevice, so preventing access of water to the interior of the molecule and helping to maintain the intersubunit interactions (Walker *et al.*, 1980). The number of salt bridges on the surface is also greater, allowing for maximum surface ion pair formation (up to sixteen extra charged residues predicted to be in salt bridges). This prevents the penetration of water into the hydrophobic interior of the molecule and helps to preserve the subunit interactions. Residues in the S-loop were also found to be more bulky in the thermophilic protein than in the mesophile (Olsen, 1983).

3.7. L-Alanine Dehydrogenase (EC 1.4.1.1)

L-alanine + H_2O + NAD^+ → pyruvate + NH_3 + NADH + H^+

L-alanine dehydrogenase from *T. thermophilus* has six subunits of identical size with a total M_r of 290,000. Its melting temperature was 86 °C and heat inactivation showed first order kinetics (Vali *et al.*, 1980). Its pH optimum was 8.0 for reduction but 10.0 for the oxidation of alanine. The K_m for pyruvate was estimated at 0.75 mM. L-cysteine and L-alanine both act as competitive inhibitors.

3.8. NADH Dehydrogenase (EC 1.6.99.3)

NADH + H^+ + acceptor → NAD^+ + reduced acceptor

NADH dehydrogenase (NADH:ferricyanide oxidoreductase) isolated from *Thermus* T351 has an M_r of 50,000 and is inhibited by NADH and ferricyanide. It apparently contains 0.05 mole FMN, 0.16 mole labile sulfur, and 2.2 mole iron per mole protein; and has an optimum pH of 3.6 and a $t\frac{1}{2}$ at 95 °C of 35 min. (Walsh *et al.*, 1983). Two NADH dehydrogenases were isolated from *T. thermophilus* HB8 of which NADH dehydrogenase 1 is a complex of 10 dissimilar polypeptides and contains noncovalently bound FMN, a nonheme iron, and acid labile sulfide. NADH dehydrogenase 2 exhibits a single band (M_r 53,000) on an SDS-PAGE gel and contains noncovalently bound FAD and no nonheme iron or acid labile sulfide. The activities of both were stable at or above 80 °C (Yagi *et al.*, 1988). Five distinct low potential iron–sulfur clusters were identified potentiometrically in the membrane particles of *T. thermophilus* HB8. Three of these clusters, one binuclear (designated [N-1H]T) and two tetranuclear ([N-2H]T and [N-3H]T) are components of the energy-coupled NADH-menaquinone oxidoreductase complex (NADH dehydrogenase I). Two very-low-potential iron-sulfur clusters (one binuclear, [N-1L]T, and one tetranuclear, [N-2L]T) were also observed in membrane particles (Meinhardt *et al.*, 1990). The NADH-binding subunit of NADH dehydrogenase 1 has been identified as a 47,000 M_r polypeptide (Xu and Yagi, 1991).

3.9. NADH Oxidase (EC 1.11.1.1)

$$O_2 + NADH + H^+ \rightarrow NAD^+ + H_2O_2$$

The flavoprotein NADH oxidase (NADH:oxidoreductase) isolated and purified from *T. aquaticus* YT1 is a homodimer with M_r of 110,000, similar to the NADH oxidase from *B. megaterium* (Cocco *et al.*, 1988). The *T. thermophilus* HB8 enzyme is a monomer with M_r 25,000 (Park *et al.*, 1992b), considerably smaller than the monomer of NADH oxidase from *Streptococcus faecalis* (Table VII). All these oxidize NADH, but both the

Table VII. The properties of NADH Oxidase from Various Sources

	Source of NADH oxidase			
	T. aquaticus[a]	*T. thermophilus*[b]	*B. megaterium*[c]	*S. faecalis*[d]
M_r Native	110,000	25,000	92,000	51,000
M_r Subunit	55,000	25,000	55,000	51,000
K_m (NADH) (μM)	39	4.14	60	41
K_m (NADPH) (μM)	—	14.0	—	—

[a]Cocco *et al.*, 1988
[b]Park *et al.*, 1992b
[c]Saeki *et al.*, 1985
[d]Schmidt *et al.*, 1986

Thermus and *B. megaterium* enzymes are two-electron donors and produce hydrogen peroxide as an end product, whereas that of *S. faecalis* is a four-electron donor producing water as an end product.

T. aquaticus NADH oxidase is highly heat stable, and still active at 95 °C and is also active over a pH range of 5.0–10.5. It has an absolute requirement for FAD of 0.7 mole/mole dimer, and has a very high specificity for NADH with little activity against other nucleotides (Cocco *et al.*, 1988).

The NADH oxidase from *T. thermophilus* is stable up to 80 °C and has a pH optimum of 5.0. Its cloned sequence provides a calculated M_r of 26,835 in close agreement with the experimental value. The usage of G+C was 70.4% overall with a preference at the third position of the codon of 81.9% (Park *et al.*, 1992a). Crystals and preliminary x-ray diffraction studies for this enzyme have been reported (Erdmann *et al.*, 1993).

3.10. Catalase

Unlike those found in other organisms, the catalase of *Thermus thermophilus* is a nonheme enzyme that utilizes an unusual dimanganese active site, which can be interconverted between four redox states. The crystal structure of this enzyme has been determined to 3 Å resolution, and was found to comprise six equivalent subunits each folded into four long antiparallel helices, a motif also found in other metalloproteins (Vainshtein *et al.*, 1984).

3.11. Superoxide Dismutase (EC 1.15.1.1)

$$O_2^- + O_2^- + 2H^+ \rightarrow O_2 + H_2O_2$$

Superoxide dismutase (SOD) catalyzes the dismutation of superoxide ions to oxygen and hydrogen peroxide. The enzymes are classified according to which redox-active metal is required for catalysis, Cu, Fe or Mn. The tetrameric manganese SOD from *T. thermophilus* HB8 has an average of 2.2 Mn ions bound per tetramer. Usually Mn superoxide dismutases are dimeric. A crystal structure of this enzyme was determined to 2.4 Å resolution and, although only a partial amino acid sequence of the *Thermus* protein was known, various residues were identified as being involved at the metal-ligand site or the active center (Stallings *et al.*, 1985). The full amino sequence was determined later, and each subunit consists of 203 residues with M_r of 23,144 compared with a tetramer M_r of 84,000 (Sato *et al.*, 1987). Using this fully determined sequence, the crystal structure was refined to 1.8 Å resolution. The monomer comprises distinct N- and

C-terminal domains. The tetramer shows 222 symmetry held together by two interfaces. Each of the metal sites is fully occupied with the Mn(III) five-coordinate in trigonal bipyramidal geometry (Ludwig *et al.*, 1991).

3.12. Cytochromes

3.12.1. Cytochrome *c*-552

Cytochrome *c*-552 purified from *T. thermophilus* HB8 has 131 amino acid residues with an M_r of 14,800 for the monoheme apoenzyme and shows sequence homology to other c-type proteins (Titani *et al.*, 1985). It is stable to heat, pH and GdmCl and was considered to be more thermostable than mesophilic cytochrome-*c* because the free energy change for denaturation is greater and has its maximum at a higher temperature (Nojima *et al.*, 1978a).

3.12.2. Cytochrome ba_3

Cytochrome ba_3, a terminal oxidase from the plasma membrane of *T. thermophilus*, is a single peptide chain of M_r about 35,000 that binds 1 heme B molecule, 1 heme A molecule and 2 copper ions as indicated by optical spectra (Zimmermann *et al.*, 1988).

3.12.3. Cytochrome c_1aa_3

Cytochrome c_1aa_3 from the periplasm of *T. thermophilus* has M_r of 93,000. It was isolated as a two subunit enzyme. The c_1 component, M_r 33,000, binds to heme c while aa_3, M_r 55,000, also contains 2 heme centers a and a_3, copper centers Cu_A and Cu_B, and an additional heme c (Yoshida *et al.*, 1984; Fee *et al.*, 1980). The nucleotide sequence of the gene for the c_1 subunit has %G+C usage overall of 67.0%, rising to 92.6% at the third position in the codon. It codes for a protein of 320 amino acid residues with a pI at 5.5. The aa_3 contains a subunit II homologous portion (amino terminal two thirds) and a cytochrome *c* homologous portion (carboxy terminal third). The subunit II sequence, like all II's, forms a transmembrane domain that anchors the subunit in the membrane, and an intermembrane domain that binds cytochrome c. The genes for typical subunit I and subunit III proteins have been sequenced (Mather *et al.*, 1991). The *caaB* gene is continuous, encoding both subunit I and III. The G+C content is 65%, slightly lower than the overall base content of *T. thermophilus*, however, at the third position this rises to 95.5%. Translation of the open reading frame predicted a polypeptide of 791 residues, with an M_r

of 89,200. This is somewhat larger than experimental values (between 54,000 to 71,000) and the 791-residue protein is thought to represent a precursor form that is processed to yield mature subunit I and III (Mather *et al.*, 1993).

3.12.4. Rieske Protein

Proteolipid complexes containing cytochromes *b* and *c*, are essential for respiratory electron transport in mitochondria, chloroplasts, and certain bacteria. The Fe/S protein of a bc$_1$ complex was first purified by Rieske *et al.* (1964). Subsequent results have shown that the Rieske protein acts as a ubiquinol-cytochrome c_1 ubisemiquinone-cytochrome *b* oxidoreductase. The Rieske Fe/S protein from *Thermus thermophilus* has an M$_r$ of 20,000 with two 2Fe-2S clusters, each of which has at least two noncysteine ligands (Fee *et al.*, 1984).

4. TRANSFERASES

4.1. tRNA Methyltransferases

tRNA (adenine-1)-methyltransferase (EC 2.1.1.36) from *T. flavus* strain 71 has M$_r$ of around 78,000, a pH optimum (at 70 °C) of 6.9 and a pH range of 6.5–7.3. Its optimum temperature for activity was 75 °C, and there was no loss of activity after 1 hour at 60 °C. It has an absolute requirement for Mg^{2+} ions and activity is stimulated by low concentrations of Na$^+$ and NH$_4^+$ (Morozov *et al.*, 1982).

A thermostable tRNA (guanosine-2'-)-methyltransferase (EC 2.1.1.34) has also been isolated from *T. thermophilus* HB27. It is a monomer of M$_r$ 20,000 with a K$_m$ for S-adenosylmethionine of 0.47 μM and for tRNA^{Phe} (from *E. coli*) of 10 nM. After 20 min at 80 °C, 90% of the original activity remained (Kumagai *et al.*, 1982).

4.2. Phosphofructokinase (EC 2.7.1.11)

ATP + D-fructose-6-phosphate → ADP + D-fructose-1,6-bisphosphate

ATP:D-fructose-6-phosphate 1-phosphotransferase (PFK) catalyses the production of fructose-1,6-bisphosphate from fructose-6-phosphate in the glycolytic pathway. PFK purified from *Thermus* X1 (Cass and Stellwagen, 1975) has increased thermal stability together with a greater stability to GdmCl, urea, and acid treatment compared with mesophilic enzymes from *E. coli* and *C. pasteurianum* (Cass and Stellwagen, 1975). The M$_r$

of the enzyme of strain X1 was 34,000, similar to that of *E. coli* and *B. stearothermophilus* (Table VIII). PFK from *Thermus* X1, *E. coli*, and *B. stearothermophilus* all exhibit K-type allosteric kinetics in which fructose 6-phosphate and ADP function as positive effectors, while phosphoenolpyruvate is a negative effector (Stellwagen and Thompson, 1979), T1$^+$, NH$_4^+$, K$^+$, Rb$^+$ or Cs$^+$ increase the activity of phosphofructokinase from X1 (Stellwagen and Thompson, 1979) in decreasing order of potency. PFK from *T. thermophilus* is also homotetrameric with M$_r$ of 36,599 (total M$_r$ 140,000). It dissociates to an inactive dimer (M$_r$ 74,000) in the presence of PEP. Tetramer regeneration is brought about by the addition of fructose-6-phosphate or MgADP or by the removal of PEP. The hyperbolic plots of velocity versus fructose-6-phosphate concentration change to sigmoidal plots on addition of PEP. Further addition of ADP causes reversion to hyperbolic plots (Xu *et al.*, 1990). A second PFK isolated from *T. thermophilus* has an M$_r$ of 132,000 with subunits of 34,500 suggesting that it is also tetrameric. The hyperbolic kinetics and molecular form of PFK2 are not affected by phosphoenolpyruvate. This PFK obeys simple Michaelis–Menten kinetics (Xu *et al.*, 1991) and is produced only when glucose is in the culture medium. The gene for PFK1 from *T. thermophilus* has been cloned in *E. coli* and its nucleotide sequence deduced. The G+C content was found to be 72.0% throughout the coding sequence and 96% for the third letter in a codon. By comparison with PFKs from other sources, it was found that residues involved in substrate binding sites are conserved. However, important differences are found at the subunit binding interface (Xu and Yagi, 1991).

Table VIII. The Properties of Phosphofructokinase from Various Sources

	M$_r$ subunit	M$_r$	pH range	T °C opt	K$_m$ (ATP) μM
X1[a]	34,000	132,000	9.5	80	59
T. caldophilus[b]	36,500	140,000	6.5–8.4	—	—
C. pasteurianum[c]	35,000	144,000	7.0–8.2	42	55
E. coli[d]	35,000	142,000	—	—	60
Rabbit[d]	38,000	—	6.0–9.2	—	42
Yeast[d]	590,000	—	6.0–9.2	—	48

[a]Cass and Stellwagen, 1975
[b]Xu *et al.*, 1990
[c]Uyeda and Kurooka, 1970
[d]Bloxham and Lardy, 1973

4.3. Phosphoglycerokinase (EC 2.7.2.3)

ADP + 1,3-diphospho-D-glycerate → ATP + 3-phospho-D-glycerate

The gene sequence for *T. thermophilus* phosphoglycerokinase (PGK), an enzyme of the glycolytic pathway, translates to an enzyme of 390 amino residues (M_r 41,791). As with other *Thermus* genes there was a strong bias in G+C content at the third position of the codon (91.3%) and 70.9% throughout the coding region (Bowen *et al.*, 1988). Crystals of this enzyme have been used to determine the three-dimensional crystal structure (Littlechild *et al.*, 1987). The physical and catalytic properties of PFK from various sources have been determined and are compared in Table IX. PGK's are monomeric enzymes with no tightly bound metal ions or prosthetic groups. Most properties of PGK from *T. thermophilus* are very similar to those from mesophilic organisms, but the isoelectric point of both the extreme thermophile (*Thermus*) and the moderate thermophile (*B. stearothermophilus*) are more acidic than the mesophilic enzymes. The PGK from *T. thermophilus* also shows the highest substrate specificity when compared with that of yeast, rabbit, and *B. stearothermophilus* (Nojima *et al.*, 1978a). PGK's are inhibited by ADP acting competitively with 3-phosphoglycerate and noncompetitively with ATP (Scopes, 1973). The reversibility of denaturation by GdmCl of both PGK from yeast and *T. thermophilus* has been studied using circular dichroism and fluorescence intensity (Nojima *et al.*, 1978b). The free energies of denaturation at 25 °C were estimated to be 11.87 ± 0.21 kcal/mole for *T. thermophilus* PGK and 5.33 ± 0.13 kcal/mole

Table IX. The Properties of Phosophoglycerokinase from Various Sources

	Source of phosphoglycerokinase			
	T. thermophilus[a]	*B. stearothermophilus*[b]	Rabbit[c]	Yeast[d]
Property	44,600	42,000	48,000	50,000
M_r				
pI	5.0	4.9	7.0	7.2
pH optimum	6.0–8.5	5.5–8.5	6.0–9.2	6.0–9.2
% helix	29	20	—	31
% sheet	11	45	—	18
K_m (ATP) mM	0.28	2.9	0.48	0.32
K_m (G3P) mM	1.79	2.2	61.22	1.30

[a]Nojima *et al.*, 1979
[b]Suzuki and Imahori, 1974
[c]Scopes, 1973
[d]Krietsch and Bücher, 1970

for yeast PGK. The van't Hoff plot of the equilibrium constant for the denaturation reaction was almost independent of temperature between 0 to 60 °C for *T. thermophilus* PGK, while that of yeast, PGK was strongly temperature dependent, as it is for other thermolabile proteins (Nojima *et al.*, 1977).

For any given temperature PGK from *T. thermophilus* HB8 is more stable and its conformational dynamics (as measured by the ability of acrylamide to quench the fluorescence of a buried tryptophan) as well as its specific activity, are lower than of the mesophilic yeast enzyme. At higher temperatures the thermodynamic stability of the thermophilic protein approaches that demonstrated by mesophilic protein at lower temperatures. These results demonstrate the constraining effect of increased stability upon conformational dynamics and enzyme activity (Varley and Pain, 1991).

4.4. RNA Polymerases (EC 2.7.7.6)

RNA polymerase purified from *T. aquaticus* has four subunits, which correspond to subunits β', β, σ and α of *E. coli* (Table X). It has an absolute requirement for Mg^{2+} or Mn^{2+} and was inhibited by KCl (Air and Harris, 1974; Fabráy *et al.*, 1992). RNA polymerase A from *T. thermophilus* can utilize various DNA's as templates and is extremely thermostable, maintaining 100% activity after three hours at 70 °C. The maximum rate of RNA synthesis was also achieved at 70 °C. Polymerase B from the same strain was only active with alternating copolymers of deoxyadenylic and deoxythymidylic acids (poly d(A-T)) as templates and also had four subunits identified as α, β, β', and σ by their M_r. Polymerase A holoenzyme comprised $(\beta\beta'\alpha_2)$ σ and polymerase B $(\beta\beta'\alpha_2)$ core enzyme lacks the σ

Table X. The Properties of RNA Polymerase from *T. aquaticus* and *T. thermophilus*

	T. aquaticus	*T. thermophilus*
Subunit M_r	165,000	170,000
β'		
β	130,000	140,000
σ	92,000	92,000
α	44,000	40,000
pH optimum	7.5–8.4	8.5
Temperature optimum (°C)	50–60	65

subunit responsible for promoter site recognition (Date *et al.*, 1975). The optimum temperature for activity was 65 °C and optimum pH of 8.5 with 18 mM MgCl$_2$, 200 mM KCl, 1 mM thermine, and 1 mM spermine (Wnendt *et al.*, 1990).

4.5. DNA Polymerases (EC 2.7.7.7)

These are more fully discussed in Chapter 8, Sections 2.4, 2.5, and 2.6.

5. HYDROLASES

5.1. Proteases

A thermostable extracellular protease from *T. caldophilus* GK24 was produced during stationary phase. It behaved as a single protein during purification but formed several bands on denaturing gel electrophoresis as did aqualysin II (Matsuzawa *et al.*, 1988). Maximum activity was at 90 °C, with a broad pH optimum; 6.6–8.0, with a peak of 7.8 with casein and 7.2 for N-carbobenzoxy-L-leucyl-L-tyrosinamide. It was insensitive to EDTA and EGTA, although this was measured at 70 °C, substantially below the optimum (Taguchi *et al.*, 1983).

Caldolysin is the trivial name for a serine protease of *Thermus* T351 isolated by Morgan *et al.* (1981). It has been purified, characterized, and patented (Cowan and Daniel, 1982; Morgan *et al.*, 1981) (Table XI). This enzyme is apparently a metalloprotease, sensitive to metal chelators EDTA and EGTA (Cowan *et al.*, 1987). Caldolysin was found to be stabilized by Ca^{2+} ions. In the presence of 10 mM Ca^{2+} the t½ of caldolysin was one hr at 90 °C, 30 hrs at 80 °C, and 193 hours at 75 °C compared with 4.8 min in its absence. Other metal ions substituted for Ca^{2+} but did not confer as much stability.

Table XI. The Properties of Some Thermus *Proteases*

Protease	Source	M$_r$	pI	pH	t½ at 85 °C (hr)
Caldolase	*Thermus* ToK3	25,000	8.9	9.5	14
Caldolysin	*Thermus* T351	21,000	8.5	8.0	30
Aqualysin	*Thermus* YT1	28,500	10.2	8.0	13.5
Ser. protease	*Thermus* Rt41A	22,500	—	10.0	—

A further extracellular alkaline proteinase isolated from *Thermus* Rt41A is a glycoprotein, containing 0.7% carbohydrate as glucose equivalents, and has four half-cystine residues present as two disulfide bonds. It was stable between pH 5.0 and 10.0 and has a maximum activity at 90 °C with 5 mM $CaCl_2$, with $t\frac{1}{2}=20$ min at 90 °C (Table XI) (Peek *et al.*, 1992). Aqualysins I and II were originally identified in culture supernatants from *T. aquaticus* YT1. They are produced during logarithmic and stationary phase respectively and are serine proteases inhibited by diisopropyl fluorophosphate. The enzymes differed in pH (10.4 for aqualysin I and 7.0 for aqualysin II) and temperature optimum (60–70 °C and 95+ °C for unpurified enzymes) (Matsuzawa *et al.*, 1988). Aqualysin II was shown to comprise several proteins, similar to the protease from *T. caldophilus* GK24 (Taguchi *et al.*, 1983). Aqualysin II also showed sensitivity to metal chelating agents, EDTA, and EGTA. Aqualysin I comprised monomers of 281 residues, with M_r 28,500. The temperature optimum of aqualysin I could be increased to 80 °C by the presence of Ca^{2+} ions. Below 70 °C activity was not affected by Ca^{2+}, indicating that the effect is on stability and not enzymic activity. At lower temperatures the enzyme became more stable to pH (5–13 at 4 °C over 22 hrs). The same is evident for protein denaturing agents (7 M urea, 6 M GdmCl, 1% SDS) (Matsuzawa *et al.*, 1988).

The nucleotide sequence for aqualysin I has been determined and the G+C content of the coding region was found to be 64.6%, increasing to 75.8% for the third position in the codon. Aqualysin I was found to comprise of 281 amino acids with a calculated M_r of 28,350. This fragment coded for the active enzyme but may have lacked the N-and C-terminal residues of the precursor protein. The amino acid sequence was found to align with a high sequence homology with the subtilisin-type serine proteases. The uses of *Thermus* proteases are discussed in Chapter 8, Sections 3.1 and 3.2.

5.2. Carbohydrases

5.2.1. α-Amylase (EC 3.1.1.1)

α-Amylase is used for producing sweeteners that contain maltotriose and maltrose. It has been produced from *Thermus* sp., especially strains AMD6, AMD22, and AMD28, and specifically hydrolyzes α-1,4-glycopyranoside bonds of amylopectin, amylose, starch, and glycogen, forming mainly maltose and maltotriose. It has optimal activity between pH 5.5–6.0 and at 70 °C. It is inhibited by Hg^{2+}, Fe^{3+}, N-dodecylbenzene sulfonate, SDS, urea, and cyclodextrin (Patent JP 1034288).

5.2.2. α-Glucosidase (EC 3.2.1.20)

The cellular locations of α- and β-glucosidases are not clear. There are some reports of extracellular production of α-glucosidase (Hudson *et al.*, 1986) as well as release by Triton. The α-glucosidase purified from *T. thermophilus* HB8 has an M_r of 67,000 and a pI at 4.5. Its temperature and pH optima were 80 °C and 5.8 and it hydrolyzes p-nitrophenyl-α-glucoside, sucrose, and maltose but not cellobiose, melibiose or soluble starch. The K_m for p-nitrophenyl-α-glucoside was 0.4 mmol/L. This enzyme is highly thermostable with 90% activity remaining after 10 hrs at 90 °C and t½=108 min at 95 °C. It is activated by Mg^{2+}, Mn^{2+}, Ca^{2+}, and Ba^{2+} and inhibited by Hg^{2+} and Cu^{2+} and both carboxyl and histidine groups are important for activity (Yang and Zhang, 1988).

5.2.3. β-Glucosidase (EC 3.2.1.21)

β-Glucosidase from *Thermus* Z-1 is induced by growth on cellobiose or laminaribinose (β-glucosides with β-1,4 or β-1,3 linkages respectively) and was purified from cell extracts without surfactant treatment. The apparent M_r was 48,000 and the enzyme was active in a broad pH range, 4.5–6.5. Its optimum temperature was 85 °C and it had a t½ at 75 °C of five days. This β-glucosidase showed optimal efficiency with cellobiose or p-nitrophenyl-β-D-glucopyranoside as substrates rather than lactose and the corresponding derivative. The K_m values for lactose and cellobiose were 36 and 2 mM respectively and for p-nitrophenyl derivatives are 6.7 and 0.21 mM. There is apparently only one catalytic site for all the substrates tested (Takase and Horikosho, 1988).

5.2.4. β-Galactosidase (EC 3.2.1.23)

β-Galactosidase was induced in *Thermus* T2 grown on lactose, galactose, and melibiose. Its temperature and pH optima were 80 °C and pH 5.0 and the estimated M_r was 570,000 (Ulrich *et al.*, 1972). It was activated by both Mn^{2+} and Fe^{2+}. β-Galactosidase isolated from *Thermus* 41A had an estimated M_r of 400,000 and pI of 4.5 and was inhibited by sulfydryl inhibitors and a number of transition metals. It was activated by EDTA and SH-containing reagents. It was stable up to 90 °C at pH 8.0, with an optimum pH of 6.0 (Cowan *et al.*, 1984).

5.2.5. Pullulanase (EC 3.2.1.41)

This enzyme specifically hydrolyzes α-(1,6) glucosidic linkages in starch and pullulan. As such it has a potential in the starch processing industry as a debranching enzyme. The ability to hydrolyze pullulan has been used as a test in some taxonomic studies on *Thermus* spp., though specific information is generally lacking. Only one pullulanase has been purified and characterized from an accepted *Thermus* (Plant *et al.*, 1986). This enzyme is associated with the outer cell membrane of *T. aquaticus* YT1 and was extractable using Triton X-100. It had a pH optimum of 6.4 and an M_r of 80–83,000. The enzyme is very thermostable, having $t^{1/2}$ of 4.5 hrs at 95 °C. At 100 °C this was shown to be pH dependent, varying from 13.5 min at pH 7.0 to 1.5 min at 4.7 (where measured activity is around 50–60% of maximum). Calcium has been implicated in protection against thermal denaturation for proteases and also appears to serve the same role in the case of pullulanase. The action of the enzyme on pullulan produces malto-triose as the ultimate product, indicating there is little or no activity against linkages other than $\alpha(1-6)$ glucosyl bonds. The pullulanase gene from *Thermus* AMD-33 has been cloned (Sashihara *et al.*, 1988) and appears to contain an endogenous promoter functional in the host organism. The *pul* gene of 2154 base pairs coded for a protein of M_r 80,237 (Sashihara *et al.*, 1993). The enzyme was produced during exponential growth. Activity was detected solely in the cells and not in the extracellular or periplasmic space. Thus the signal sequences for this enzyme do not appear to function in *E. coli*, in contrast to the case of aqualysin I. Otherwise, the recombinant enzyme behaves as the parent. This enzyme shows a relatively sharp pH optimum of 5.6, with a temperature optimum at 70 °C. Maltotriose is the sole product from pullulan. Enzymes capable of hydrolyzing pullulan have been purified and characterized from *Thermus* AMD33 (Nakamura *et al.*, 1989). The two enzymes found differed in pI (4.2 and 4.3 respectively) but were otherwise nearly identical. The K_m values of 1.33 and 0.77 mg·ml^{-1} for pullulan and 1.45 and 1.41 mg·ml^{-1} for short-chain amylose were reported for enzymes I and II. By contrast with other pullulanases the K_m is rather high, but the ratio between values for pullulan and glycogen is low. Pullulanase I had a M_r of 135,000, a pI of 4.2, a pH optimum of 5.5–6.0 and a temperature optimum at 70 °C. The temperature optimum was shown to increase in the presence of 3 mM $CaCl_2$ to 75 °C with a similar increase in thermostability. Mercuric and ferric ions inhibited activity as did SDS and urea. Mercuribenzoate and iodoacetate showed no effect at 0.1 and 1 mM respectively. However, there is some doubt as to whether *Thermus* AMD33 is a genuine *Thermus* since it has a G+C content of only 55% (normally >60% for *Thermus* sp.) and a temperature optimum for growth of 60 °C, which is rather low for the genus (Williams, 1991).

5.3. Alkaline Phosphatase (EC 3.1.3.1)

A repressible alkaline phosphate from *T. aquaticus*, either dimeric or tetrameric, had a subunit M_r of 51,000 and an apparent native M_r of

143,000. It had a pI at 8.4, a pH optimum of 9.2, and an optimum temperature range of 75–80 °C. The K_m was estimated at 0.8 mM for p-nitrophenyl phosphate (Yeh and Trela, 1976). This alkaline phosphatase was also a phosphodiesterase and a monoesterase. At pH 7.2, the diesterase activity was maximal at 80–85 °C, a higher temperature and a lower pH than for monoesterase activity. Phosphate monoesters were inhibitors of both activities, though greater for the diesterase. This may be due to the low pH at which the diesterase was measured, which would be expected to change the enzyme conformation significantly (Smile et al., 1977). Alkaline phosphatase activity also occurs, apparently constitutively, in all strains of Thermus ruber tested (Egorova and Loginova, 1984; Cossar and Sharp, 1989). The enzyme from T. ruber strain 16063 isolated from Iceland was partially purified by detergent treatment for optimal recovery from cell extracts (implying a membrane association). The optima were at pH 8.9 (in diethanolamine buffer) and 67 °C respectively, and the subunit M_r was 40–50,000 Da. This enzyme was strongly inhibited by Tris and all activity was lost on anion exchange chromatography using this buffer (Debora Johnson).

5.4. Ribonucleases (EC 3.1.4.22/23)

The gene encoding ribonuclease H from T. thermophilus HB8 shows 69.9% of G+C overall and gives rise to a protein sequence of 164 amino acids with a calculated M_r of 18,335. The amino acid sequence shows a 56% similarity to E. coli ribonuclease HI but little or no identity to E. coli ribonuclease HII. It is more stable than E. coli ribonuclease HI with 90% activity remaining after 10 min at 65 °C (the E. coli enzyme loses 95% of its activity under the same conditions) (Itaya and Kondo, 1991). A corrected sequence of 166 amino acids had a 59% similarity to the E. coli enzyme. The pI was 10.5 and there was a broad pH optimum of 8.5–9.5. The maximum activity was obtained at 70 °C with Mg^{2+} at 1 mM required for activity (Kanaya and Itaya, 1992).

5.5. EFTu

The elongation factor EFTu, which elongates polypeptide chains during protein biosynthesis has been isolated from both T. aquaticus YT1 and T. thermophilus HB8. EFTu from T. thermophilus was found to have an M_r 142,000 (Arai et al., 1978) and was stable toward heat and to protein denaturants such as 5.5 M urea and 1.5 GdmCl (Nakamura et al., 1978). T. thermophilus HB8 grown in the presence of [^{32}P] orthophosphate was found to produce phosphorylated EFTu (Lippmann et al., 1993). Two genes for this protein have been isolated from *T. thermophilus*. *TufA* has been cloned and sequenced and has a G+C usage of 62.9% overall, rising to 84.5% at the third position in a codon. It codes for a protein of 405 amino acids with a calculated M_r of 44,685. It has a high sequence homology with *E. coli* and *Saccharomyces cerevisae* (65–70%) (Kushiro *et al.*, 1987). *TufA* has also been sequenced from *Thermus* strain B and the G+C content at the third position of the codon is 89.5% with an overall usage of 60.7%. The derived protein sequence differs from *tufA* and *tufB*-encoded sequences from *T. thermophilus* in ten out of the 405 amino acids. Both have an additional loop of ten amino acids (182–191) compared with the elongation factor of *E. coli* (Voss *et al.*, 1992).

The *tufB* gene sequence from *T. thermophilus* codes for a protein of 406 amino acids and differs from *tufA* by four amino acids and at ten positions out of 1221 nucleotide residues. The G+C content was again high, 62.5% overall, and 84.8% at the third position. The *tufB* product showed a 70.9% homology to the corresponding sequence of the *tufB* product of *E. coli* (Satoh *et al.*, 1991). EFTu from both *T. thermophilus* and *T. aquaticus* have been crystallized (Reshetnikova *et al.*, 1991; Lippmann *et al.*, 1988).

5.6. Asparaginase (EC 3.5.1.1)

$$\text{Asparagine} + H_2O \rightarrow \text{Aspartate} + NH_3$$

Both L- and D-asparaginases have been isolated from *T. aquaticus* and their properties compared (Table XII). L-asparaginase was highly stereospecific, with no activity at all on D-asparagine. Its K_m of 8.6 mM is relatively high compared with those of mesophilic microbial L-asparaginases. Several amino acids, including L-aspartate, L- and D-lysine, and L- and D-serine inhibit this enzyme. It has a $t^{1/2}$ at 85 °C of 40 min (Curran *et al.*, 1985). D-asparaginase was also isolated from *Thermus* T351 as a dimer with subunit M_r of 30,000. It showed a pH optimum at 9.5, although it was active between the range of 8.5–10.5. It has six disulfide bonds and a histidine was implicated at the active site. It was found to be activated by oxo acids and inhibited by 2-oxoglutarate, pyruvate, and oxaloacetate. The

Table XII. The Properties of D- and L-Asparaginase from *Thermus aquaticus*

	M_r	pI	Optimum pH	K_m (mM)	$t^{1/2}$ at 85 °C (min)
L-Asparaginase	80,000	4.6	9.5	8.6	40
D-Asparaginase	60,000	4.8	9.5	2.2	25

K_m for D-asparaginase was 2.2 mM and it was found to be thermostable with 90% activity remaining after 10 hrs at 75 °C, and t½ at 85 °C of 25 min (Guy and Daniel, 1982).

5.7. Inorganic Pyrophosphatase (EC 3.6.1.1)

$$\text{Pyrophosphate} + H_2O \to 2\text{-orthophosphate}$$

The purification and characterization of inorganic pyrophosphatase from *T. aquaticus* produced a tetramer of M_r 84,000, with a monomer M_r of 21,000. It has a pI at 5.7, a pH optimum of 8.3, and a temperature optimum of 80 °C. The K_m for pyrophosphate (PPi) was 0.6 mM and only PPi is used as a substrate. It has an absolute requirement for a divalent cation with Mg^{2+} being preferred (Verhoeven *et al.*, 1986).

5.8. ATPase (EC 3.6.1.3)

$$\text{ATP} + H_2O \to \text{ADP} + \text{orthophosphate}$$

ATPase from *T. thermophilus* has a native M_r of 360,000 and is made up of four subunits of 66,000 (α), 55,000 (β), 30,000 (τ), and 12,000 (δ). It shows low hydrolytic activity (0.07 μmole Pi released mg^{-1}/min^{-1}) suggesting that it belongs to the V-type ATPase family. The α subunit is of a similar molecular size to that of the catalytic subunits of V-type ATPases and the N-terminal sequences of the α and β subunits show homology with other V-type sequences. The V-type ATPases are found in eukaryotic endomembrane vacuolar vesicles but, so far in prokaryotes, the only ones found are in the plasma membranes of archaea. Since *T. thermophilus* is a bacterium, it was not expected that a V-type ATPase would be found (Yokoyama *et al.*, 1990). The genes, *aptAB*, coding for the α and β subunits of *T. thermophilus* HB8 have been isolated and cloned. The deduced amino acid sequence of α (583 amino acids) and β (478 amino acids) show a high similarity to other V-type ATPases (Tsutsumi *et al.*, 1991).

5.9. DNA Endonucleases

DNA endonucleases (restriction endonucleases or site-specific endonucleases) recognize a distinct DNA sequence or set of sequences and usually occur together with a modification methylase with the same sequence specificity. Type I enzymes have subunits with both methylase and endonuclease activity, whereas type II are small molecules with endonuclease activity only. The cloning and sequencing of the genes from *T. thermophilus* HB8 for the restriction–modification system has been reported (recognition sequence T↓CGA). The endonuclease nucleotide sequence translated to a single chain of 263 amino acids, while that of the modification gene translated to 428 amino acids. A high degree of amino acid conservation between this and the *T. aquaticus* restriction–modification system suggests a common origin for them. However, codon usage and G+C content of these genes were much lower than for other *Thermus* genes, suggesting that they were only recently introduced into the genus *Thermus* (Barany *et al.*, 1992). TflI from *Thermus flavus* is an isoschizomer of Taq I.

The DNA methylase from *T. thermophilus* HB8 is a single polypeptide of M_r 41–44,000. It has a maximum activity at pH 7.4 and 70 °C. Enzyme activity is inhibited by 0.2 M NaCl and 2 mM $HgCl_2$. It carries out the transfer of methyl groups from S-adenyl-L-methionine to a double stranded DNA and obeys Michaelis–Menten kinetics with a K_m for S-adenyl-L-methionine and lambda phage DNA of 0.8 μM and 10 μg/ml, respectively (Sato *et al.*, 1980). *T. aquaticus* has at least two restriction–modification systems, *Taq*I recognizes double stranded DNA at the palindromic sequence TCGA and cleaves both strands between the T and C, leaving a 2 base 5′ overlap. It is stable up to 70 °C (Sato *et al.*, 1977). The *Taq*I restriction–modification system has been cloned and sequenced. The methylase was 1089 base pairs (363 amino acids, M_r 40,576), and the endonuclease, 702 base pairs (234 amino acids, M_r 27,523) with a separation of 132 base pairs between the two genes. The G+C content of each gene was 59% for the methylase and 52% for the endonuclease (Slatko *et al.*, 1987).

*Tsp*EI has been isolated from *Thermus* strain E. It has optimum activity at 65 °C and an optimum pH at 25 °C of 7.8–8.4. It is stable up to 90 °C and it recognizes the DNA sequence AATT, the same as *Eco*RI. Since *Thermus* DNA contains 65% G+C, and this figure rises when considering the third position of the codon, the recognition sequence of *Tsp*EI is relatively rare in the coding sequence of its own genome. It may therefore have a role protecting against exogenous DNA (Raven *et al.*, 1993). An isoschizomer, Tsp509I, has been described (Roberts, 1993).

A type II endonuclease, *Tth*111I has been isolated from *Thermus thermophilus* strain 111. It has optimum activity between 60–70 °C and with 6–10 mM $MgCl_2$ and 50–100 mM NaCl. It recognizes the DNA sequence 5′-TGACN↓NNGTC-3′ (Shinomiya *et al.*, 1980a). A second type II endonuclease (*Tth*111II) has also been isolated. It has optimum activity between 65–70 °C, with activity up to 80 °C. It requires either Mg^{2+} or Mn^{2+} for endonuclease activity and recognizes the DNA sequence 5′-CAAPu-CA(N)$_{11}$-3′, cleaving after the 11th N base (Shinomiya and Sato, 1980b). *Taq*II has also been isolated from *T. aquaticus* YT1 and it recognizes the DNA sequences GACCGA and CACCCA (Barker *et al.*, 1984).

*Tsp*45I has been purified from *Thermus* strain 45. It has optimum activity at 65 °C and has a low requirement for inorganic ions. It has an M_r by gel filtration of 80,000 and gives a single band on SDS-PAGE estimated at 38,000. It cleaves the recognition sequence 5'-↓GTSAC-3' (Raven *et al.*, 1993). Other new restriction enzymes that have been described in *Thermus* strains are TspRI (↓CASTG) and TfiII (G↓ATWC). Isoschizomers of enzymes from mesopiles include TruI (GGWCC, AvaII), TruII (GATC, MboI), Tru201I (R↓GATCY, XhoII) TspZnI (GGCC, HaeIII), TaqXI (CCWGG, AeuII), and Tsp49I (A↓CGT, MaeII) (Roberts, 1993).

6. LYASES

6.1. Anthranilate Synthetase (EC 4.1.3.27)

Chorismate + L-glutamine → anthranilate + pyruvate + L-glutamate

Anthranilate synthetase I and II have been cloned and sequenced from *T. thermophilus* HB8 (Sato *et al.*, 1988). A comparison of *trpE* (anthranilate synthetase I) and *trpG* (anthranilate synthetase II) genes and proteins can be seen in Table XIII.

6.2. Enolase (EC 4.2.1.11)

2-phospho-D-glycerate → phosphoenolpyruvate + H_2O

Enolase (2-phospho-D-glycerate hydrolyase) is an enzyme of the glycolytic pathway, which catalyzes the reaction from 2-phosphoglycerate to phosphoenolpyruvate. The enolases from both *Thermus X1* and *T. aquaticus* have also been purified and characterized (Stellwagen *et al.*, 1973; Barnes and Stellwagen, 1973) and their properties compared with enolases from other mesophilic sources (Table XIV). The two *Thermus* enzymes were found to be very similar to enolases from rabbit muscle, salmon, trout, *E. coli*, and yeast in all but two properties (Barnes and Stellwagen, 1973). First, the enolases from *Thermus* are much more thermostable and have a

Table XIII. The Properties of Anthranilate Synthetase I and II

	% G+C overall	% G+C 3rd	M_r native	M_r subunit	Optimum (°C)
trpE	69.1	93.0	110,000	50,000	60
trpG	68.9	94.3	20,000	20,000	—

Table XIV. The Properties of Enolase from Various Sources

Property	Source of enolase			
	Yeast	Rabbit	*X1*	*T. aquaticus*
M_r native	88,000	82,000	382,000	328,000
M_r subunit	44,000	44,000	48,000	44,000
% α-helix	25	30	33	32
% β-sheet	37	38	29	16
K_m (2-PGA) (μM)	150	90	110	110
E_{act} (Kcal/mole)	11.8	16.6	15.4	15.8
Temperature optimum (°C)	49	57	70	89

Adapted from Stellwagen & Barnes, 1978.

higher optimum temperature, with the X1 enzyme having an optimum temperature for catalysis 16 °C lower than enolase from *T. aquaticus*. Secondly, each *Thermus* enolase is a globular protein made up of eight identical subunits. The M_r of each subunit is around 44,000 (Stellwagen *et al.*, 1973; Barnes and Stellwagen, 1973) as determined by circular dichroism measurements. However, the mesophilic native enzymes have an M_r of around 88,000 and exist as dimers (Wold, 1971). Comparison of the amino acid composition of the enolases of rabbit, yeast, and the two *Thermus* species by Stellwagen and Barnes (1978) suggested that the four proteins were related, although the reduced β-sheet content of the *Thermus* species may be an indication of their increased thermal stability. The catalytic data obtained showed no major differences between the thermophilic and mesophilic enzymes and it is therefore thought that the catalytic sites in the four enzymes would also be similar and that the four enolases are members of a closely related, if not homologous, series of proteins.

6.3. Fumarase (EC 4.2.1.2)

$$\text{Fumarate} + H_2O \rightarrow \text{L-malate}$$

Fumarase (L-malate hydro-lysase) catalyses the reversible hydration of fumarate to L-malate in the tricarboxylic cycle. It has been purified from the thermophilic bacteria *Thermus* X1 and compared with fumarase from the moderate thermophile *B. stearothermophilus* (Cook and Ramaley, 1976). The fumarase from *Thermus* was found to be slightly more thermostable with an optimum catalytic reaction temperature of 83 °C compared with 81 °C. Apart from thermostability, the catalytic properties of these fumarases were extremely similar to each other and to those of the pig heart fumarase

(Table XV) (Bock and Alberty, 1953). Both enzymes showed similar saturation kinetics as a function of temperature.

6.4. Tryptophan Synthetase (EC 4.2.1.20)

$$\text{L-serine} + \text{indole} \rightarrow \text{L-tryptophan} + H_2O$$

The *trpA* gene sequence of *T. thermophilus* translated to a 404-amino-acid sequence, while that of *trpB* gave 271 residues. The G+C usage of genes A and B was 69.3 and 70.2%, respectively, and 90.0 and 92.3% at the third nucleotide. The deduced amino acids were compared with those from *E. coli*, *S. typhimurium*, and *Saccharomyces cerevisiae* (Table XVI, Koyama and Furukawa, 1990).

7. ISOMERASES

7.1. D-Xylose Isomerase (EC 5.3.1.5)

D-xylose isomerase catalyzes the reversible isomerization of D-xylose to D-xylulose, which is the first step of xylose metabolism in the pentose phosphate cycle. D-xylose is important industrially because the enzyme also interconverts glucose and fructose and is used to produce high fructose corn syrup. D-xylose isomerase purified from *Thermus* has a native M_r of 196,000, a subunit M_r of 50,000, and a tetrameric structure like mesophilic xylose isomerases. It has a pI at 5.5 and an optimum pH range between 5.5–8.5 (Lehmacher and Bisswanger, 1990a). Besides D-xylose and D-glucose it converts D-ribose and D-arabinose as substrates, though with lower affinities. Activation of the enzyme depends on divalent cations, with Mn^{2+} having the greatest effect. It is strongly inhibited by Cu^{2+}, and weakly inhibited by Ni^{2+}, Fe^{2+}, and Ca^{2+} (Lehmacher and Bisswanger, 1990b).

Table XV. The Properties of Fumarases from Various Sources

	Thermus X1[a]	*B. stearothermophilus*[a]	Pig heart[b]
M_r	180,000	180,000	194,000
Temperature optimum (°C)	83	81	—
pH optimum	7.2	6.9	6.0–8.0

[a]Cook and Ramaley, 1976
[b]Massey, 1953

**Table XVI. The Similarity of Microbial
Tryptophan Synthetases of and Other Sources**

% Similarity between *Thermus thermophilus* enzyme and		
E. coli	*S. typhimurium*	*S. cerevisiae*
A	54.8	54.8	59.4
B	28.7	28.7	42.6

The xylose isomerase gene from *T. thermophilus* cloned in *E. coli* codes for a polypeptide of three hundred and eighty-seven amino acid residues with a calculated M_r of 44,000. Comparison of the sequence with other xylose isomerase sequences shows that the amino acids involved in substrate binding and isomerization are conserved. The thermostability of *Thermus* isomerases was attributed to substitution of amino acids that might be involved in irreversible inactivation, and also from increased hydrophobicity (Dekker *et al.*, 1991).

8. LIGASES

8.1. Aminoacyl tRNA Synthetase

Amino acid + ATP + tRNA → Amino acyl-tRNA + AMP + Pyrophosphate

In procaryotes there are only 20 aminoacyl tRNA synthetases, each one recognizing a specific amino acid. Aminoacyl tRNA synthetases range from monomeric structures to tetramers. The majority perform a two stage catalysis; first the aminoacylation of ATP with the elimination of pyrophosphate to give aminoacyl adenylate, then the transfer of the amino acid to the 3' end of its cognate tRNA chain (Schimmel and Söll, 1979). Properties of *Thermus* aminoacyl tRNA synthetases are summarized in Table XVII. Methionyl-tRNA synthetase of *T. thermophilus* HB8 can be cleaved by limited digestion into four domains, T1 (29,000), T2 (23,000), T3 (14,500), and T4 (7,500). T1 is a small aminoacylation unit, T2 is enzymatically inactive and interacts with tRNA[Metf], T3 forms a dimer and is responsible for the dimer assembly, and T4 is the flexible tail (Kohda *et al.*, 1987). The gene from *T. thermophilus* HB8 has been expressed in *E. coli* and translates to an amino acid sequence of 616 residues with a calculated M_r of 70,737. It has little sequence homology (25%) with methionyl-tRNA

Table XVII. Properties of Various tRNA Synthetases from *T. thermophilus*

	EC number	Number of subunits	M_r	Other remarks
Tyrosyl[a]	6.1.1.1	2 α_2	100,000	
Threonyl[b]	6.1.1.3	2 α_2	2×74,000	Crystals obtained
Isoleucyl[c]	6.1.1.5	1	108,000	2 Zn^{2+} tightly bound
Methionyl[c]	6.1.1.10	2	150,000	2 Zn^{2+} tightly bound
Seryl[b]	6.1.1.11	2	46,000	Crystals of enzyme and with its cognate tRNA
Glutamyl[d]	6.1.1.17	1	50,000	
Valyl[c]	6.1.1.19	1	129,000	2 Zn^{2+} tightly bound
Phenylalanyl[e]	6.1.1.20	4 $\alpha_2\beta_2$	264,000	α 40,000 β92,000

[a]Yaremchuk *et al.*, 1990
[b]Garber *et al.*, 1990; Yaremchuk *et al.*, 1990
[c]Kohda *et al.*, 1984
[d]Hara-Yokoyama *et al.*, 1984
[e]Bobkova *et al.*, 1991

synthetase from *E. coli*. The codon usage for the third letter of a codon was 94% G+C (Nureki *et al.*, 1991).

Glutamyl-tRNA synthetase purified from *T. thermophilus* HB8 has an optimum temperature for aminoacylation of tRNA^{Glu} of 65 °C with an optimum pH range of 8.0–9.0 in the presence of Mg^{2+}. It is very thermostable with 70% activity remaining after nine hrs at 65 °C. The K_m determined in the presence of 50 mM KCl 10 mM $MgCl_2$, pH 8.0 at 65 °C was 230 μM for ATP, 70 μM for L-glutamate and 0.6 μM *T. thermophilus* tRNA^{Glu} (Hara-Yokoyama *et al.*, 1984). The gene for glutamyl-tRNA synthetase produces a protein of 468 amino acids, with a calculated M_r of 53,901. The G+C usage was again found to be high (94%) at the third position of the codon, the amino acid sequence showed a 35–45% identity with other bacterial sequences and the binding site for ATP and for the 3' terminus of tRNA^{Glu} are highly conserved. Crystals of this protein were successfully grown (Nureki *et al.*, 1991).

Phenylalanyl-tRNA synthetase from *T. thermophilus* HB8 had an optimum temperature of aminoacylation close to 80 °C. The heterotetrameric enzyme, which has been crystallized (Ankilova *et al.*, 1988), contained α-subunits, primarily isolated as α_2-dimers that tended to aggregate, and β_2-dimer subunits. The pI values are compared with those of *E. coli* phenylalanyl-tRNA synthetase in Table XVIII (Bobkova *et al.*, 1991). The *E. coli* enzyme loses activity above 43 °C, both in the [32]PPi-ATP exchange

**Table XVIII. The pIs of the Subunits of
Phenylanalyl-tRNA Synthetase**

	pI	
	T. thermophilus	*E. coli*
α	5.20	6.05
β	5.13	4.85
native	5.00	4.75

reaction and in the overall aminoacylation, whereas that from *T. thermophilus* is stable up to 90 °C (Bobkova *et al.*, 1992). The genes for phenylalanyl-tRNA synthetase have been cloned into *E. coli* and sequenced. The genes *PheS* and *PheT* encode for the α- and β-subunits with molecular weights of 39,000 and 87,000 respectively. Three conserved sequence motifs typical for a class II tRNA synthetase was found to occur in the α-subunit (Keller *et al.*, 1992). In the β-subunit clusters of hydrophilic amino acids and a leucine zipper motif were identified (Kreutzer *et al.*, 1992). Phenylalanyl-tRNA synthetase from *T. thermophilus* can attach more than one molecule of phenylalanine per tRNA^{Phe}. The 'hyperaminoacylated' tRNA^{Phe} was found to be the bis-2',3'-O-phenylalanyl-tRNA^{Phe} and its formation was typical for the thermophilic enzyme but found not to occur with *E. coli* phenylalanyl-tRNA synthetase under the same conditions (Stepanov *et al.*, 1992).

8.2. Succinyl-CoA Synthetase (EC 6.2.1.4/5 GDP/ADP)

The nucleotide sequence of the succinyl-CoA synthetase α subunit from *Thermus* B translates to a protein of 288 amino acids, which shows similarity to other known succinyl-CoA synthetases (Nicholls *et al.*, 1988). The gene encoding for both the α and β subunits of succinyl-CoA, *scsA*, and *scsB* have been found in an open reading frame upstream of the *mdh* gene in *T. flavus*. A high G+C content was found throughout these genes and at the third position of the codon. The amino acid sequence of the α subunit showed a 57% similarity to that of the *E. coli* α subunit and the similarities between the β-subunits was also reported to be high (Nishiyama *et al.*, 1991).

8.3. DNA Ligase

DNA ligase catalyzes the formation of a covalent phosphate link between nicked strands of DNA, using NAD (or ATP) as a cofactor. There are three reversible steps (Barany and Gelfand, 1991). The first involves the formation of a high-energy enzyme intermediate by transfer of the adenyl group from the cofactor to the ε-amino group of the lysine residue. This is followed by transfer of the adenyl group to the 5′ phosphate of one DNA strand and attack of this activated 5′ end by a 3′-hydroxyl group on the adjacent strand. This forms the phosphodiester link between the two DNA strands and eliminates AMP. The enzyme from *T. thermophilus* HB8 is a monomer of M_r of 79,000 with an absolute requirement for either Mg^{2+} or Mn^{2+} and NAD as a cofactor. It has a K_m for NAD of 1.85×10^{-8} M and a pH optimum of 7.6. It is activated by K^+ and NH_4^+. The nick-closing activity of this enzyme was active over a range of 15–85 °C, with an optimum at 65–72 °C (Takahashi *et al.*, 1984). This thermostable DNA ligase was found to be able to discriminate between a mismatched and complement DNA helix and to retain its activity after multiple thermal cycles. Its nucleotide sequence was determined and the codon usage of G+C overall was 66.0%, rising to 91.4% for the third codon. It translates to a protein of 676 amino acid residues, which show a 47% identity to *E. coli* DNA ligase (Barany and Gelfand, 1991).

8.3.1. The Ligase Chain Reaction

The ligase chain reaction (LCR) is able to detect a single base mutation. The method uses both strands of genomic DNA as targets for oligonucleotide hybridization. Four oligonucleotides are used, two of which lie end by end and are complementary to each DNA strand. The target DNA is first heat-denatured to separate the two strands, and then the four oligonucleotides are hybridized to the target DNA near their melting temperature. Thermostable DNA ligase will then covalently link the adjacent oligonucleotides, but only if there is complete match at the junction. The products of the ligation generated in the first round joined the DNA strands as targets for successive cycles of denaturation, annealing, and ligation. Thus the amplification of ligated oligonucleotides that absolutely match the original target sequence continues exponentially. A single-base mismatch at the oligonucleotide junction will not support amplification, since ligation will not occur.

Thermostable DNA ligases are necessary in the LCR since temperatures of 90–100 °C are used in 30- to 60-second cycles to complete denaturation of double stranded DNA at each round. *Taq* ligase retains activity after 20 or 30 repeated one min exposures to 94 °C (Barany, 1991a,b).

9. MISCELLANEOUS

9.1. Chaperonins

Molecular chaperones are a ubiquitous family of cellular proteins that mediate the correct folding of other polypeptides, and sometimes in their assembly into oligomeric structures, but which are not themselves components of these final structures (Ellis, 1990). Chaperonin isolated from *T. thermophilus* consisted of two proteins of M_r 58,000 (cpn60) and 10,000 (cpn10), with the M_r of the native protein being greater than 1,000,000 and showing a seven-fold symmetry from the top view and a 'football'-like structure from the side view under electron microscopy. It has a weak ATPase activity, inhibited by sulfite and activated by bicarbonate. *In vivo* chaperonin can promote the ATP-dependent refolding of several GdmCl-denatured enzymes (Taguchi *et al.*, 1991). It can also protect several heat labile enzymes from irreversible heat denaturation in the presence of MgATP (Taguchi and Yoshida, 1993). A stable complex cpn60/cpn10 'holo-chaperonin' from *T. thermophilus* has been isolated and crystallized (Lissin *et al.*, 1992).

REFERENCES

Air, G. M., and Harris, J. I., 1974, DNA-dependent RNA polymerase from the thermophilic bacterium *Thermus aquaticus*, *FEBS Lett.* **38**:277–281.

Ankilova, V. N., Reshetnikova, L. S., Chernaya, M. M., and Lavrik, O. I., 1988, Phenylalanyl-tRNA synthetase from *Thermus thermophilus* HB8: Purification and properties of the crystallising enzyme, *FEBS Lett.* **227**:9–13.

Arai, K-I., Yoshimi, O., Arai, N., Nakamura, S., Henneke, C., Oshima, T., and Kaziro, Y., 1978, Studies on polypeptide-chain-elongation factors from an extreme thermophile, *Thermus thermophilus* HB8, *Eur. J. Biochem.* **92**:509–519. (Abstract)

Argos, P., Rossmann, M. G., Grau, U. M., Zuber, H., Frank, G., and Tratschin, J. D., 1979, Thermal stability and protein structure, *Biochemistry* **18**:5698–5703.

Atkinson, T., 1976, Thermostable enzymes, *J. Appl. Biotechnol.* **26**:39–40.

Barany, F., 1991a, Genetic disease detection and DNA amplification using cloned thermostable ligase, *Proc. Natl. Acad. Sci. USA* **88**:189–193.

Barany, F., 1991b, The ligase chain reaction in a PCR world, *PCR Methods Appl.* **1**:5–16.

Barany, F., Danzitz, M., Zebala, J., and Mayer, A., 1992, Cloning and sequencing of genes encoding the TthHB81 restriction and modification enzymes: Comparison with the isoschizomeric TaqI enzymes, *Gene* **112**:3–12.

Barany, F., and Gelfand, D. H., 1991, Cloning, overexpression and nucleotide sequence of a thermostable DNA ligase-encoding gene, *Gene* **109**:1–11.

Barker, D., Hoff, M., Oliphant, A., and White, R., 1984, A second type II restriction endonuclease from *Thermus aquaticus* with an unusual sequence specificity, *Nucleic Acids Res.* **12**:5567–5581.

Barnes, L. D., and Stellwagen, E., 1973, Enolase from the thermophile *Thermus X-1*, *Biochemistry* **12**:1559–1565.

Biesecker, G., Harris, J. I., Thierry, J. C., Walker, J. E., and Wonacoff, A. J., 1977, Sequence and structure of D-glyceraldehyde 3-phosphate dehydrogenase from *Bacillus stearothermophilus*, *Nature* **266**:323–333.

Biffen, J. H. F., and Williams, R. A. D., 1976, Purification and properties of malate dehydrogenase from *Thermus aquaticus*, in: *Enzymes and proteins from thermophilic micro-organisms* (H. Zuber, ed.), Birkhauser Verlag, Basel and Stuttgart, pp. 157–166.

Birktoft, J. J., Rhodes, G., and Banaszak, L. J., 1989, Refined crystal structure of cytoplasmic malate dehydrogenase at 2.5 Å resolution, *Biochemistry* **28**:6065–6981.

Bloxham, D. P., and Lardy, H. A., 1973, Phosphofructokinase, in: *The Enzymes*, 3rd ed. (P. D. Boyer, ed.), Academic Press, New York, San Francisco, London, pp. 239–278.

Blundell, T., Sibanda, B. L., Sternberg, M. J. E., and Thornton, J. M., 1987, Knowledge-based prediction of protein structures and the design of novel molecules, *Nature* **326**:347–352.

Bobkova, E. V., Mashanov Golikov, A. V., Wolfson, A., Ankilova, V. N., and Lavrik, O. I., 1991, Comparative study of subunits of phenylalanyl-tRNA synthetase from *Escherichia coli* and *Thermus thermophilus*, *FEBS Lett.* **290**:95–98.

Bobkova, E. V., Stepanov, V. G., and Lavrik, O. I., 1992, A comparative study of the relationship between thermostability and function of phenylalanyl-tRNA synthetases from *Escherichia coli* and *Thermus thermophilus*, *FEBS Lett.* **302**:54–56.

Bock, R. M., and Alberty, R. A., 1953, Studies of the enzyme fumerase. I. Kinetics and equilibrium, *J. Am. Chem. Soc.* **75**: 1921–1925.

Bowen, D., Littlechild, J. A., Fothergill, J. E., Watson, H. C., and Hall, L., 1988, Nucleotide sequence of the phosphoglycerate kinase gene from the extreme thermophile *Thermus thermophilus*. Comparison of the deduced amino acid sequence with that of the mesophilic yeast phosphoglycerate kinase, *Biochem. J.* **254**:509–517.

Brock, T. D., and Freeze, H., 1969, *Thermus aquaticus* gen.n. and sp.n.; a nonsporulating extreme thermophile, *J. Bacteriol.* **98**:289–297.

Burley, S. K., and Petsko, G. A., 1988, Weakly polar interactions in proteins, *Adv. Prot. Chem.* **39**:125–189.

Burley, S. K., and Petsko, G. A., 1989, Electrostatic interactions in aromatic oligopeptides contribute to protein stability, *Tibtech* **7**:354–359.

Cass, K. H., and Stellwagen, E., 1975, A thermostable phosphofructokinase from an extreme thermophile *Thermus X-1*, *Arch. Biochem. Biophys.* **171**:682–689.

Cocco, D., Rinaldi, A., Savini, I., Cooper, J. M., and Bannister, J. V., 1988, NADH oxidase from the extreme thermophile *Thermus aquaticus* YT-1. Purification and characterisation. *Eur. J. Biochem.* **174**:267–271.

Cook, W. R., and Ramaley, R. F., 1976, Purification and catalytic properties of "thermophilic" fumerase from *Bacillus stearothermophilus* NU-10 and *Thermus X1*, in: *Enzymes and proteins from thermophilic micro-organisms* (H. Zuber, ed.), Birkhäuser, Verlag, Basel and Stuttgart, pp. 207–222.

Cossar, J. D., and Sharp, R. J., 1989, Alkaline phosphatase from *Thermus ruber*, in: *FEMS Symposium 39. Microbiology of extreme environments and its potential for biotechnology* (R. A. D. Williams, M. Costa, J. Duarte, and W. Boone, eds.), Elsevier, London, p. 384.

Cowan, D. A., and Daniel, R. M., 1982, Purification and some properties of an extracellular protease (caldolysin) from an extreme thermophile, *Biochim. Biophys. Acta* **705:**293–305.

Cowan, D. A., Daniel, R. M., Maltn, A. M., and Morgan, H. W., 1984, Some properties of a β-galactosidase from an extremely thermophilic bacterium, *Biotechnol. Bioeng.* **26:**1141–1145.

Cowan, D. A., Daniel, R. M., and Morgan, H. W., 1987, A comparison of extracellular serine proteases from four strains of *Thermus aquaticus, FEMS Microbiol. Lett.* **43:**155–159.

Creighton, T. E., 1988, Disulphide bonds and protein stability, *Bioessays* **8:**57–63.

Curran, M. P., Daniel, R. M., Guy, G. R., and Morgan, H. W., 1985, A specific L-asparaginase from *Thermus aquaticus, Arch. Biochem. Biophys.* **241:**571–576.

Daniel, R. M., Cowan, D. A., Morgan, H. W., and Curran, M. P., 1982, A correlation between protein thermostability and resistance to proteolysis, *Biochem. J.* **207:**641–644.

Date, T., Suzuki, K., and Imahori, K., 1975, Purification and some properties of DNA-dependent RNA polymerase from an extreme thermophile *Thermus thermophilus* HB8, *J. Biochem.* (Tokyo) **78:**845–858.

Dekker, K., Yamagata, H., Sakaguchi, K., and Udaka, S., 1991, Xylose (glucose) isomerase gene from the thermophile *Thermus thermophilus*: Cloning, sequencing and comparison with other thermostable xylose isomerase, *J. Bacteriol.* **173:**3078–3083.

Devanathan, T., Akagi, J. M., Hersh, R. T., and Himes, R. H., 1969, Ferredoxin from two thermophilic *Clostridia, J. Biol. Chem.* **244:**2846–2853.

Edlin, J. D., and Sundaram, T. K., 1989, Isocitrate dehydrogenase from thermophilic and mesophilic bacteria. Isolation and some characteristics, *Eur. J. Biochem.* **21:**1203–1210.

Egorova, L. A., and Loginova, L. G., 1984, Selection of an alkaline phosphatase-producing thermophilic bacterium of the genus *Thermus, Microbiologiya* **532:**242–245.

Eguchi, H., Wakagi, T., and Oshima, T., 1989, A highly stable NADP-dependent isocitrate dehydrogenase from *Thermus thermophilus* HB8: Purification and general properties, *Biochim. Biophys. Acta* **990:**133–137.

Ellis, R. J., 1990, The molecular chaperone concept, *Seminars in Cell Biology* **1:**1–9.

Erdmann, H., Hecht, H-J., Park, H-J., Sprinzl, M., Schomberg, D., and Schmid, R. D., 1993, Crystallization and preliminary X-ray diffraction studies of NADH oxidase from *Thermus thermophilus* HB8, *J. Mol. Biol.* **230:**1086–1088.

Fabráy, M., Sümegi, J., and Veneatianer, P., 1992, Purification and some properties of the RNA polymerase of an extremely thermophilic bacterium: *Thermus aquaticus* T2, *Biochim. Biophys. Acta* **435:**228–235.

Fee, J. A., Choc, M. G., Findling, K. L., Lorence, R., and Yoshida, T., 1980, Properties of a copper-containing cytochrome c_1aa_3 complex: A terminal oxidase of the extreme thermophile *Thermus thermophilus* HB8, *Proc. Natl. Acad. Sci. USA* **77:**147–151.

Fee, J. A., Findling, K. L., Yoshida, T., Hille, R., Tarr, G. E., Hearshen, D. O., Dunham, W. R., Day, E. P., Kent, T. A., and Münck, E., 1984, Purification and characterization of the Rieske iron-sulfur protein from *Thermus thermophilus, J. Biol. Chem.* **259:**124–133.

Garber, M. B., Yaremchuk, A. D., Tukalo, M. A., Egorova, S. P., Berthet Colominas, C., and Leberman, R., 1990, Crystals of seryl-tRNA synthetase from *Thermus thermophilus*. Preliminary crystallographic data, *J. Mol. Biol.* **213:**631–632.

Ghosh, D., O'Donnell, S., Furey, W. S., Robbins, A. H., and Stout, C. D., 1982, Iron-sulphur clusters and protein structure of *Azotobacter* ferredoxin at 2.0 Å resolution, *J. Mol. Biol.* **158:**73–109.

Guy, G. R., and Daniel, R. M., 1982, The purification and some properties of a stereospecific D-asparaginase from an extremely thermophilic bacterium, *Thermus aquaticus, Biochem. J.* **787:**790.

Hara-Yokoyama, M., Yokoyama, S., and Miyazawa, T., 1984, Purification and characterization of glutamyl-tRNA synthetase from an extreme thermophile, *Thermus thermophilus* HB8, *J. Biochem.* (Tokyo) **96**:1599–1607.

Hecht, R. M., Garza, A., Lee, Y.-M., Miller, M. D., and Pisegna, M. A., 1989, Nucleotide sequence of the glyceraldehyde-3-phosphate dehydrogenase from *Thermus aquaticus* YT-1, *Nucleic Acids Res.* **17**:10123.

Hocking, J. D., and Harris, J. I., 1976, Glyceraldehyde-3-phosphate dehydrogenase from an extreme thermophile, *Thermus aquaticus*, *Experimentia Suppl.* **26**:121–133.

Hocking, J. D., and Harris, J. I., 1980, D-glyceraldehyde 3-phosphate dehydrogenase: Amino-acid sequence of the enzyme from the extreme thermophile *Thermus aquaticus*, *Eur. J. Biochem.* **108**:567–579.

Hon-nami, K., and Oshima, T., 1980, Cytochrome oxidase from an extreme thermophile, *Thermus thermophilus* HB8, *Biochem. Biophys. Res. Commun.* **92**:1023–1029.

Howard, J. B., Lorsbach, T., and Que, L., 1976, Iron-sulpher clusters and cysteine distribution in a ferredoxin from *Azotobacter vinelandii*, *Biochem. Biophys. Res. Commun.* **70**:582–588.

Hudson, J. A., Morgan, H. W., and Daniel, R. M., 1986, A numerical classification of some *Thermus* isolates, *J. Gen. Microbiol.* **132**:195–200.

Iijima, S., Saiki, T., and Beppu, T., 1980, Physicochemical and catalytic properties of thermostable malate dehydrogenase from an extreme thermophile *Thermus flavus* AT-62, *Biochim. Biophys. Acta* **613**:1–9.

Iijima, S., Saiki, T., and Beppu, T., 1984, Reversible denaturation of thermophilic malate dehydrogenase by guanidinium chloride and acid, *J. Biochem.* (Tokyo) **95**:3–1281.

Imada, K., Sato, M., Tanaka, N., Katsube, Y., Matsuura, Y., and Oshima, T., 1991, Three-dimensional structure of a highly thermostable enzyme, 3-isopropylmalate dehydrogenase of *Thermus thermophilus* at 2.2 Å resolution, *J. Mol. Biol.* **222**: 725–738.

Itaya, M., and Kondo, K., 1991, Molecular cloning of a ribonuclease H (RNase HI) gene from an extreme thermophile *Thermus thermophilus* HB8: a thermostable RNase H can functionally replace the *Escherichia coli* enzyme in vivo, *Nucleic Acids Res.* **19**:4443.

Jaenicke, R., 1981, Enzymes under extremes of physical conditions, *Ann. Rev. Biophys. Bioeng.* **10**:1–67.

Johnson, D. J., 1993, Purification, characterisation and genetic studies of an alkaline phosphates from the thermophile *T. ruber* (strain 16063), Ph.D thesis, University of Manchester.

Johnson, M. K., Bennett, D. E., Fee, J. A., and Sweeney, W. V., 1987, Spectroscopic studies of the seven-iron-containing ferredoxins from *Azotobacter vinelandii* and *Thermus thermophilus*, *Biochim. Biophys. Acta.* **911**:81–94.

Kagawa, Y., Nojima, H., Nukiwa, N., Ishizuka, M., Nakajima, T., Yashuhara, T., Tanaka, T., and Oshima, T., 1984, High guanine plus cytosine content in the third letter of codons of an extreme thermophile, *J. Biol. Chem.* **259**:2956–2960.

Kanaya, S., and Itaya, M., 1992, Expression, purification and characterization of a recombinant ribonuclease H from *Thermus thermophilus* HB8, *J. Biol. Chem.* **267**:10184–10192.

Katsube, Y., Tanaka, N., Takenaka, A., Yamada, T., and Oshima, T., 1988, Crystallization and preliminary X-ray data for 3-isopropylmalate dehydrogenase of *Thermus thermophilus*, *J. Biochem.* (Tokyo) **104**:679–680.

Keller, B., Kast, P., and Hennecke, H., 1992, Cloning and sequence analysis of the phenylalanyl-tRNA synthetase genes (*pheST*) from *Thermus thermophilus*, *FEBS Lett.* **301**:83–88.

Kellis, J. T. Jr., Nyberg, K., Sali, D., and Fersht, A. R., 1988, Contribution of hydrophobic interactions to protein stability, *Nature (London)* **333**:784–786.

Kelly, C. A., Sarfaty, S., Nishiyama, M., Beppu, T., and Birktoft, J. J., 1991, Preliminary

X-ray diffraction analysis of a crystallizable mutant of malate dehydrogenase from the thermophile *Thermus flavus*, *J. Mol. Biol.* **221**:383–385.

Kelly, C. A., Nishiyama, M., Ohnishi, Y., Beppu, T., and Birktoft, J. J., 1993, Determinants of protein thermostability observed in the 1.9-Å crystal structure of malate dehydrogenase from the thermophilic bacterium *Thermus flavus*, *Biochemistry* **32**:3913–3922.

Kirino, H., and Oshima, T., 1991, Molecular cloning and nucleotide sequence of 3-isopropylmalate dehydrogenase gene (*leuB*) from an extreme thermophile *Thermus aquaticus* YT-1, *J. Biochem.* (Tokyo) **109**:852–857.

Kohda, D., Yokoyama, S., and Miyazawa, T., 1984, Thermostable valyl-tRNA, isoleucyl-tRNA and methionyl-tRNA synthetases from an extreme thermophile *Thermus thermophilus* HB8: protein structure and Zn^{2+} binding, *FEBS Lett.* **174**:20–23.

Kohda, D., Yokoyama, S., and Miyazawa, T., 1987, Functions of isolated domains of methionyl-tRNA synthetase from an extreme thermophile, *Thermus thermophilus* HB8, *J. Biol. Chem.* **262**:558–563.

Koide, S., Iwata, S., Matsuzawa, H., and Ohta, T., 1991, Crystallization of an allosteric lactate dehydrogenase from *Thermus caldophilus* and preliminary crystallographic data, *J. Biochem.* **109**:6–7.

Koyama, Y., and Furukawa, K., 1990, Cloning and sequence analysis of tryptophan synthetase genes of an extreme thermophile, *Thermus thermophilus* HB27: Plasmid transfer from replica-plated *Escherichia coli* recombinant colonies to competent *Thermus thermophilus* cells, *J. Bacteriol.* **172**:3490–3495.

Kreutzer, R., Kruft, V., Bobkova, E. V., Lavrik, O. I., and Sprinzl, M., 1992, Structure of the phenylalanyl-tRNA synthetase genes from *Thermus thermophilus* HB8 and their expression in *Escherichia coli*, *Nucleic Acids Res.* **20**:4173–4178.

Krietsch, W. K. G., and Bücher, T., 1970, 3-phosphoglycerate kinase from rabbit sceletal muscle and yeast, *Eur. J. Biochem.* **17**:568–580.

Kumagai, I., Watanabe, K., and Oshima, T., 1982, A thermostable tRNA (guanosine-2'-)-methyltransferase from *Thermus thermophilus* HB27 and the effect of ribose methylation on the conformational stability of tRNA, *J. Biol. Chem.* **257**:7388–7395.

Kunai, K., Machida, M., Matsuzawa, H., and Ohta, T., 1986, Nucleotide sequence and characteristics of the gene for L-lactate dehydrogenase of *Thermus caldophilus* GK24 and the deduced amino-acid sequence of the enzyme, *Eur. J. Biochem.* **160**:433–440.

Kushiro, M., Shimizu, M., and Tomita, K., 1987, Molecular cloning and sequence determination of the *tuf* gene coding for the elongation factur Tu of *Thermus thermophilus* HB8, *Eur. J. Biochem.* **170**:93–98.

Kwon, S. T., Terada, I., Matsuzawa, H., and Ohta, T., 1988b, Nucleotide sequence of the gene for aqualysin I (a thermophilic alkaline serine protease) of *Thermus aquaticus* YT-1 and characteristics of the deduced primary structure of the enzyme, *Eur. J. Biochem.* **173**:11–497.

Lehmacher, A., and Bisswanger, H., 1990a, Isolation and characterisation of an extremely thermostable D-xylose isomerase from *Thermus aquaticus* HB8, *J. Gen. Microbiol.* **136**:679–686.

Lehmacher, A., and Bisswanger, H., 1990b, Comparative kinetics of D-xylose and D-glucose isomers activities of D-xylose isomerase from *Thermus aquaticus* HB8, *Biol. Chem. Hoppe Sayler* **371**:527–536.

Lippmann, C., Betzel, C., Dauter, Z., Wilson, K., and Erdmann, V. A., 1988, Crystallization and preliminary X-ray diffraction studies of intact EF-Tu from *Thermus aquaticus* YT-1, *FEBS Lett.* **240**:139–142.

Lippmann, C., Lindschau, C., Vijgenboom, E., Schröder, W., Bosch, L., and Erdmann, V.

A., 1993, Cytochrome oxidase from *Thermus thermophilus* strain HB8, *J. Biochem.* (Tokyo) **85**:1509–1517.

Lissin, N. M., Sedelnikova, S. E., and Ryazantsev, S. N., 1992, Crystallization of the cpn60/cpn10 complex ('holo-chaperonin') from *Thermus thermophilus, FEBS Lett.* **311**:22–24.

Littlechild, J. A., Davies, G. J., Gamblin, S. J., and Watson, H. C., 1987, Phosphoglycerate kinase from the extreme thermophile *Thermus thermophilus.* Crystallization and preliminary X-ray data, *FEBS Lett.* **225**:123–126.

Loewenthal, R., Sancho, J., and Fersht, A. R., 1992, Histidine-aromatic interactions in barnase, *FEBS Lett.* **224**:759–770.

Love, D. R., Fisher, R., and Bergquist, P. L., 1988, Sequence structure and expression of a cloned β-glucosidase from an extreme thermophile, *Mol. Gen. Genet.* **213**: 84–92.

Ludwig, M. L., Metzger, A. L., Pattridge, K. A., and Stallings, W. C., 1991, Manganese superoxide dismutase from *Thermus thermophilus.* A structural model refined at 1.8 Å resolution, *J. Mol. Biol.* **219**:335–358.

Machida, M., Matsuzawa, H., and Ohta, T., 1985a, Fructose 1,6-bisphosphate-dependent L-lactate dehydrogenase from *Thermus aquaticus* YT-1, an extrene thermophile: activation by citrate and modification reagents and comparison with *Thermus caldophilus* GK24 L-lactate dehydrogenase, *J. Biochem.* (Tokyo) **97**:899–909.

Machida, M., Yokoyama, S., Matsuzawa, H., Miyazawa, T., and Ohta, T., 1985b, Allosteric effect of fructose 1,6-bisphosphate on the conformation of NAD+ as bound to L-lactate dehydrogenase from *Thermus caldophilus* GK24, *J. Biol. Chem.* **260**:16143–16147.

Massey, V., 1953, Studies on fumerase, *Biochem. J.* **55**:172–177.

Mather, M. W., Springer, P., and Fee, J. A., 1991, Cytochrome oxidase genes from *Thermus thermophilus.* Nucleotide sequence and analysis of the deduced primary structure of subunit II of cytochrome caa_3, *J. Biol. Chem.* **266**:5025–5035.

Mather, M. W., Springer, P., Hensel, S., Buse, G., and Fee, J. A., 1993, Cytochrome oxidase genes from *Thermus thermophilus, J. Biol. Chem.* **268**:5395–5408.

Matsumura, M., Becktel, W. J., Levitt, M., and Matthewson, B. W., 1988, Stabilization of phage T4 lysozyme by engineering disulphide bonds, *Proc. Natl. Acad. Sci. USA* **86**:6562–6566.

Matsuzawa, M., Tokugawa, K., Hamaoki, M., Mizoguchi, M., Taguchi, H., Terada, I., Kwon, S-T., and Ohta, T., 1988, Purification and characterization of Aqualysin I (a thermophilic alkaline serine protease) produced by *Thermus aquaticus* YT-1, *Eur. J. Biochem.* **171**:441–447.

Matthews, B. W., Nicholson, H., and Becktel, W. J., 1987, Enhanced protein thermostability from site-directed mutations that decrease the entropy of unfolding, *Proc. Natl. Acad. Sci. USA* **84**:6663–6667.

Meinhardt, S. W., Wang, D. C., Hon-nami, K., Yagi, T., Oshima, T., and Ohnishi, T., 1990, Studies on the NADH-menaquinone oxidoreductase segment of the respiratory chain in *Thermus thermophilus* HB-8, *J. Biol. Chem.* **265**:1360–1368.

Merkler, D. J., Farrington, C. K., and Wedler, F. C., 1981, Protein thermostability, *Int. J. Pept. Protein Res.* **18**:430–442.

Miyazaki, K., Eguchi, H., Yamagishi, A., Wakagi, T., and Oshima, T., 1992, Molecular cloning of the isocitrate dehydrogenase gene of an extreme thermophile, *Thermus thermophilus* HB8, *Appl. Environ. Microbiol.* **58**:93–98.

Morgan, H. W., Daniel, R. M., Cowan, D. A., and Hickey, C. W., 1981, *Thermostable microorganism and proteolytic enzyme prepared therefrom, and processes for the preparation of the micro-organism and the proteolytic enzyme,* European patent number 0 024 182.

Morozov, I. A., Gambaryan, A. S., Lvova, T. N., Nedospasov, A. A., and Venkstern, T. V.,

1982, Purification and characterization of tRNA (adenine-1-)-methyl transferase from *Thermus flavus* strain 71, *Eur. J. Biochem.* **129**:429–436.

Nakamura, N., Sashihara, N., Nagayama, H., and Horikoshi, K., 1989, Characterisation of pullulanase and α-amylase activities of *Thermus* sp. AMD-33, *Starch* **41**:112–117.

Nakamura, S., Ohta, T., Arai, K-I., Arai, N., Oshima, T., and Kaziro, Y., 1978, Studies on polypeptide-chain-elongation factors from an extreme thermophile, *Thermus thermophilus* HB8, *Eur. J. Biochem.* **92**:533–543. (Abstract)

Nicholls, D. J., Sundaram, T. K., Atkinson, T., and Minton, N. P., 1988, Nucleotide sequence of the succinyl-CoA synthetase alpha-subunit from *Thermus aquaticus* B, *Nucleic Acids Res.* **16**:9858.

Nicholls, D. J., Sundaram, T. K., Atkinson, T., and Minton, N. P., 1989, Cloning and nucleotide sequences of the *mdh* and *sucD* genes from *Thermus aquaticus* B, *FEMS Microbiol. Lett.* **70**:7–14.

Nishiyama, M., Matsubara, N., Yamamoto, K., Iijima, S., Uozumi, T., and Beppu, T., 1986, Nucleotide sequence of the malate dehydrogenase gene of *Thermus flavus* and its mutation directing an increase in enzyme activity, *J. Biol. Chem.* **261**:14178–14183.

Nishiyama, M., Horinouchi, S., and Beppu, T., 1991, Characterization of an operon encoding succinyl-CoA synthetase and malate dehydrogenase from *Thermus flavus* AT-62 and its expression in *Escherichia coli*, *Mol. Gen. Genet.* **226**:1–9.

Nojima, H., Ikai, A., Oshima, T., and Noda, H., 1977, Reversible thermal unfolding of thermostable phosphoglycerate kinase, *J. Mol. Biol.* **116**:429–442.

Nojima, H., Hon-nami, K., Oshima, T., and Noda, H., 1978a, Reversible thermal unfolding of thermostable cytochrome *c*-552, *J. Mol. Biol.* **122**:33–42.

Nojima, H., Ikai, A., Hon-nami, K., and Oshima, T., 1978b, Thermodynamic studies on reversible denaturation of thermostable proteins from an extreme thermophile, in: *Biochemistry of Thermophily (Papers and seminars)* (S. M. Freidman, ed.), Academic Press, New York, pp. 305–323.

Nojima, H., Oshima, T., and Noda, H., 1979, Purification and properties of phosphoglycerate kinase from *Thermus thermophilus* strain HB8, *J. Biochem.* (Tokyo) **85**:1509–1517.

Nosoh, Y., and Sakiguchi, T., 1990, Protein engineering for thermostability, *Trends Biotechnol.* **8**:16–20.

Nureki, O., Muramatsu, T., Suzuki, K., Kohda, D., Matsuzawa, H., Ohta, T., Miyazawa, T., and Yokoyama, S., 1991, Methionyl-tRNA synthetase gene from an extreme thermophile, *Thermus thermophilus* HB8. Molecular cloning, primary-structure analysis, expression in *Escherichia coli*, and site-directed mutagenesis, *J. Biol. Chem.* **266**:3268–3277.

Olsen, K. W., 1983, Structural basis for the thermal stability of glyceraldehyde-3-phosphate dehydrogenase, *Int. J. Pept. Protein Res.* **22**:469–475.

Ono, M., Matsuzawa, H., and Ohta, T., 1990, Nucleotide sequence and characteristics of the gene for L-lactate dehydrogenase of *Thermus aquaticus* YT-1 and the deduced amino acid sequence of the enzyme, *J. Biochem.* (Tokyo) **107**:21–26.

Oshima, T., Sakaki, Y., Wakayama, N., Watanabe, K., Ohashi, Z., and Nishimura, S., 1976, Biochemical studies of an extreme thermophile *Thermus thermophilus*: thermal stability of cell constituents and a bacteriophage, *Experimentia Suppl.* **26**:317–330.

Pakula, A. A., and Sauer, R. T., 1990, Reverse hydrophobic effects relieved by amino-acid substitutions at a protein surface, *Nature* **344**:363–364.

Park, H. J., Kreutzer, R., Reiser, C. O., and Sprinzl, M., 1992a, Molecular cloning and nucleotide sequence of the gene encoding a H_2O_2-forming NADH oxidase from the extreme thermophile *Thermus thermophilus* HB8 and its expression in *Escherichia coli*, *Eur. J. Biochem.* **205**:875–879.

Park, H. J., Reiser, C. O., Kondruweit, S., Erdmann, H., Schmid, R. D., and Sprinzl, M.,

1992b, Purification and characterization of a NADH oxidase from the thermophile *Thermus thermophilus* HB8, *Eur. J. Biochem.* **205**:881–885.

Peek, K., Daniel, R. M., Monk, C., Parker, L., and Coolbear, T., 1992, Purification and characterization of a thermostable proteinase isolated from *Thermus* sp. strain Rt41A, *Eur. J. Biochem.* **207**:1035–1044.

Perutz, M. F., and Raidt, H., 1975, Stereochemical basis of heat stability in bacterial ferredoxin and in haemoglobin A2, *Nature* **255**: 256–259.

Plant, A. R., Morgan, H. W., and Daniel, R. M., 1986, A highly thermostable pullulanase from *Thermus aquaticus* YT-1, *Enzyme Microb. Technol.* **8**:668–672.

Qaw, F. S., and Brewer, J. M., 1986, Arginyl residues and thermal stability in proteins, *Mol. Cell. Biochem.* **71**:121–127.

Ramaley, R. F., and Hudock, M. O., 1973, Purification and properties of isocitrate dehydrogenase (NADP) from *Thermus aquaticus* YT-1, *Bacillus substilis*-168 and *Chlamydomonas reinhardti* y^2, *Biochim. Biophys. Acta* **315**:22–36.

Raven, N. D. H., Kelly, C. D., Carter, N. D., Eastlake, P., Brown, C., and Williams, R. A. D., 1993, A new restriction endonuclease, *TspEI*, from the genus *Thermus* that generates cohesive termini compatible with those of *Eco*RI, Gene **131**:83–86.

Reiske, J. S., MacLennan, D. H., and Coleman, R., 1964, Isolation and properties of an iron-protein from the (reduced coenzyme Q)—cytochrome C reductase complex of the respiratory chain, *Biochem. Biophys. Res. Commun.* **15**:338–344.

Reshetnikova, L. S., Reiser, C. O., Schirmer, N. K., Berchtold, H., Storm, R., Hilgenfeld, R., and Sprinzl, M., 1991, Crystals of intact elongation factor Tu from *Thermus thermophilus* diffracting to high resolution, *J. Mol. Biol.* **221**:375–377.

Roberts, R., 1993, REBASE, *Nucleic Acids Res.* **21**:3125–3137.

Rupley, J. A., Gratton, E., and Careri, G., 1983, Water and globular proteins, *Trends Biochem. Sci.* **8**:18–22.

Saeki, Y., Nozaki, M., and Matsumoto, K., 1985, Purification and some properties of NADH oxidase from *Bacillus megaterium, J. Biochem.* (Tokyo) **98**:1433–1440.

Saiki, T., Kimura, R., and Arima, K., 1972, Isolation and characterisation of extremely thermophilic bacteria from hot springs, *Agric. Biol. Chem.* **36**:2357–2366.

Saiki, T., Mahmud, I., Matsubara, N., Taya, K., and Arima, K., 1978, Purification and some properties of NADP+ specific isocitrate dehydrogenase from an extreme thermophile, *Thermus flavus* AT-62, in: *Enzymes and proteins from thermophilic micro-organisms* (H. Zuber, ed.), Birkhauser Verlag, Basel/Stuttgart, pp. 287–303.

Sashihara, N., Nakamura, N., Nagayama, H., and Horikoshi, K., 1988, Cloning and expression of the thermostable pullulanase gene from *Thermus* sp. strain AMD-33 in *Escherichia coli, FEMS Microbiol. Lett.* **49**:385–388.

Sashihara, N., Nakamura, N., and Horikoshi, K., 1993, Subcloning and nucleotide sequencing of the pul gene of *Thermus* sp. AMD-33 thermostable enzyme gene cloning and characterization, *Starch* **45**: 144–150.

Sato, S., Nakada, Y., Kanaya, S., and Tanaka, T., 1988, Molecular cloning and nucleotide sequence of *Thermus thermophilus* HB8 *trp*E and *trp*G, *Biochim. Biophys. Acta* **950**:3–312.

Sato, S., Hutchinson, III, C. A., and Harris, J. I., 1977, A thermostable sequence-specific endonuclease from *Thermus aquaticus, Proc. Natl. Acad. Sci. USA* **74**:542–546.

Sato, S., Nakazawa, K., and Shinomiya, T., 1980, A DNA methylase from *Thermus thermophilus* HB8, *J. Biochem.* (Tokyo) **88**:737–747.

Sato, S., Nakazawa, K., Hon-nami, K., and Oshima, T., 1981, Purification, some properties and amino acid sequence of *Thermus thermophilus* HB8 ferredoxin, *Biochim. Biophys. Acta* **668**:277–289.

Sato, S., Nakada, Y., and Nakazawa Tomizawa, K., 1987, Amino-acid sequence of a tetrameric, manganese superoxide dismutase from *Thermus thermophilus* HB8, *Biochim. Biophys. Acta* **912:**178–184.

Sato, S., and Harris, J. I., 1977, Superoxide dismutase from *Thermus aquaticus, Eur. J. Biochem.* **73:**373–381.

Satoh, M., Tanaka, T., Kushiro, A., Hakoshima, T., and Tomita, K., 1991, Molecular cloning, nucleotide sequence and expression of the *TufB* gene encoding elongation factor Tu from *Thermus thermophilus* HB8, *FEBS Lett.* **288:**98–100.

Schimmel, P. R., and Söll, D., 1979, Aminoacyl-tRNA synthestases: General features and recognition of transfer RNAs, *Ann. Rev. Biochem.* **48:**601–648.

Schmidt, H. L., Stöcklein, W., Danzer, J., Kirch, P., and Limbach, B., 1986, Isolation and properties of an H_2O-forming NADH oxidase from *Streptococcus faecalis, Eur. J. Biochem.* **156:**149–155.

Schroeder, G., Matsuzawa, H., and Ohta, T., 1988, Involvement of the conserved histidine-188 residue in the L-lactate dehydrogenase from *Thermus caldophilus* GK24 in allosteric regulation by fructose 1,6-bisphosphate, *Biochem. Biophys. Res. Commun.* **152:**1236–1241.

Scopes, E. K., 1973, 3-phosphoglycerol kinase, in: *The Enzymes,* 3rd ed. (P. D. Boyer, ed.), Academic Press, New York, pp. 335–351.

Seideler, L., Peter, M. E., Meissner, F., and Sprinzl, M., 1987, Sequence and identification of the nucleotide binding site for the elongation factor Tu from *Thermus thermophilus* HB8, *Nucleic Acids Res.* **15:**9263–9277.

Shinomiya, T., Kobayashi, M., and Sato, S., 1980a, A second site specific endonuclease from *Thermus thermophilus* 111, Tth 111 II, *Nucleic acids Res.* **8:**3275–3285.

Shinomiya, T., and Sato, S., 1980b, A site specific endonuclease from *Thermus thermophilus* 111, Tth 111 I, *Nucleic Acids Res.* **8:**43–56.

Singleton, R. Jr., Kimmel, J. R., and Amelunxen, R. E., 1969, The amino acid composition and other properties of thermophilic glyceraldehyde 3-phosphate dehydrogenase from *Bacillus stearothermophilus, J. Biol. Chem.* **224:**1623–1630.

Slatko, B. E., Benner, J. S., Jager-Quinton, T., Moran, L. S., Simcox, T. G., Van Cott, E. M., and Wilson, G. M., 1987, Cloning, sequencing and expression of the *Taq*I restriction-modification system, *Nucleic Acids Res.* **15:**9781–9796.

Smile, D. H., Donohue, M., Yeh, M-F., Kenkel, T., and Trela, J. M., 1977, Repressible alkaline phosphatase from *Thermus aquaticus, J. Biol. Chem.* **252:**3399–3401.

Stallings, W. C., Pattridge, K. A., Strong, R. K., and Ludwig, M. L., 1985, The structure of manganese superoxide dismutase from *Thermus thermophilus* HB8 at 2.4-Å resolution, *J. Biol. Chem.* **260:**16424–16432.

Stellwagen, E., Cronlund, M. M., and Barnes, L. D., 1973, A thermostable enolase from the extreme thermophile *Thermus aquaticus* YT-1, *J. Biochem.* (Tokyo) **12:**1552–1559.

Stellwagen, E., and Barnes, L. D., 1978, Analysis of the thermostabilityof enolase, in: *Biochemistry of thermophily (Papers and seminars)* (S. M. Freidman, ed.), Academic Press, New York, pp. 223–227.

Stellwagen, E., and Thompson, S. T., 1979, The activation of phosphofructokinase by monovalent cations, *Biochim. Biophys. Acta* **569:**6–12.

Stellwagen, E., and Wilgus, H., 1978, Thermostabilityof proteins, in: *Biochemistry of Thermophily (Papers and Seminars)* (S. M. Freidman, ed.), Academic Press, New York, pp. 223–232.

Stepanov, V. G., Moor, N. A., Ankilova, V. N., and Lavrik, O. I., 1992, Phenylalanyl-tRNA synthetase from *Thermus thermophilus* can attach two molecules of phenylalanine to tRNA (Phe), *FEBS Lett.* **311:**192–194.

Stewart, D. E., and Weiner, P. K., 1987, *J. Molec. Graphics* **5:**133–140.

Suzuki, K., and Imahori, K., 1974, Phosphoglycerate kinase of *Bacillus stearothermophilus*, *J. Biochem.* **76**:771–782.

Taguchi, H., Yamashita, M., Matsuzawa, H., and Ohta, T., 1982, Heat-stable and fructose 1,6-bisphosphate-activated L-lactate dehydrogenase from an extremely thermophilic bacterium, *J. Biochem.* **91**:1343–1348.

Taguchi, H., Haraski, M., Matsuzawa, H., and Ohta, T., 1983, Heat stable extracellular proteolytic enzyme produced by *Thermus caldophilus* strain GK24, an extremely thermophilic bacterium, *J. Biochem.* (Tokyo) **93**: 7–13.

Taguchi, H., Matsuzawa, H., and Ohta, T., 1984, L-Lactate dehydrogenase from *Thermus caldophilus* GK24, an extremely thermophilic bacterium. Densensitization to fructose 1,6-bisphosphate in the activated state by arginine-specific chemical modification and the N-terminal amino acid sequence, *Eur. J. Biochem.* **145**:283–290.

Taguchi, H., Konishi, J., Ishii, N., and Yoshida, M., 1991, A chaperonin from a thermophilic bacteria, *Thermus thermophilus*, that controls refolding of several thermostable enzymes, *J. Biol. Chem.* **266**:22411–22418.

Taguchi, H., and Yoshida, M., 1993, Chaperonin from *Thermus thermophilus* can protect several enzymes from irreversible heat denaturation by capturing denaturation intermediate, *J. Biol. Chem.* **268**:5371–5375.

Takahashi, H., Yamaguchi, E., and Uchida, T., 1984, Thermophilic DNA ligase, *J. Biol. Chem.* **259**:10041–10047.

Takase, M., and Horikoshi, K., 1988, A thermostable β-glucosidase isolated from a bacterium species of the genus *Thermus*, *Appl. Microbiol. Biotechnol.* **2955**:55–60.

Tanaka, T., Kawano, N., and Oshima, T., 1981, Cloning of 3-isopropylmalate dehydrogenase gene of an extreme thermophile and partial purification of the gene product, *J. Biochem.* **89**:677–682.

Titani, K., Ericsson, L. H., Hon-nami, K., and Miyazawa, T., 1985, Amino acid sequence of cytochrome c-552 from *Thermus thermophilus* HB8, *Biochem. Biophys. Res. Commun.* **128**:781–787.

Tsutsumi, S., Denda, Y., Yokoyama, K., Oshima, T., Date, T., and Yoshida, M., 1991, Molecular cloning of genes encoding the major two subunits of a eubacterial V-type ATPase from *Thermus thermophilus*, *Biochim. Biophys. Acta* **1098**:13–20.

Ulrich, J. T., McFeters, G. A., and Temple, K. L., 1972, Induction and characterization of β-galactosidase in an extreme thermophile, *J. Bacteriol.* **110**:691–698.

Uyeda, K., and Kurooka, S., 1970, Crystallization and properties of phosphofructokinase from *Clostridia pasteurianum*, *J. Biol. Chem.* **245**:3315–3324.

Vainshtein, B. K., Melik-Adamyan, V. R., Barynin, V. V., and Vagin, A. A., 1984, in: *Progress in Bioorganic Chemistry and Molecular Biology* (Y. Ovchinnikov, ed.), Elsevier, New York, pp. 1117–1132.

Vali, Z., Kilar, F., Lakatos, S., Venyaminov, S. A., and Zavodszky, F., 1980, L-alanine dehydrogenase from *Thermus thermophilus*, *Biochim. Biophys. Acta* **615**:34–47.

Varley, P. G., and Pain, R. H., 1991, Relation between stability, dynamics and enzyme activity in 3-phosphoglycerate kinases from yeast and *Thermus thermophilus*, *J. Mol. Biol.* **220**:531–538.

Verhoeven, J. A., Schenck, K. M., Meyer, R. R., and Trela, J. M., 1986, Purification and characterization of an inorganic pyrophosphatase from the extreme thermophile *Thermus aquaticus*, *J. Bacteriol.* **168**:318–321.

Vihinen, M., 1987, Relationship of protein flexibility to thermostability, *Protein. Eng.* **1**:477–480.

Voss, R. H., Hartmann, R. K., Lippmann, C., Alexander, C., Jahn, O., and Erdmann, V. A., 1992, Sequence of the *tufA* gene encoding factor EF-Tu from *Thermus aquaticus* and overproduction of the protein in *Escherichia coli*, *Eur. J. Biochem.* **207**:839–846.

Walker, J. E., Wonacott, A. J., and Harris, J. I., 1980, Heat stability of a tetrameric enzyme, D-glyceraldehyde-3-phosphate dehydrogenase, *Eur. J. Biochem.* **108**:581–586.

Walsh, A. T., Daniel, R. M., and Morgan, H. W., 1983, A soluble NADH dehydrogenase (NADH:ferricyanide oxidoreductase) from *Thermus aquaticus* strain T351, *Biochem. J.* **209**:427–437.

Williams, R. A. D., 1992, The genus *Thermus*, in: *Thermophilic Eubacteria* (J. Kristjansson, ed.), CRC Press, Boca Raton, FL, pp. 51–62.

Wnendt, S., Hartmann, R. K., Ulbrich, N., and Erdmann, V. A., 1990, Isolation and physical properties of the DNA-directed RNA polymerase from *Thermus thermophilus* HB8, *Eur. J. Biochem.* **191**:467–472.

Wold, F., 1971, Enolase, in: *The Enzymes*, 3rd ed. (P. D. Boyer, ed.), Academic Press, New York, pp. 499–538.

Xu, J., Oshima, T., and Yoshida, M., 1990, Tetramer-dimer conversion of phosphofructokinase from *Thermus thermophilus* induced by its allosteric effectors, *J. Mol. Biol.* **215**:597–606.

Xu, J., Oshima, T., and Yoshida, M., 1991, Phosphoenolpyruvate-insensitive phosphofructokinase isozyme from *Thermus thermophilus* HB8, *J. Biochem.* (Tokyo) **109**:199–203.

Xu, X. M., and Yagi, T., 1991, Identification of the NADH-binding subunit of energy-transducing NADH-quinone oxidoreductase (NDH-1) of *Thermus thermophilus* HB-8, *Biochem. Biophys. Res. Commun.* **174**:667–672.

Yagi, T., Hon-nami, K., and Ohnishi, T., 1988, Purification and characterization of two types of NADH-quinone reductase from *Thermus thermophilus* HB-8, *Biochemistry* **27**:2008–2013.

Yamada, T., Akutsu, N., Miyazaki, K., Kakinuma, K., Yoshida, M., and Oshima, T., 1990, Purification, catalytic properties, and thermal stability of threo-Ds-3-isopropylmalate dehydrogenase coded by *leuB* gene from an extreme thermophile, *Thermus thermophilus* strain HB8, *J. Biochem.* (Tokyo) **108**:449–456.

Yang, S-J., and Zhang, S-Z., 1988, Purification and characterization of α-glucosidase from an extreme thermophile, *Thermus thermophilus* HB8, *Ann. N.Y. Acad. Sci.* **542**:210–212.

Yang, S., Cai, M., Liu, J., Gu, Y., Wang, Z., and Zhang, S., 1991, Studies on the thermostable L-lactate dehydrogenase from thermophilic bacteria, *Wei-Sheng-Wu-Hsueh-Pao* **31**:438–443.

Yaremchuk, A. D., Tukalo, M. A., Egorova, S. P., Konovalenko, A. V., and Matsuka, G. K., 1990, [Isolation of tyrosyl-tRNA-synthetase from *Thermus thermophilus* HB-27], *Ukr. Biokhim. Zh.* **62**:97–99.

Yeh, M-F., and Trela, J. M., 1976, Purification and characterization of a repressible alkaline phosphatase from *Thermus aquaticus*, *J. Biol. Chem.* **251**:3134–3139.

Yokoyama, K., Oshima, T., and Yoshida, M., 1990, *Thermus thermophilus* membrane-associated ATPase. Indication of a eubacterial V-type ATPase, *Eur. J. Biochem.* **265**:21946–21950.

Yoshida, T., Larence, R. M., Choc, M. G., Tarr, G. E., Findling, K. L., and Fee, J. A., 1984, Respiratory proteins from the extremely thermophilic aerobic bacteria *Thermus thermophilus*, *J. Biol. Chem.* **259**:112–123.

Zimmermann, B. H., Nitsche, C. I., Fee, J. A., Rusnak, F., and Munck, E., 1988, Properties of a copper-containing cytochrome ba_3: a second terminal oxidase from the extreme thermophile *Thermus thermophilus*, *Proc. Natl. Acad. Sci. USA* **85**:5779–5783.

The Cell Walls and Lipids of *Thermus*

5

MILTON S. DA COSTA

1. INTRODUCTION

The cell walls and membranes, being extensive and extremely well-organized structures, are important in maintaining the integrity of the cell within its growth temperature range. The properties of these structures contribute to the ability of microorganisms to grow at high temperatures. Whereas the cell wall has primarily a structural role, the cell membrane controls the exchange of substances between the internal and the external environments. The lipid fraction of the membrane must maintain an appropriate fluidity for membrane protein function under changing environmental conditions. The biochemical and biophysical characteristics of these structures are, therefore, of great interest in our understanding of life at high temperatures. The composition of cell walls and membranes is important in assessing the taxonomic and phylogenetic relationships of microorganisms, and complements phenotypic and molecular-genetic information. The bacteria of the genus *Thermus,* due to their simple growth requirements, have contributed a significant proportion of information on the biochemistry and physiology of thermophiles.

2. CELL WALL STRUCTURE AND COMPOSITION

Most *Thermus* strains in culture form pleomorphic rod-shaped cells and short filaments, and stain gram-negative (Brock, 1978). Electron mi-

---

MILTON S. DA COSTA • Departamento de Zoologia, Universidade de Coimbra, 3049 Coimbra Cedex, Portugal.

Thermus Species, edited by Richard Sharp and Ralph Williams. Plenum Press, New York, 1995.

croscopy of cross-sections of the cell envelope show a thin dense layer, presumably representing the peptidoglycan, closely adherent to the cell membrane. An outer layer, in most strains, has a characteristic corrugated "cobble stone" or "annelid" appearance with more or less regularly spaced indentations closely connected to the peptydoglycan (Brock and Edwards, 1970; Hensel *et al.*, 1986; Jackson *et al.*, 1973; Pask-Hughes and Williams, 1978). *Thermus filiformis*, uniquely, has a stable filamentous morphology in culture, and rod shaped cells are never observed (Hudson *et al.*, 1987). It also has an extra wall layer, external to the corrugated layer, which runs interrupted over zones of septum formation. The peptidoglycan of all strains examined contains ornithine (Pask-Hughes and Williams, 1978; Hensel *et al.*, 1986; Williams, 1989), which is also found in the peptidoglycan of the bacteria of the distantly related genus *Deinococcus*, which stains gram-positive (Embley *et al.*, 1987). The primary structure of the peptidoglycan of *T. ruber* and strain H3 (Hensel *et al.*, 1986), is identical to that of some strains of *Deinococcus* (Schleifer and Kandler, 1972). The external corrugated layer is probably not chemically or structurally equivalent to the outer membrane of other gram-negative bacteria; neither heptose nor ketodeoxyoctulosonate were detected in lipopolysaccharide preparations of several strains (Pask-Hughes and Williams, 1978). A crystalline layer of hexagonal symmetry was identified in the cell envelope of *T. thermophilus* HB8 (Berenguer *et al.*, 1988; Faraldo *et al.*, 1988). This is probably analogous to the crystalline surface layers (S-layers) of other bacteria (Sleytr and Messner 1988). This crystalline layer is formed of one protein (P100) whose oligomeric complexes are extremely resistant to thermal denaturation in the presence of Ca^{2+}. The crystalline layer proteins appear to be associated with the outer layer of the cell envelope. The complex cell wall of *Deinococcus radiodurans* strain Stark resembles, in some ways, the envelope of *Thermus* strains. There is an outer "membrane" and an hexagonal surface protein array (Lancy and Murray, 1978; Thompson and Murray, 1981; Thompson *et al.*, 1982). The formation of multicell aggregates, commonly described as "rotund bodies", is observed in many strains of *Thermus*, both the yellow (higher temperature), and red-pigmented (lower temperature) strains (Brock and Edwards, 1970; Kraepelin and Gravenstein, 1960; Pask-Hughes and Williams, 1978). These structures are normally large, comprising several cells bound together by a common outer layer of the cell envelope which that has partially peeled off each cell. This network of cells, enclosing a large inner space devoid of any visible structures, is also seen in other bacteria, such as the anaerobic thermophilic bacterium *Dictyoglomus thermophilum* (Saiki *et al.*, 1985), as well as some mesophiles such as *Thiobacillius thiooxidans* (Mahoney and Edwards, 1966). Moreover, rotund bodies are not consistent morphological features of the strains observed, indicating that these structures are formed under conditions leading to reduced growth rate.

3. LIPID COMPOSITION

3.1. Respiratory Quinones

Fully saturated menaquinone 8 (MK-8) is the predominant respiratory quinone in all the *Thermus* strains examined (Collins and Jones, 1981; Hensel *et al.*, 1986; Williams, 1989); surprisingly, these menaquinones are very rare in gram-negative bacteria. However, the presence of MK-8 together with ornithine in the peptidoglycan, reinforces the phylogenetic relationship between *Thermus* and the gram-positive genus *Deinococcus* (Hensel *et al.*, 1986; Weisburg *et al.*, 1989).

3.2. Polar Lipids

The polar lipid patterns of all strains of the genus on single and two-dimensional thin layer chromatography (TLC) are very similar, although some components exhibit small, but reproducible, differences in migration. In the high temperature strains, one glycolipid (GL1) and one phospholipid (PL2) are major components of the polar lipid fraction (Figure 1). At the optimum growth temperature, two additional minor glycolipids and two minor phospholipids, one of which contains a free amino group, are also present in most strains examined. These include *T. aquaticus* YT1, *T. thermophilus* HB8, and several strains from mainland Portugal and the island of S. Miguel. Other minor glycolipids and phospholipids are also detected in a few strains, such as strain LFC1, which has a ninhydrin-positive glycolipid, and strain FQ3, with one extra phospholipid and three additional glycolipids (Donato *et al.*, 1990).

By contrast to the high-temperature strains, *T. ruber*, "*T. rubens*" and several red-colored isolates from mainland Portugal, the island of S. Miguel, Iceland, New Zealand, and Yellowstone National Park, U.S.A., have two prominent glycolipids (GL1a and GL1b) migrating close to each other at the position usually occupied by GL1. A predominant phospholipid, corresponding to PL2, is also found in these strains, although the minor components are barely detectable or absent (Donato *et al.*, 1991). At the optimum growth temperature, the major glycolipid and major phospholipid of the high-temperature strains examined comprise between 83 and 96% of the polar lipid carbohydrate and 77 to 92% of the polar lipid phosphorus. In *T. ruber* and *T. rubens* the two prominent glycolipids (GL1a

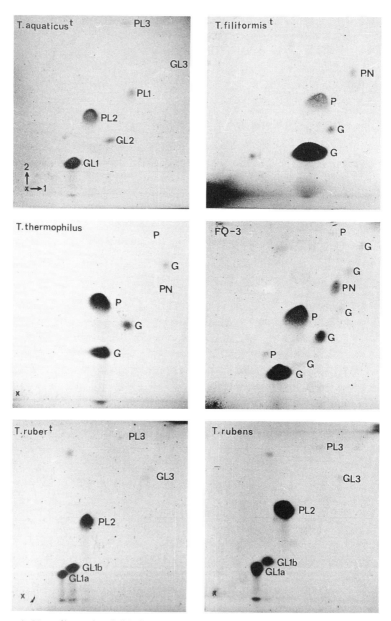

Figure 1. Two dimensional thin-layer chromatography of the polar lipids from *Thermus* strains. GL_1, glycolipid 1; GL_{1a}, glycolipid 1a; GL_{1b}, glycolipid lb; GL_2, glycolipid 2; GL_3, glycolipid 3; PL_1, phospholipid 1; PL_2, phospholipid 2; PL_3, phospholipid 3.

and GL1b) together account for over 90% of the polar lipid carbohydrate and the major phospholipid is practically the only phospholipid present during growth at 60 °C (Prado *et al.*, 1988; Donato *et al.*, 1991) (Table I). The structure of the major glycolipid of *T. thermophilus* HB8 was established by Oshima and Yamakawa (1974), who found it to contain two galactose residues, one glucose residue, and glucosamine. These authors also reported the terminal galactose to be in the furanose form, and iso-C17(15 methylhexadecanoic acid) to be the sole fatty acid amide-linked to glucosamine. The precise structure of the major glycolipid of this strain described is galactofuranosyl-(1→2)-galactopyranosyl-(1→6)-glucosaminyl(15-methylhexadecanoyl)-(1→2)-glucopyranosyl-diacylglycerol. The structures of the major glycolipid of several strains of *Thermus*, described in some detail, while their precise structure is unknown, are also generically diglycosyl-(Nacyl)-glycosaminyl glucosyl-diacylglycerols (Table II). In all cases, the polar head group is composed of three hexose residues and one hexosamine residue, which is amide-linked to a fatty acid. The ratio of glucose to galactose varies among the strains, but in most strains glucose is the only hexose present. Glucosamine is found in the major glycolipid of all strains examined with the exception of *T. aquaticus* YT1, where galactosamine is present (Pask-Hughes and Shaw, 1982, Prado *et al.*, 1988). Both prominent glycolipids of *T. ruber* and *T. rubens* have identical polar head groups composed of three glucose residues and one glucosamine. The difference in migration of these two glycolipids on thin layer chromatography must be due to unknown differences in the fine structure of the polar head group or in the fatty acid composition (Table II). The structure of the major phospholipid is unknown but it is reported to

Table I. Relative Proportions of Glycolipids (GL) and Phospholipids (PL) in Strains of the Genus *Thermus* at the Optimum Growth Temperature[a]

Strains	Polar lipids							
	GL_1	GL_{1a}	GL_{1b}	GL_2	GL_3	PL_1	PL_2	PL_3
T. aquaticus YT-1	94.6	—	—	3.4	tr	7.1	90.6	1.8
Thermus CG-2	96.4	—	—	2.1	tr	16.6	76.8	2.0
Thermus VI-13	83.2	—	—	12.3	1.5	10.7	84.2	4.0
Thermus SPS-11	94.8	—	—	3.0	tr	6.0	90.7	1.2
Thermus SPS-17	95.8	—	—	2.9	tr	5.6	91.5	0.9
T. ruber	—	41.8	57.1	tr	tr	1.4	93.3	3.4
"*T. rubens*"	—	75.8	22.4	tr	tr	—	98.7	1.6

[a] Results from Prado *et al.* (1988) and Donato *et al.* (1991); tr = trace.

Table II. Composition of the Polar Head Group of the
Major-Glycolipid (GL_1) of High-Temperature Strains of *Thermus*
and the Two Prominent Glycolipids (GL_{1a} and GL_{1b}) of *T. tuber*
and "*T. rubens*"

	Composition (ratio)[a]				
Strain	Glu	Gal	GluNH	GalNH	Gly
T. aquaticus YT-1	2.8	—	—	1.0	0.9
T. thermophilus HB-8	1.0	2.0	1.0	—	1.0
Thermus H	2.8	—	1.0	—	0.8[b]
Thermus J	2.0	1.1	1.0	—	0.9[b]
Thermus NH	1.0	2.1	1.0	—	0.9[b]
Thermus SPS-11	2.8	—	1.0	—	1.0
Thermus SPS-17	3.0	—	1.0	—	0.9
Thermus CG-2	2.0	1.0	1.0	—	0.9
Thermus VI-13	2.8	—	1.0	—	1.1
T. ruber (GL1a)	3.1	—	1.0	—	1.2
(GL1b)	3.3	—	1.0	—	0.9
T. rubens (GL1a)	3.2	—	1.0	—	0.9
(GL1b)	2.9	—	1.0	—	0.9

[a]Glu, glucose; Gal, galactose; GluNH, glucosamine; GalNH, galactosamine; Gly, glycerol.
[b]Results from Pask-Hughes and Shaw (1982).

contain one phosphorus and three fatty acids per molecule, and glucosamine is detected after acid hydrolysis (Pask-Hughes and Shaw, 1982). The minor phospholipids and glycolipids have not been identified, and like the major components may be unusual and restricted to the genus *Thermus*.

3.3. Fatty Acids

The membrane lipids of many thermophilic bacteria, including *Thermus*, contain large amounts of iso- and anteiso-branched chain fatty acids. The higher melting iso-branched fatty acids are more abundant in thermophiles than anteiso-fatty acids, which have considerably lower melting points. Exceptions to the predominance of iso-and anteiso-branched chain of fatty acids are found in some thermophilic bacteria, and illustrate the different strategies which have evolved to produce stable and functional membranes at high temperatures. Thus the lipids of *Thermomicrobium roseum* have high large proportions of internally branched fatty acids, primarily 12-methyl C18 fatty acid either ester-linked to long chain diols or amide-linked to amino sugars (Langworthy and Pond, 1986), and in the thermoacidophile *Bacillus acidocaldarius* ω-cyclohexyl fatty acids predominate (de Rosa *et al.*, 1972; Oshima and Ariga, 1975). Other notable exceptions are the presence of "diabolic acid" (15,16-dimethyltriacontanedioic acid), originally described in a mesophilic *Butyrivibrio* spp. (Klein *et al.*, 1979), and a novel glycerol ether lipid (15,16-dimethyl-30-glyceryloxytriacontanoic acid) in addition to normal-chain fatty acids in *Thermotoga maritima* (de Rosa *et al.*, 1989). Branched chain fatty acids are not restricted to thermophiles, and are present in high relative proportions in many mesophilic bacteria (Kaneda, 1991). Therefore, branched-chain fatty acids in thermophilic bacteria reflects phylogenetic relationships more than a requirement for growth at high temperatures.

It was initially demonstrated that, at the optimum growth temperature, iso-C17 (15-methylhexadecanoic acid) is the predominant fatty acid in *T. aquaticus* YT1, *T. thermophilus* HB8, "*T. flavus*," and *Thermus* strains X1, H, J, and NH, followed by smaller relative proportions of iso-C15 (13-methyltetradecanoic acid). Due to the very high growth temperatures, anteiso-C15 (12-methyltetradecanoic acid) and anteiso-C17 (14-methylhexadecanoic acid) were found in smaller relative proportions than the odd numbered iso-branched fatty acids (Hensel *et al.*, 1986, Jackson *et al.*, 1973; Oshima, 1978, Oshima and Yamakawa, 1974; Pask-Hughes and Shaw, 1982; Ray *et al.*, 1971a). Even-numbered *iso* and straight-chain fatty acids are only found as minor components. Monounsaturated fatty acids are rare or vestigial at elevated growth temperatures but, recently, a monounsaturated anteiso-C17:1 was identified in several *Thermus* strains grown at low temperatures (Nordstrom and Laakso, 1992). Comparative studies of a large number of *Thermus* strains indicate that the relative proportions of the predominant fatty acids are more variable than originally expected (Donato *et al.*, 1990). At optimum growth temperatures of 70–73 °C, many high-temperature isolates have, for example, higher relative proportions of the iso-C15 than of iso-C17 fatty acid. Some strains also contain high proportions of the iso-C16 (14-pentadecanoic acid) fatty acid. The *T. ruber*-like strains have a high iso-C15 to iso-C17 fatty acid ratio, perhaps due to the lower growth temperature of these strains. Nevertheless, some high-temperature strains also have similar iso-C15/iso-C17 ratios, suggesting that control of membrane fluidity depends on factors intrinsic to each strain, rather than on the growth temperature alone (Donato *et al.* 1991; Hensel *et al.*, 1986). *T. filiformis* provides an interesting example of strain-specificity of fatty acid composition, in which it appears to be unique. Under the conditions tested, this strain contained very large proportions of anteiso-C15 and anteiso-C17 fatty acids at the optimum growth temperature (70 °C) (Donato *et al.*, 1990).

3.4. Carotenoids

Due the presence of carotenoids, most high-temperature *Thermus* strains, including *T. filiformis*, are yellow-pigmented; nevertheless, some are colorless. The strains isolated from natural habitats exposed to sunlight are generally pigmented, while those form man-made thermal environments lacking illumination, such as hot water systems and hot water tanks, are frequently nonpigmented. About 87% of *Thermus* strains isolated by Brock and Boylen (1973) and 70% of the strains isolated by Stramer and Starzyk (1981) from domestic and commercial hot-water tanks were nonpigmented. In contrast, the latter authors reported that 70% of the isolates from steel mill discharges and nuclear power plant cooling water were pigmented. However, the yellow-pigmented strains from these man-made thermal environments may not have originated in these nonilluminated sites. Carotenoids of *Thermus* strains are presumed to have a photoprotective function (Brock, 1978; Brock, 1981), implying that pigmentation is selected for by exposure to sunlight, but it must be stressed that in some bacteria, e.g. *Acholeplasma laidlawi*, carotenoids may reinforce the lipid bilayer, and so have a structural role equivalent to the sterols in eucaryotes (Rottem and Markowitz, 1979). It has been suggested that pigmentation in some strains may be an unstable characteristic. One study reported the frequent isolation of colorless mutants from *T. aquaticus* YT1 in continuous culture (Cometta *et al.*, 1982), but this has never been observed in many years of continuous culture of several *Thermus* species in other laboratories (R. Sharp, personal communication). With strain YS45, the intensity of pigmentation varied directly with the degree of illumination of chemostats where growth was limited by the carbon source (Cossar and Sharp, 1989) but was never lost completely. Several halotolerant *Thermus* isolates also consistently produce nonpigmented colonies (C. Manaia and O. Nunes, personal communication). All low temperature isolates related to *T. ruber* are to date red-pigmented, but nonpigmented strains have not been sought (Hensel *et al.*, 1986; Loginova *et al.*, 1984; Sharp and Williams, 1988). Absorption spectra of acetone extracts of pigmented high-temperature strains demonstrate that the yellow-pigmented strains exhibit absorption maxima at about 450 and 480 nm (Pask-Hughes and Williams, 1977), while the red-pigmented strains have absorption maxima at longer wavelengths (480 and 510 nm) (Hensel *et al.*, 1986). The carotenoid pigments have only been partially characterized and include phytoene and β-carotene; however other major carotenoid components have not been identified (Ray *et al.*, 1971b).

4. TEMPERATURE-INDUCED ALTERATIONS IN LIPIDS

4.1. Polar Lipids

Within their growth temperature ranges, bacteria show changes in membrane lipid composition in response to changes in the growth temperature. This "homeoviscous adaptation" is a prerequisite for membrane fluidity throughout the temperature range, allowing the regulation of membrane-bound enzymes and transport systems (Russell, 1984). Alterations in fatty acid composition significantly change the membrane fluidity, although changes in the relative proportions of glycolipids and phospholipids also appear to be involved in "homeoviscous adaptation" in *Thermus* and other thermophiles (Hasegawa *et al.*, 1980). Temperature-related alterations in the complex lipids composition in a small number of strains indicates that the growth temperature has a profound effect on the fatty acid composition and the complex lipids. It was initially noted that the carotenoids, phospholipids, and glycolipids increased in proportion with growth temperature (Ray *et al.*, 1971b). Increase in the growth temperature of strains SPS11 (nonpigmented) and SPS17 (yellow-pigmented) from 50 °C to their maximum growth temperature was associated with an increase in the glycolipid content, but not the phospholipid content. The carotenoids of strain SPS17 increased up to the optimum growth temperature, but decreased at 78 °C; no alteration was observed in the carotenoid content of strain SPS11, which was unable to grow at 78 °C (Prado *et al.*, 1988). The growth temperature also affects the polar lipid components in some *Thermus* strains. Oshima (1978) reported a large increase in the major glycolipid of *T. thermophilus* as the growth temperature was increased from 45 °C to 80 °C. The relative concentration of the major glycolipid (GL1) of strain SPS17 increases with a concomitant decrease in two minor glycolipids (GL2 and GL3) as the growth temperature was raised from 50 °C to 78 °C (Prado *et al.*, 1988). Similar alterations were not found, however, in strain SPS11. Unlike the major glycolipid, the relative concentration of major phospholipid (PL2) decreased in strains SPS11 and SPS17 (Prado *et al.*, 1988) as the growth temperature was raised, while the relative proportion of a minor phospholipid increased.

**Table III. Fatty Acid Composition of *Thermus* Strains
at the Optimum Growth Temperature[a]**

Strains	Growth temp. (°C)	Fatty acids (%)[b]								
		iC14	iC15	aC15	nC15	iC16	nC16	iC17	aC17	nC17
High-temperature										
T. aquaticus YT-1	73	—	26.2	3.8	—	7.1	7.4	46.6	7.4	1.0
"T. Thermphilus"	73	tr	30.1	7.6	tr	4.4	1.0	44.9	10.5	—
T. filiformis	73	tr	5.6	22.5	—	8.9	1.7	9.2	48.6	1.0
SPS-17	73	—	43.3	6.8	2.0	3.0	tr	36.9	5.3	1.0
VI-5b2	70	tr	60.5	8.2	3.5	2.4	—	22.1	2.1	—
VI-13	73	—	50.6	6.6	tr	tr	1.1	34.2	3.3	2.0
LFC-1	73	—	25.1	10.2	tr	4.9	tr	41.9	14.7	—
LFF-1	73	tr	60.7	11.2	2.6	2.3	tr	19.1	2.3	tr
RQ-1	73	1.1	13.6	11.0	—	15.9	2.6	30.2	22.7	tr
Low-temperature										
T. ruber	60	tr	66.5	7.8	1.7	1.6	1.4	13.8	2.3	tr
"T. rubens"	60	tr	56.4	5.0	1.4	1.5	2.0	20.5	3.6	—
16501	60	tr	62.6	13.1	1.5	1.5	2.3	13.5	2.0	tr
16503	60	tr	55.2	8.0	1.0	1.7	1.8	18.9	4.7	tr
16106	60	tr	58.5	8.5	1.5	1.7	1.2	15.8	3.1	tr
16294	60	tr	62.8	6.4	2.1	2.6	1.7	15.2	2.7	tr
15058	60	1.5	46.9	8.0	2.4	7.4	2.9	18.4	3.6	tr
SPS-R4	60	tr	62.0	8.8	1.9	2.4	1.7	13.7	3.5	tr

[a]Results from Donato *et al.* (1990) and Donato *et al.* (1991).
[b]Abbreviation for fatty acids; i, *iso*-branched; a, *anteiso*-branched; n, straight chain; tr, trace amounts.

4.2. Fatty Acids

The effect of the growth temperature on the fatty acid composition has been studied in a few high-temperature strains of *Thermus*. In general, an increase in the growth temperature of these organisms results in an increase in the *iso*-branched fatty acids and a decrease in the *anteiso*-branched fatty acids (Oshima, 1978; Prado *et al.*, 1988; Ray *et al.*, 1971a). For example, as the growth temperature of *T. thermophilus* HB8 is raised from 49 °C to 82 °C there is a pronounced increase in the iso-C17/iso-C15 ratio and a very large decrease in the relative proportions of anteiso-C15 and iso-C16 fatty acids, while the anteiso-C17 fatty acid remains virtually unchanged (Oshima, 1978). Increasing the growth temperature of *T. filiformis*, by contrast to other strains examined, resulted in small decreases in iso-C15, iso-C17, and anteiso-C15 fatty acids with pronounced increases in the relative proportions of iso-C16 and n-C16 fatty acids. At all growth temperatures, however, the anteiso- to iso-fatty acid ratio was very high (Donato *et al.*, 1990). In addition to temperature-induced alterations in the proportions of saturated iso-/anteiso-fatty acids, the monounsaturated anteiso-C17:1 was not detected in high temperature *Thermus* strains growing at 70 °C, but was present in cells grown at 40–45 °C. This fatty acid reached particularly high relative proportions in two *T. ruber*-like strains grown at 35 °C (Nordstrom and Laakso, 1992).

5. FINAL CONSIDERATIONS

Despite the progress made in understanding the composition of the cell walls and lipids of the bacteria of the genus *Thermus*, there are still large gaps in our knowledge of the structural and biophysical properties of these cell components. The biophysical properties of the membranes have only been dealt with superficially (Pinheiro *et al.*, 1987; Prado *et al.*, 1990). In addition, the apolar lipid fraction has been practically ignored and further progress in this area will be important in understanding membrane structure at high growth temperatures. No single cell component can be considered more important than others for life at high temperatures, since all of them must have functional stability at temperatures which support growth. The structure of the cell wall and membrane components will, nevertheless, provide insight into the mechanisms which contribute to growth at temperatures which are alien to our view of life.

ACKNOWLEDGMENTS. The results from the author's laboratory were obtained by Adelina Prado, Eduardo Seleiro, and M. Manuel Donato. This work was supported by the Junta Nacional de Investigacao Cientifica e Tecnologica (JNICT/BIO/87/79; BIO/908/90), and Instituto Nacional de Investigacao Cientifica (Centro de Biologia Celular).

REFERENCES

Berenguer, J., Faraldo, M. L., and Pedro, M. A., 1988, Ca^{2+}-stabilized oligomeric protein complexes are major components of the cell envelope of *"Thermus thermophilus"* HB8, *J. Bacteriol.* **170:**2441–2447.

Brock, T. D., 1978, *Thermophilic Microorganisms and Life at High Temperatures*, Springer-Verlag, New York.

Brock, T. D., 1981, Extreme Thermophiles of the Genera *Thermus* and *Sulfolobus*, in: *The Prokaryotes: A Handbook on Habits, Isolation and Identification of Bacteria*, Volume 1 (M. P. Starr, H. Stolp, H. G. Truper, A. Balows, and H. G. Schlegel, eds.), Springer-Verlag, New York, pp. 978–984.

Brock, T. D., and Boylen, K. L., 1973, Presence of thermophilic bacteria in laundry and domestic hot water heaters, *Appl. Microbiol.* **23:**72–76.

Brock, T. D., and Edwards, M. R., 1970, Fine structure of *Thermus aquaticus*, an extreme thermophile, *J. Bacteriol.* **104:**509–517.

Collins, M. D., and Jones, D., 1981, Distribution of isoprenoid quinone structural types in bacteria and their taxonomic implications, *Microbiol. Rev.* **45:**316–354.

Cometta, S., Sonnleitner, B., and Fietcher, A., 1982, The growth behaviour of *Thermus aquaticus* in continuous culture, *Eur. J. Appl. Microbiol. Biotechnol.* **15:**69–74.

Cossar, D., and Sharp, R. J., 1989, Loss of pigmentation in *Thermus* spp., in: *Microbiology of Extreme Environments and its Potential for Biotechnology* (M. S. da Costa, J. S. Duarte, and R. A. D. Williams, eds.), **385**. Elsevier, London.

de Rosa, M., Gambacorta, A., Minale, L., and BúLock, J. D., 1972, The formation of ω-cyclohexyl fatty acids from shikimate in an acidophilic thermophilic *Bacillus, Biochem. J.* **128:**751–754.

de Rosa, M., Gambacorta, A., Huber, R., Lanzotti, V., Nicolaus, B., Stetter, K. O., and Trincone, A., 1989, Lipid structure in Thermotoga maritima, in: *Microbiology of Extreme Environments and its Potential for Biotechnology* (M. S. da Costa, J. C. Duarte, and R. A. D. Williams, eds.), Elsevier, London, pp. 167–173.

Donato, M. M., Seleiro, E. A., and da Costa, M. S., 1990, Polar lipid and fatty acid composition of strains of the genus *Thermus, System. Appl. Microbiol.* **13:**234–239.

Donato, M. M., Seleiro, E. A., and da Costa, M. S., 1991, Polar lipid and fatty acid composition of strains of *Thermus ruber, System. Appl. Microbiol.* **14:**235–239.

Embley, T. M., O'Donnell, A. G., Wait, R., and Rostron, I., 1987, Lipid and cell wall amino acid composition in the classification of members of the genus *Deinococcus, System. Appl. Microbiol.* **10:**20–27.

Faraldo, M. L., Pedro, M. A., and Berenguer, J., 1988, Purification, composition, and Ca^{2+}-binding properties of the monomeric protein of the S-layer of *Thermus thermophilus, FEBS Letters* **235:**117–121.

Hasegawa, Y., Kawada, N., and Nosoh, Y., 1980, Change in chemical composition of the membrane of *Bacillus caldotenax* after shifting growth temperature, *Arch. Microbiol.* **126:**103–108.

Hensel, R., Demharter, W., Kandler, O., Kroppenstedt, R. M., and Stakebrandt, E., 1986, Chemotaxonomic and molecular-genetic studies of the genus *Thermus*. Evidence for a phylogenetic relationship of *Thermus aquaticus* and *Thermus ruber* to the genus *Deinococcus, Int. J. System. Bacteriol.* **36:**444–453.

Hudson, J. A., Morgan, H. W., and Daniel, R. M., 1987, *Thermus filiformis* sp. nov., a filamentous caldoacitive bacterium, *Int. J. System. Bacteriol.* **37:**431–436.

Jackson, T. J., Ramaley, R. F., and Meinschein, W. G., 1973, Fatty acids of a nonpigmented, thermophilic bacterium similar to *Thermus aquaticus, Arch. Microbiol.* **88:**127–133.

Kaneda, T., 1991, Iso- and anteiso-fatty acids in bacteria: biosynthesis, function and taxonomic significance, *Bacteriol. Rev.* **55:**288–302.

Klein, A., Hazelwood, G. P., Kempt, P., and Dawson, M. C., 1979, A new series of long-chain dicarboxylic acids with vicinal dimethyl branching as major components of the lipids of *Butyrivibrio* spp., *Biochem. J.* **183:**691–700.

Kraepelin, G., and Gravenstein, H. U., 1980, Experimentelle induktion von "rotund bodies" bei *Thermus aquaticus, Zeitschr. Algm. Mikrobiol.* **20:**33–45.

Lancy, P., and Murray, R. G. E., 1977, The envelope of *Micrococcus radiodurans*: isolation, purification and preliminary analysis of the cell wall layers, *Can. J. Microbiol.* **24:**162–176.

Langworthy, T. A., and Pond, J. L., 1986, Membranes and lipids of thermophiles, in:

Thermophiles: General, Molecular and Applied Microbiology (T. D. Brock, ed.), John Wiley, New York, pp. 107–135.

Loginova, L. G., Egorova, R. S., Golovacheva, R. S., and Seregina, L. M., 1984, *Thermus ruber* sp. nov., nom. rev, *Int. J. System. Bacteriol.* **34**:498–499.

Mahoney, R. P., and Edwards, M. R., 1966, Fine structure of *Thiobacillus thiooxidans*, *J. Bacteriol.* **92**:487–495.

Oshima, M., 1978, Structure and function of membrane lipids in thermophilic bacteria, in: *Biochemistry of Thermophily* (S. M. Friedman, ed.), Academic Press, New York, pp. 1–10.

Oshima, M., and Ariga, T., 1975, ω-cyclohexyl fatty acids in acidophilic thermophilic bacteria, *J. Biol. Chem.* **250**:6963–6968.

Oshima, M., and Yamakawa, T., 1974, Chemical structure of a novel glycolipid from an extreme thermophile, *Flavobacterium thermophilum*. *Biochem.* **13**:1140–1146.

Nordstrom, K. M., and Laakso, S. V., 1992, Effect of growth temperature on fatty acid composition of ten *Thermus* strains, *Appl. Environ. Microbiol.* **58**:1656–1660.

Pask-Hughes, R. A., and Shaw, N., 1982, Glycolipids from some thermophilic bacteria belonging to the genus *Thermus*, *J. Bacteriol.* **149**:54–58.

Pask-Hughes, R. A., and Williams, R. A. D. 1977, Yellow-pigmented strains of *Thermus* spp. from Icelandic hot springs, *J. Gen. Microbiol.* **102**:375–383.

Pask-Hughes, R. A., and Williams, R. A. D., 1978, Cell envelope components of strains belonging to the genus *Thermus*, *J. Gen. Microbiol.* **107**:65–72.

Pinheiro, T. J., Vaz, W. C., Geraldes, C. F., Prado, A., and da Costa, M. S., 1987, A ^{31}P-NMR study on multilamellar liposomes formed from the lipids of a thermophilic bacterium, *Biochem. Biophys. Res. Comm.* **148**:397–402.

Prado, A., da Costa, M. S., and Madeira, V. M. C., 1988, Effect of growth temperature on the lipid composition of two strains of *Thermus* sp., *J. Gen. Microbiol.* **134**:1653–1660.

Prado, A., Costa, M. S., and Madeira, V. M. C., 1990, Liposomes from a thermophilic eubacterium, Thermus sp: Fluorescence polarization and permeability properties, *Int. J. Biochem.* **22**:1497–1502.

Ray, P. H., White, D. C., and Brock, T. D., 1971a, Effect of the temperature on the fatty acid composition of *Thermus aquaticus*, *J. Bacteriol.* **106**:25–30.

Ray, P. H., White, D. C., and Brock, T. D., 1971b, Effect of the growth temperature on the lipid composition of *Thermus aquaticus*, *J. Bacteriol.* **108**:221–225.

Rottem, S., and Markowitz, O., 1979, Carotenoids act as reinforcers of the *Acholeoplasma laidlawi* lipid bilayer, *J. Bacteriol.* **140**:944–948.

Russell, N. J., 1984, Mechanisms of thermal adaptations in bacteria: blueprints for survival, *TIBS*, **9**:108–112.

Saiki, T., Kobayashi, Y., Kawagoe, K., and Beppu, T., 1985, *Dictyoglomus thermophilum* gen. nov., sp. nov., a chemoorganotrophic, anaerobic, thermophilic bacterium, *Int. J. System. Bacteriol.* **35**:253–259.

Schleifer, K. H., and Kandler, O., 1972, Peptidoglycan types of bacterial cell walls and their taxonomic implications, *Bacteriol. Rev.* **36**:407–477.

Sharp, R., and Williams, R. A. D., 1988, Properties of *Thermus ruber* strains isolated from Icelandic hot springs and DNA: DNA homology of *Thermus ruber* and *Thermus aquaticus*, *Appl. Environ. Microbiol.* **54**:2049–2053.

Sleytr, U. B., and Messner, P., 1988, Crystalline surface layers in prokaryotes, *J. Bacteriol.* **170**:2891–2897.

Stramer, S. L., and Starzyk, M. J., 1981, The occurrence and survival of *Thermus aquaticus*, *Microbios* **32**:99–110.

Thompson, B. G., and Murray, R. G. E., 1981, Isolation and characterization of the plasma membrane and the outer membrane of *Deinococcus radiodurans* strain Sark, *Can. J. Microbiol.* **27:**729–734.

Thompson, B. G., Murray, R. G. E., and Boyce, J. F., 1982, The association of the surface array and the outer membrane of *Deinococcus radiodurans, Can. J. Microbiol.* **28:**1081–1088.

Weisburg, W. C., Giovannoni, J. C., and Woese, C. R., 1989, The *Deinococcus-Thermus* phylum and the effect of rRNA composition on the phylogenetic tree construction, *System. Appl. Microbiol.* **11:**128–134.

Williams, R. A. D., 1989, Biochemical taxonomy of the genus *Thermus,* in: *Microbiology of Extreme Environments and its Potential for Biotechnology* (M. S. da Costa, J. C. Duarte, and R. A. D. Williams, eds.), Elsevier, London, pp. 82–97.

Genetics of *Thermus*

6

Plasmids, Bacteriophage, Potential Vectors, Gene Transfer Systems

NEIL D. H. RAVEN

1. INTRODUCTION

In a review of the applied genetics of aerobic thermophiles, Imanaka and Aiba (1986) identified six potential advantages of a thermophilic cloning system compared with a mesophilic system.

1. The host–vector system would be nonpathogenic because thermophiles do not grow at the appropriate temperature.
2. The mode of expression and the gene products of cloned genes from mesophiles and/or thermophiles can be comparatively examined at elevated temperatures in thermophiles. A genetic system would contribute to the elucidation of the molecular basis of thermophily.
3. The production of thermostable enzymes in thermophiles will be enhanced by cloning into plasmid vectors (providing they are efficiently expressed) because of the gene dosage effect.
4. The amount of energy required for cooling during cultivation of thermophiles in a large-scale fermenter is less than that for mesophiles.
5. The cultivation time is shortened because of accelerated growth rates of thermophiles.
6. The probability of contamination diminishes because of higher cultivation temperatures (>55 °C).

NEIL D. H. RAVEN • Centre for Applied Microbiology and Research, Porton Down, Salisbury, Wiltshire SP4 0JG, United Kingdom.

Thermus Species, edited by Richard Sharp and Ralph Williams. Plenum Press, New York, 1995.

A further advantage is the ability to select, *in vivo*, thermostable variants of cloned genes encoding thermolabile enzymes. This would greatly facilitate both the systematic investigation of the mechanisms involved in thermostabilizing enzymes and in the enhancement of the thermostability of industrially important enzymes.

Although in the last ten years more extremely thermophilic microorganisms have been reported (Stetter *et al.*, 1990), the two genera discussed by Imanaka and Aiba (1986), *Bacillus* and *Thermus*, remain the best candidates for the development of a general thermophilic host–vector cloning system. Both genera grow aerobically at temperatures up to 85 °C, rapidly reaching high cell densities on conventional media, and form colonies on solid media at 70 °C within one to two days. The only additional requirements for the growth of these organisms compared with *Escherichia coli* are incubators capable of accurately maintaining high temperatures.

Until recently it seemed that a viable high-temperature cloning system would be more readily attainable for *Bacillus* than *Thermus*. This chapter will describe developments that reversed this situation. The review will also cover methods for the isolation of plasmids from *Thermus*, their curing, their potential for use in cloning vectors, the transformation of *Thermus*, the development of *Thermus* vectors, and their selectable phenotypes. *Thermus* bacteriophage isolation and cultivation, methods for the isolation of phage DNA, and the mapping of one phage will be detailed. Finally the potential applications of gene transfer systems for *Thermus* will be discussed and specific examples cited.

2. *THERMUS* PLASMIDS

2.1. Isolation of Plasmids from *Thermus*

The first paper to describe the isolation of plasmids from *Thermus* (Hishinuma *et al.*, 1978) examined eight strains by centrifugation of cell lysates containing tritiated adenosine labeled DNA in cesium chloride–ethidium bromide equilibrium density gradients. Satellite peaks to the chromosomal DNA in four of the strains were found to be plasmid DNA. Eberhard *et al.* (1981) used a large-scale isolation procedure to screen *Thermus thermophilus* HB8 and prepared a restriction map of the plasmid pTT1 (pTT8 of Hishinuma *et al.*, 1978) for 11 restriction enzymes. They prepared cleared lysates by high-speed centrifugation in an ultracentrifuge followed by a series of phenol extractions, Biogel A50M chromatography, and a series of acid phenol extractions. Using the same procedure,

Vasquez *et al.*, (1981) characterized plasmid pTF62, isolated from *Thermus flavus* AT62, and later demonstrated the presence of a much larger plasmid pVV8 in *Thermus thermophilus* HB8 (Vasquez *et al.*, 1983). Koh (1985) observed plasmids in two of three screened Icelandic *Thermus* strains, one in strain B_1, and two in strain B_2, using a cleared lysate procedure, followed by polyethylene glycol precipitation and caesium chloride–ethidium bromide centrifugation. Munster *et al.* (1985) examined 56 *Thermus* strains by the single colony lysis procedure of Eckhardt (1978) and showed that more than 60% carried plasmid DNA. Raven and Williams (1985) isolated and characterized two plasmids from one of these strains, *Thermus* YS045, using an alkaline denaturation procedure based upon that of Birnboim and Doly (1979) and its large-scale version (Marko *et al.*, 1982). Subsequently, Raven and Williams (1989), in a more extensive survey of *Thermus* strains (Munster *et al.*, 1985), confirmed the general applicability of the alkaline denaturation procedure for the isolation of plasmids from *Thermus*. Becker *et al.* (1986) also used an alkaline denaturation procedure and detected four plasmids in *Thermus aquaticus* YT1. Kröger et al. (1988) used a procedure again involving alkaline denaturation with *Thermus aquaticus* YT1. Further purification involved Sephacryl S-1000 gel filtration, followed by ethanol precipitation and desalting by Sephadex G50 gel filtration. Five plasmids were detected, but variation in the number of plasmids from colony to colony was observed.

Recent studies also appear to have used alkaline denaturation for the isolation of *Thermus* plasmids (Mather and Fee, 1990; Lasa *et al.*, 1992b). An exception has been the use by Denman *et al.* (1991) of thermal denaturation (also referred to in passing by Kröger *et al.*, 1988) to detect plasmids in *Thermus* strains from the Australian Artesian Basin. Exposure to 95 °C for 45 sec in the STET buffer of Holmes and Quigley (1981) was sufficient to release plasmid DNA and allowed direct electrophoresis of the clarified extract. If this procedure is applicable to other *Thermus* strains (especially those able to grow at 85 °C) then its rapidity would lend itself to screening for recombinant clones prepared with *Thermus* host–vector systems. The plasmid DNA produced, however, is insufficiently pure to allow restriction endonuclease digestion without further purification. A combination of this procedure and an alkaline denaturation step yields plasmid DNA, which is both substantially free of chromosomal DNA and also digestible by restriction endonucleases (Raven, unpublished observations).

2.2. Restriction Endonuclease Mapping

Thermus strains are an abundant source of plasmids, in contrast to the thermophilic *Bacillus* strains (T_{max} > 75 °C), where plasmids are rare.

Although a large number of *Thermus* plasmids has been reported, coherent circular restriction endonuclease maps have been published for only four. Plasmids range in size from more than 70 Kb for pVV8 from *Thermus thermophilus* HB8 (Vasquez *et al.*, 1983), to 1.5 Kb for pNZ1200 from *Thermus* sp. T4–1A (Berquist *et al.*, 1987) (Table I).

2.3. Plasmid-Encoded Phenotypes

Investigators have attempted to ascribe phenotypic functions to plasmids by isolating cured derivatives of the wild type strain and comparing their respective phenotypes. Novobiocin (Vasquez *et al.*, 1981, 1983; Becker *et al.*, 1986), freeze-drying (Raven and Williams, 1989), and sparging with 100% oxygen (Mather and Fee, 1990) have been used successfully to cure *Thermus* strains. Other agents that did not effect curing were acridine orange, ethidium bromide, sodium dodecylsulfate, and rifampicin (Vasquez *et al.*, 1983, 1984).

The potentially plasmid-encoded phenotypes that have been principally sought are antibiotic resistance and heavy metal tolerance. These markers can be directly selected and lend themselves to the development of plasmid vectors. Other phenotypes screened for include bacteriocin production (Hishinuma *et al.*, 1978), restriction endonuclease production (Vasquez *et al.*, 1981, 1983), bacteriophage sensitivity (Vasquez *et al.*, 1983), amylase, protease, and alkaline phosphatase production (Becker *et al.*,

Table I. Restriction Endonuclease-Mapped Plasmids

Plasmid	Source	Size (Md)	Size (Kb)	Number of restriction enzymes used	Restriction enzymes reported to have single sites	Reference
pTT8	*T. thermophilus* HB8	6.0	9.2	11	BclI, BglII, BstEII, SacII[a,b]	Eberhard *et al.*, (1981)
pTF62	*T. flavus* AT62	6.8	10.5	23	BclI[a]	Vasquez *et al.*, (1981)
pTYS45-1	*Thermus* sp.	3.7	5.9	16	BamHI, PstI, SphI	Raven and Williams (1985)
pTYS14-1	*Thermus* sp.	1.3	2.0	36	BamHI, HincII, HindIII, NotI, PstI, XbaI	Raven and Willams (1989)

[a]Both pTT8 and pTF62 were subsequently reported to have a single HpaI site (Vasquez *et al.*, 1984)
[b]pTT8 was also reported to have a single SalI site (Koyama, 1992)

1986); and rotund body formation (Becker *et al.*, 1986). The only property established as plasmid-encoded is the phenomenon of cell aggregation in *T. thermophilus* HB8 (Mather and Fee, 1990) that is associated with the large plasmid pVV8. Although the aggregation phenotype could not be directly selected, collecting only the cells adhering to the walls of a cured shake flask culture enriched the phenotype. Reintroduction of pVV8 into cured cells reinstated the aggregation phenotype, but the plasmid was no longer stably maintained and was eventually lost in the absence of further enrichment. Mather and Fee (1990) suggested two possible causes for this; the curing procedure induced a genetic change affecting the stability of pVV8 in *Thermus thermophilus* HB8, or only a proportion of the population obtained from the enrichment procedure contained pVV8, with the plasmid-free cells out-competing the plasmid bearing cells. The aggregation phenotype is not an ideal plasmid-encoded marker, however, even if it were on a smaller replicon and gave stable transformants.

The absence of directly selectable markers on *Thermus* plasmids has been a major disadvantage. By contrast the moderately thermophilic *Bacilli* have several antibiotic resistance-encoding plasmids (Bingham *et al.*, 1979; Imanaka *et al.*, 1981). The replication functions of cryptic *Thermus* plasmids are, however, suitable for the development of high-temperature vectors since they are inherently stable up to 85 °C.

3. TRANSFORMATION

3.1. *Bacillus stearothermophilus* Host–Vector Cloning System

Prior to 1986, there were no published reports of the transformation of *Thermus*, and the thermophilic *Bacilli* appeared to be more promising for the development of a high-temperature cloning system. Imanaka *et al.*, (1982) were able to transform *Bacillus stearothermophilus* CU21 (derived from the *B. stearothermophilus* type strain IFO12550) with plasmids pUB110 (from the mesophile *Staphylococcus aureus*) (Gryczan *et al.*, 1978) and pTB19 and its derivatives (from an uncharacterized thermophilic *Bacillus* sp) (Imanaka *et al.*, 1981). Transformation frequencies in excess of 10^7 μg DNA were obtained with the plasmid construct pTB90 using kanamycin or tetracycline resistance selection. Although unstable at temperatures above 60 °C in the absence of phenotypic selection, plasmid pTB90 was shown to allow growth of *B. stearothermophilus* CU21 up to 65 °C in the presence of either kanamycin or tetracycline. A viable host–vector cloning system capable of operating at elevated temperatures is, therefore, available for *Bacillus stearothermophilus*, but has three inherent limitations: (1) the conditions to achieve high transformation frequencies are very precise, (2) theprocedure is complex and time-consuming, requiring preparation of protoplasts followed by their regeneration over a period of five days, and (3) the maximum growth temperature of *Bacillus stearothermophilus* CU21 is only 70 °C, therefore, selections at a higher temperature are not possible.

This transformation system demonstrated the direct selection of thermostable mutants of enzymes. Thermostable variants of the kanamycin nucleotidyl transferase gene encoded by plasmid pUB110 have been produced. Matsumura and Aiba (1985) subjected the gene to hydroxylamine mutagenesis before recloning it into a vector plasmid, pTB922. Two separate single base pair substitutions permitted growth on kanamycin-containing medium at 61 °C compared with 55 °C for the wild type enzyme. Independently, Liao *et al.* (1986), using the same gene on a plasmid construct pBST2, isolated mutants containing both substitutions that allowed growth at 70 °C. Single-step variants with higher thermostability arose spontaneously at low frequencies and at higher frequencies by passage through an *E. coli* mutator strain. Sequential mutants could be produced either by plating the *Bacillus stearothermophilus* host (strain 1174) at progressively higher temperatures or by similar rounds of selection and temperature increase in a chemostat.

3.2. *Thermus* Transformation

The discovery by Koyama *et al.*, (1986) that *Thermus* strains are naturally transformable (i.e., DNA can be taken up and incorporated directly under normal physiological conditions) renders *Thermus* the prime candidate for the development of an extremely thermophilic host–vector cloning system. Koyama *et al.* (1986) showed that *T. thermophilus* HB27, could be transformed by chromosomal DNA at exceptionally high frequencies (10^{-1}–10^{-2}) and by the cryptic plasmid pTT8 at similarly high frequencies. Transformation was achieved by adding DNA directly to growing *Thermus* cells. Streptomycin resistance could be transferred, and amino acid auxotrophs transformed to prototrophy. No treatment was required to induce competence, although higher transformation frequencies were achieved by supplementation of the basal medium with divalent cations (Ca^{2+} and Mg^{2+}). Competence was maintained throughout the growth phase, and transformation occurred over a wide range of temperatures. Colony formation occurred within 36–48 hours even on minimal medium.

Given the simplicity and efficiency of this system, it is surprising that it had not been reported previously. This may be explained by the misinterpretation that *Thermus* is a classical gram-negative organism. Natural competence in Gram-negative bacteria normally had some complicating factor, e.g., piliated cells may be required, as in *Neisseria gonorrhoeae*, or specific base sequences required, as in *Haemophilus influenzae*. Plasmid transformation, the principal requirement, generally occurs at frequencies, several orders of magnitude lower than chromosomal DNA transformation.

It is now known from evolutionary studies that *Thermus* descended from an early bacterial lineage which predates both Gram-negative and Gram-positive cell types (Woese, 1987); this explains the mixture of Gram-negative and Gram-positive characteristics possessed by *Thermus*. The fortuitously simple and efficient transformation of *Thermus* may, therefore, be a primitive feature.

4. *THERMUS* VECTORS

4.1. TrpB Complementation Vectors

A plasmid vector for *Thermus* was first described by Koyama *et al.*, (1989) (Table II). The plasmid pYK105 was constructed from the tryptophan synthetase genes of *Thermus* strain T2 (Ulrich *et al.*, 1972), *E. coli* plasmid pUC13 and plasmid pTT8 from *T. thermophilus* HB8. The vector had a total size of 16.7 Kb and replicated in both *T. thermophilus* HB27, where tryptophan auxtrophy (*trp*B5) could be complemented, and in *E. coli*, where ampicillin resistance was expressed.

Initially Koyama *et al.* (1989) attempted to complement the *T. thermophilus* HB27 *trp*B5 mutant using the trytophan synthetase genes (*trp*BA) from prototrophic *T. thermophilus* HB27. Using a chromosomal library in *E. coli* MC1009, colonies expressing the *trp*B gene were detected by direct plasmid transfer into *Thermus* cells (Koyama and Furukawa, 1990) modified from a method developed by Van Randen and Venema (1984). The *E. coli* clones were replicated onto minimal agar coated with viable cells of the *T. thermophilus* HB27 *trp*B5 mutant. After 48 hrs at 70 °C several prototrophic *Thermus* colonies had developed. A clone was selected that carried a plasmid designated pKA2, which contained the *T. thermophilus* HB27 tryptophan synthetase genes. Although pKA2 replicated stably in *E. coli* and also transformed *T. thermophilus* HB27, the cloned *trp* genes were not suitable as markers in *T. thermophilus* HB27 because of the high frequency of recombination with their chromosomal counterparts. *Thermus* strain T2 *trp* genes were chosen because *Thermus* T2 chromomsomal DNA

Table II. *Thermus* Vectors

Plasmid	*Thermus* host	Size	Shuttle vector	Selection in *Thermus*	Cloning sites	Insertional inactivation	Reference
pYK105	HB27 *trp*	16.7 Kb	+	Trp complementation	EcoRI	+ (in *E. coli*)	Koyama *et al.* (1989)
pYK109	HB27 *trp*	14.2 Kb	+	Trp complementation	EcoRI	+ (in *E. coli*)	Koyama *et al.* (1990a)
pYK9	KB27 *lac*	16.8 Kb	–	Blue colonies (X-gal)	BglII	+ (replacement)	Koyama *et al.* (1990b)
pMKM001	HB8	8.2 Kb	–	Kanamycin resistance	?	–	Mather and Fee (1992)
pLU1	HB27	7.2 Kb	+	Kanamycin resistance	?	–	Lasa *et al.* (1992b)
pLU2	HB27	12.0 Kb	+	Kanamycin resistance	?	–	Lasa *et al.* (1992b)
pLU3	HB27	10.5 Kb	+	Kanamycin resistance	?	–	Lasa *et al.* (1992b)
pLU4	HB27	9.9 Kb	+	Kanamycin resistance	?	–	Lasa *et al.* (1992b)
pMY1	HB27	8.4 Kb	+	Kanamycin resistance	?	–	Lasa *et al.* (1992b)
pMY2	HB27	7.7 Kb	+	Kanamycin resistance	?	–	Lasa *et al.* (1992b)
pMY3	HB27	13.1 Kb	+	Kanamycin resistance	?	–	Lasa *et al.* (1992b)
pMY1.1	HB27	8.4 Kb	+	Kanamycin resistance	EcoRI	+ (in *E. coli*, by replacement)	Lasa *et al.* (1992b)
pYK179	HB27	14.8 Kb	–	Kanamycin resistance Blue colonies (X-gal)	?	+ (replacement)	Koyama (1992)
pYK189	HB27	9.1 Kb	–	Kanamycin resistance	EcoRI, HindIII, KpnI, PstI, SphI, XbaI	–	Koyama, unpublished
pYK130	HB27	12.1 Kb	–	Kanamycin resistance	SalI	–	Koyama, unpublished
pYK141	HB27	11.5 Kb	–	Kanamycin resistance	EcoRI	–	Koyama, unpublished

transformed *T. thermophilus* HB27 *trp*B5 to Trp⁺ at a very low frequency (10^{-5}), implying a low level of homology.

Thermus T2 *trp* genes were cloned into pUC13 and recombinants containing the *Thermus* T2 *trp* genes selected by colony hybridization using *T. thermophilus* HB27 *trp* genes as a probe. A plasmid (pKA207) was found to carry the *Thermus* T2 tryptophan synthetase genes on a 6.8 Kb insert. Three BglII sites external to the *trp* genes allowed BglII digested pKA208 to be annealed to BglII linearized pTT8. The ligation mixture was used to transform *T. thermophilus* HB27 *trp*B5 and a Trp⁺ transformant obtained, which yielded pYK105.

A smaller *Thermus–E. coli* shuttle vector was produced from a derivative of pKA207 containing only the *Thermus* T2 *trp*B gene (Koyama *et al.*, 1990a) The *trp*B gene was first subcloned in pUC13 on a 1.8 Kb BglII-SacI fragment to generate a 4.5 Kb plasmid pKA216. This plasmid was then cleaved with BamHI and ligated to pTT8 digested with BclI. A recombinant plasmid pYK109 (14.2 Kb) was obtained from a Trp⁺ transformant. Plasmid pYK109 could transform *E. coli* MC1061 to ampicillin resistance and *T. thermophilus* HB27 *trp*B5 to Trp⁺ at a frequency of 10^6 transformants/µg DNA.

A practical demonstration of the use of pYK105 was provided by Touhara *et al.*, 1991. Aqualysin I had been previously cloned and expressed in *E. coli* under the control of an inducible *E. coli* consensus promoter (Terada *et al.*, 1990). Although *E. coli* yielded mature aqualysin I after treatment at 65 °C, secretion of the enzyme did not occur as in *Thermus aquaticus* YT1. Direct cloning of the aqualysin I gene under the control of its own promoter was unsuccessful. Touhara *et al.* (1991), produced a plasmid pYL005 (16.8 Kb) in which the pUC13 segment of pYK105 had been replaced by an EcoRI-XbaI fragment containing the 5′ terminal portion of the aqualysin I gene. Plasmid pUC18 was cloned into the XbaI site of pYL005 to generate pYL055 (19.5 Kb), which was selected after transformation of *E. coli*. The pUC18 portion of pYL055 was replaced by the 3′ terminal segment of the aqualysin I gene by XbaI cleavage followed by religation. This construct was then introduced into a *T. thermophilus* HB27 *trp*B mutant in which constitutive expression of the gene was considered unlikely to be toxic. Approximately half of the Trp⁺ colonies formed clearing zones on minimal agar plates containing 1% skimmed milk, indicating successful secretion of mature aqualysin I. One of the aqualysin I expression plasmids was characterized and designated pNK006 (20.3 Kb). Although expression was no higher than in *Thermus aquaticus* YT1, the *Thermus thermophilus* HB27 *trp*B5/pNK006 host–vector system provides the possibility of increasing aqualysin I production in the future.

4.2. β-Galactosidase Expression Vectors

Koyama *et al.* (1990b) reported a further *Thermus* vector pYK9 (16.8 Kb) that utilized β-galactosidase expression for screening transformants. Cells expressing β-galactosidase activity are able to hydrolyze the chromogenic substrate 5-bromo-4-chloro-3-indolyl-β-galactopyranoside (X-gal) to give a nondiffusing blue pigment. In *E. coli*, β-galactosidase positive transformants produce blue colonies visible among the background of white, untransformed colonies. In yellow *Thermus* strains, equally distinctive blue-green colonies are produced. Koyama *et. al.*, (1990b) used *Thermus* strain T2 as the source of DNA because it produces a thermostable β-galactosidase (Ulrich *et al.*, 1972). Mbo I partial digests (4–10 Kb) of *Thermus* T2 chromosomal DNA were ligated to BamHI-digested consensus promoter plasmid vector pDR540 and transformed into *E. coli* MC1061, a β-galactosidase deleted host. Colonies appearing at 37 °C on plates containing X-gal were further incubated at 70 °C for 2 hrs. Although this caused the death of the *E. coli* cells, it allowed thermophilic β-galactosidase activity to be visualized. Recombinant plasmids carrying *Thermus* T2 β-galactosidase genes could be recovered from the dead cells. One of these plasmids (pOS76) expressed both α- and β-galactosidase activities. A 6.7 Kb fragment from a SacI partial digest of pOS76, subcloned into SacI digested pUC18, produced the plasmid pOS101 that also directed both α- and β-galactosidase production. Plasmid pOS101 was modified by inserting synthetic DNA linkers containing BglII sites on both sides of the insert DNA. The resultant plasmid pOS161 (9.8 Kb) carried the *Thermus* T2 α- and β-galactosidase genes as a 7.1 Kb gene cartridge bounded by BglII sites. A β-galactosidase minus mutant of *T. thermophilus* HB27 was produced by N-methyl-N'-nitro-N-nitrosoguanidine mutagenesis and transformed using DNA resulting from the ligation of the gene cartridge to BglII-cleaved pTT8 and spread on agar plates containing X-gal (50 µgml^{-1}). Plasmid pYK9 was isolated from a blue colony after incubation at 70 °C. Expression of the cloned *Thermus* T2 β-galactosidase occurred in *T. thermophilus* HB27 although the enzyme was produced constitutively.

4.3. Kanamycin Resistance Vectors

Selection in *Thermus* using β-galactosidase expression has the drawback that nontransformed cells also form colonies, while tryptophan synthetase complementation can only be used with the appropriate auxotrophic host. Since appreciable homology exists between the plasmid-borne markers and their inactivated chromosomal counterparts, a significant background of wild type colonies without plasmids are produced by recombination. These obstacles were overcome by Mather and Fee (1992), who showed that the thermostable kanamycin nucleotidyl transferase (KNTase) mutant produced by Matsumura and Aiba (1985) could be cloned and expressed in *Thermus*.

4.3.1. Plasmid pMKM001

The KNTase gene on a 930bp EcoRI fragment was inserted into the *E. coli* vector pKK223-3 (Pharmacia Biotech, St. Albans, United Kingdom). This "gene cartridge" or "gene cassette" contained the entire KNTase gene preceded by only 165bp of DNA, allowing possible expression of the gene by insertion downstream of a native *Thermus* promoter. The gene cartridge was isolated from the remainder of the plasmid by EcoRI cleavage then rendered blunt-ended by S1 nuclease digestion and "polished" with Klenow fragment. *T. thermophilus* plasmid pTT8 was randomly cleaved by partial digestion with DNaseI in the presence of manganese ions and rendered blunt-ended by S1 nuclease and Klenow fragment treatment. This mixture of fragments was then ligated to the KNTase fragment and used to transform competent *T. thermophilus* HB8. Transformants were selected by growth at 60 °C on plates containing 100μgml^{-1} kanamycin. A new 8.2 Kb plasmid designated pMKM001 was able to transform *T. thermophilus* HB8 to kan^R, but several problems were encountered in its use.

The amount of pMKM001 recoverable by alkaline lysis was always less than pTT8 from an equivalent quantity of HB8 cells. Attempts to purify covalently closed circular pMKM001 from agarose gels resulted in the linearization of the plasmid. Recovery by electroelution gave a mixture of closed circular and open circular forms. Mather and Fee (1992) suggested that this apparent fragility of pMKM001 could be due to the large G+C difference between the KNTase insert (42% G+C) and the remainder of the plasmid (~68% G+C).

Transformation of *T. thermophilus* HB8 derivatives apparently lacking pTT8 resulted in kan^R transformants containing pMKM001 and a second plasmid, the same size as pTT8, suggesting that some apparently cured *Thermus* strains may contain the plasmid (or plasmids) as chromosomal integrants. Plasmid pMKM001 may be able to mobilize pTT8 from the chromosome by recombinational rescue. When *T. thermophilus* HB27 was transformed, only mixtures of pMKM001 and pTT8 were able to generate kan^R transformants. It was concluded that the 2 Kb of DNA deleted from pTT8 during the construction of pMKM001 contained regions important for the maintenance of the plasmid and that these functions were being provided in *trans* by pTT8. Overall transformation efficiencies were relatively low (600 transformants per μg of plasmid DNA at 200μgml⁻¹ kanamycin, sufficient to completely suppress spontaneous resistant mutants).

4.3.2. plU and pMY Series Vectors

Plasmids have subsequently been produced that use KNTase selection but are independent replicons. Lasa *et al.* (1992b) described seven vectors based on the KNTase gene cartridge of Matsumura and Aiba (1985) and replication origins provided by cryptic plasmids from *Thermus aquaticus* YVII-51B (Brock and Freeze, 1969) and *Thermus* sp. T2 (Ulrich *et al.*, 1972) Initially, KNTase expression in *Thermus* was demonstrated by insertion of the KNTase gene cartridge downstream of the DNA sequence encoding the promoter of the *slp*A gene of *T. thermophilus* HB8. This promoter was chosen since the *slp*A gene encodes the major component of the S-layer of *T. thermophilus* HB8 (Faraldo *et al.*, 1991). This is one of its most abundant proteins, therefore, the *slp*A gene promoter was expected to be an extremely strong promoter. A single copy of the KNTase gene under the control of the *slp*A gene promoter and integrated into the chromosome of *T. thermophilus* HB8 (Lasa *et al.*, 1992a) resulted in a minimal inhibitory concentration for kanamycin of about 20μg ml⁻¹ in plate assays. In a multicopy plasmid, higher levels of kanamycin resistance would be expected.

This hybrid gene was constructed *in vitro* by ligating a 0.18 Kb HindIII-NdeI fragment containing the *slp*A promoter to a 0.76 Kb NdeI-EcoRI fragment containing the KNTase gene coding region. The nucleotide sequence encoding the amino terminal region of the KNTase gene was modified by PCR to contain the recognition sequence for NdeI. The forward oligonucleotide contained an NdeI recognition sequence followed by nucleotides complementary to the amino terminal region of the KNTase gene. The reverse oligonucleotide was complementary to the promoter region of the β-galactosidase gene of pUC9 into which the KNTase gene had been previously cloned. PCR of the KNTase gene with these oligonucleotides produced the appropriate sized fragment that was then phosphorylated and blunt-end cloned into the SmaI site of pBluescript vector KS to create plasmid pKT0. The 0.76 Kb Nde I-Eco RI fragment was isolated and purified from pKT0 and ligated in the presence of the 0.18 Kb HindIII-NdeI fragment and HindIII, EcoRI digested pBluescript KS. The ligation mixture was used to transform the *E. coli* host strain JM109 and recombinants were selected for amp^R and kan^R on minimal agar plates. A plasmid designated pKT1 which was recovered was used to produce large quantities of the *slp*A-KNTase gene cartridge by either BamHI or PstI restriction enzyme digestion.

Lasa *et al.* (1992b) ligated the purified gene cartridges to total plasmid preparations from both *T. aquaticus* YVII-51B and *Thermus* sp.T2 cleaved with BamHI or PstI. Ligation mixtures were used to transform *T. thermophilus* HB27 and resistant colonies were selected on 20μgml⁻¹ kanamycin plates by growth at 70 °C for 48 hrs. The BamHI-*Thermus* sp.T2 and PstI-*T. aquaticus* YVII-51B plasmid DNA combinations yielded very few colonies. The BamHI-*T. aquaticus* YVII-51B and PstI-*Thermus* sp.T2 plasmid DNA combinations, however, yielded many colonies from which total plasmid DNA was recovered and maintained as two separate pools. Each mixture was digested with EcoRI, which cuts *Thermus* DNA infrequently and has a single site within the gene cartridge external to the KNTase gene sequence. Cloning into the EcoRI site of pUC9 or pUC19 yielded *E. coli* colonies resistant to both ampicillin (100μgml⁻¹) and kanamycin (30μgml⁻¹). Examination of these colonies revealed seven different plasmids that were capable of autonomous replication in both *E. coli* and *Thermus*. The four derived from *T. aquaticus* YVII-51B and the three from *Thermus* sp.T2 were designated pLU1 to pLU4 and pMY1 to pMY3 respectively. Only two plasmids, pMY1 and pMY2, appeared to have *Thermus* plasmid sequences in common (from restriction endonuclease digests and cross-hybridizations). The remaining plasmids, therefore, were considered to have distinct replication origins, each deriving from a different native *Thermus* plasmid. These plasmids were capable of allowing growth of *T. thermophilus* HB27 in the presence of 200μgml⁻¹ kanamycin.

4.3.3. Plasmid Stability and Transformation

The seven plasmids were tested for stability in *T. thermophilus* HB27 in the absence of kanamycin. Plasmid-bearing cultures grown on nonselective medium were plated out and probed with the kanamycin resistance gene. Only plasmids pMY1 and pMY2 were found to be unstable. This instability did not correlate with plasmid size because pLU1 is smaller than either pMY1 or pMY2.

Stable plasmids pLU1, pLU2, and pLU4 have low copy numbers similar to those of pMY1 and pMY2. Both pMY plasmids, however, hybridized to a 16Kbp PstI fragment from plasmid preparations of *Thermus* sp. T2. This suggested that they arose as deletion products from a larger plasmid. The loss of some of this plasmid might explain their instability.

Plasmid DNA purified from a methylase deficient *E. coli* strain, transformed *T. thermophilus* HB27 at frequencies of 10²–10⁴ transformant colonies per µg DNA. There was no apparent correlation between efficiencies and either plasmid size or stability. At least tenfold higher transformation efficiencies were observed for the pMY plasmids isolated from *Thermus* in comparison to *E. coli*. For pMY1 (8.4Kbp) a frequency of 10^5 transformant *Thermus* per µg DNA could be obtained.

4.3.4. Plasmid pTCM1

To demonstrate the utility of these *Thermus–E. coli* shuttle vectors Lasa *et al.* (1992b) used a derivative of pMY1, pMY1.1, in which restriction sites had been deleted from the polylinker, to clone and express a thermostable cellulose gene (*cel* A) from *Clostridium thermocellum* in both *E. coli* and *Thermus*. A synthetic gene was constructed that fused the promoter and amino terminal region (33 amino acids) of the S-layer gene (*slp*A) from *T. thermophilus* HB8 to the *cel*A gene deleted of its promoter and amino acid terminal region (28 amino acids). This construct was assembled in pUC9 to give plasmid pPSC1 and cloned in *E. coli*. Surprisingly extracellular cellulose was observed, indicating that the signal peptide of the S-layer gene from *T. thermophilus* HB8 is functional in this mesophile.

To express the fusion protein in *T. thermophilus* HB27 the pUC9-derived region of pMY1.1 was removed by EcoRI cleavage and replaced by EcoRI linearized pPSC1. Plasmid pTCM1 (10.7Kbp) was isolated from cellulose positive colonies after transformation of competent *Thermus* cells, and directed expression and secretion of cellulose in *E. coli*, confirming its bifunctionality.

4.3.5. Replacement Vector pYK179

Koyama (1992) reported the development of *Thermus* replacement vector pYK179, which permits the detection of cloned fragments by color selection and allows transcription of genes downstream of a Kan^R gene expressed from a strong promoter. The promoter was selected by inserting a library of *T. aquaticus* YT1 DNA (prepared by MboI digestion) into the BamHI site of the polylinker of pUC119 into which had been previously cloned the thermostable kanamycin resistance gene of Liao *et al.*, (1986). SalI-linearized pTT8 was then also ligated into the interrupted polylinker and the mixture of plasmids produced used to transform *T. thermophilus* HB27. Transformants were selected on plates containing 200µgml^{-1} kanamycin, since only clones containing a strong promoter and strongly expressing the KNTase would be able to grow. One of these transformants pYK167 (14.9Kbp) contained a 12Kbp insert of *T. aquaticus* YT1 DNA.

Plasmid pYK179 was constructed by replacing the pUC119-derived component of pYK167 with the previously cloned β-galactosidase gene from *Thermus* sp.T2 (Koyama *et al.*, 1990b) Since pYK179-bearing colonies hydrolyze X-gal, they form blue-green colonies on kanamycin X-gal agar. Where the β-galactosidase gene has been replaced, therefore, colonies return to their usual yellow color while retaining kan^R resistance. Since replacement DNA can be inserted at an EcoRI site immediately downstream of the kan^R gene, read-through transcription of cloned genes from the strong promoter upstream can also take place.

4.3.6. Insertion Vector pYK189

In a derivative of pYK179 (Koyama, 1993), the β-galactosidase gene and surrounding restriction sites have been replaced by the polylinker of pUC119. Additionally some redundant regions and unwanted restriction sites derived from pTT8 have been deleted allowing DNA to be inserted into seven different single restriction sites in the polylinker. The plasmid, designated pYK189 (9.1Kbp) also allows potential expression of the cloned genes since the polylinker is located immediately downstream of the kanamycin resistance gene and its *T. aquaticus* YT1-derived promoter (Figure 1).

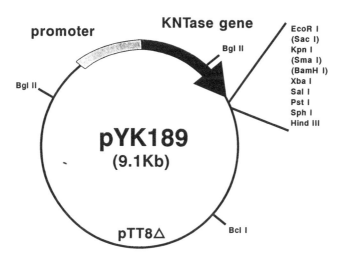

Figure 1. Restriction map of plasmid pYK189.

4.3.7. Promoter Probe Vector pYK130 and Expression Vector pK141

Further *Thermus* vectors with specific applications (Koyama, unpublished) have been derived from the *trp*B complementation vector pYK109 (Koyama *et al.*, 1990a) (Figure 2). Plasmid pYK128 was produced by deleting pUC13 and the XbaI-SalI fragment from pYK109 while leaving the EcoRI site and SalI site intact. The kan^R gene of Liao *et al.* (1986) was engineered to produce a promoter-less gene cartridge flanked by EcoRI sites which was then introduced into the EcoRI site of pYK128 to generate pYK130 (12.1Kbp). Cloning into the SalI site of pYK130 and plating on high concentrations of kanamycin allows the selection of fragments containing strong promoters. Plasmid pYK134 was produced by this means, resulting in the location of such a promoter on a 261 bp DNA fragment from an unidentified thermophilic bacterium. The use of pYK130 allows the rapid identification of *Thermus* promoters and facilitates the determination of consensus sequences and their optimal spacing. Plasmid pYK141 (11.5 Kbp) was produced by EcoRI cleavage of pYK134 to release the promoter-less kanamycin resistance gene cartridge and religation of the remaining large fragment. This plasmid, therefore, contains a strong promoter immediately upstream of a single EcoRI restriction site into which introduced genes can be positioned to maximize their potential for expression.

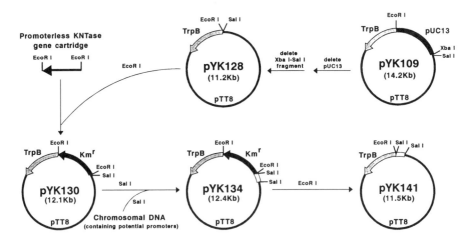

Figure 2. Construction of plasmids pYK130 and pYK141.

5. *THERMUS* BACTERIOPHAGES

5.1. Bacteriophage φYS40

5.1.1. Isolation and Host Range

Sakaki and Oshima (1975) reported the first bacteriophages infectious to *Thermus*. Several bacteriophages were isolated from Mine and Atagawa hot springs in Japan by plating with *Thermus thermophilus* HB8 on double-layer agar plates. After overnight incubation at 65 °C, plaques were picked from the soft upper agar layer and purified by several rounds of reinfection and plating from single plaques. The phage was considered to be virulent since it formed clear plaques on a lawn of *Thermus thermophilus* HB8. It also infected five other *Thermus* strains of Japanese origin, although turbid plaques were produced on three of the strains. Plating efficiencies ranged from 0.01 to 1.0, but the one non-Japanese strain, *Thermus aquatics* YT1, was resistant to infection.

5.1.2. Growth Characteristics and Purification

Phage φYS40 was found to have a latent period of 80 min and a burst size of 80 at its optimal growth temperature 65 °C. The phage grew over the range 56–78 °C, within the growth range of its host (50–80 °C) Thermostability at 80 °C was greatest in either the hot spring water from which it was isolated or the growth medium. High titre lysates were prepared by growing *Thermus thermophilus* HB8 to high cell density (2 × 10^{9}ml^{-1}) and infecting at a multiplicity of 0.5. After incubation at 65 °C for 10 min to allow adsorption of the phage to the host, the culture was diluted tenfold with growth medium and incubation with shaking continued for several hours. Titres in the range 1–5 × 10^{10} pfu ml^{-1} were achieved.

Lysed cultures were treated with DNase and RNase to digest nucleic acids and cleared of cell debris by centrifugation at 5000g for 10 min. Phage φYS40 was concentrated by precipitation with 10% polyethylene glycol 6000 in the presence of 0.5M NaCl at 4 °C overnight. After centrifugation, the precipitate was resuspended in buffer (10mM Tris HCl, pH7.5; 10mM MgCl$_2$) and residual insoluble material removed by centrifugation at 5000g for a further 10 min. The phage could not be purified by CsCl equilibrium centrifugation. Aliquots of the crude phage suspension were, therefore, loaded onto a 10–30% sucrose gradient and centrifuged in a Beckman SW27 rotor for 20 min at 20,000 rpm and phage recovered from the middle of the gradient.

5.1.3. Phage Morphology and Nucleic Acids

Electron micrographs of φYS40 prepared by negative staining with uranyl acetate revealed a hexagonal head (indicating an icosahedral structure three-dimensionally) with a diameter of 125nm. The tail was 178nm long, 27nm wide, and terminated in a base plate and tail fibers. The phage appeared to be larger than either phage lambda or T4. Phage nucleic acid, purified by phenol extraction and column chromatography (Mandel and Hershey, 1960), exhibited absorption and circular dichroism spectra typical of double-stranded DNA. This was confirmed by its resistance to RNase and S1 nuclease, and its sensitivity to DNase. Its G+C content was determined by chemical analysis, DNA melting temperature and buoyant density in CsCl. Since no unusual bases were detected by chemical analysis it was concluded that the genome of φYS40 contained DNA with no significant modification. All three procedures gave approximately 35% G+C, very different from the host *Thermus thermophilus* HB8 (reported to have a G+C content from 65–69%). Sakaki and Oshima (1975) concluded that stabilizing cofactors might be required by φYS40 DNA to prevent the melting of AT-rich regions. The M_r of the phage DNA was determined by its sedimentation velocity relative to phage T4 DNA in a 5–20% sucrose gradient. The φYS40 DNA was labeled with ^{32}P, while T4 DNA was labeled with ^{3}H, and the positions of φYS40 and T4 DNA were determined by liquid scintillation counting of fractions from the gradient. The M_r for φYS40 DNA was determined as 1.36×10^8D relative to phage T4 DNA at 1.3×10^8D, corresponding to a genome size for φYS40 of approximately 210 kb pairs. The genome size of phage T4 has now been more accurately established to be 166.5 kb (Kutter *et al.*, 1990), giving a revised value for φYS40 DNA of approximately 175 kb.

5.2. Other *Thermus* Bacteriophages

There have been few other reports of *Thermus* bacteriophage, which is surprising given their potential as cloning vectors and as a source of thermostable enzymes, including DNA polymerases, DNA ligases, RNA polymerases, polynucleotide kinases, and lysozymes. Although techniques for isolating and manipulating large DNA molecules *in vitro* have improved (e.g., by embedding in agarose blocks or beads), the φYS40 genome is probably too large to permit its routine use as a vector. Two other *Thermus* phages have smaller genomes than φYS40. Berquist *et al.*, (1987) reported the isolation by Patel (1985) of a *Thermus* phage from the North Island thermal region of New Zealand. The phage, called W28P, had a morphology like *E. coli* phage lambda with a double-stranded DNA genome of approximately 35 kb. The DNA appeared to be able to form an open circular configuration and, therefore, like phage lambda DNA might also have cohesive ends. The phage was limited by its ability only to infect the strain on which it was isolated, *Thermus* sp. W28P.

Raven (1990) reported the isolation of a lytic bacteriophage able to infect *Thermus thermophilus* HB8 and other widely used *Thermus* strains. This phage has been subsequently characterized in detail (Raven *et al.*, manuscript in preparation) and found to have a genome size of 81 kb.

5.2.1. *Thermus* Bacteriophage φYB10

Bacteriophages designated φYB1-φYB10 were isolated from biomass samples collected from hot springs in the Potts Basin area of Yellowstone National Park (Sharp and Munster, unpublished). Electron microscopy revealed similar morphologies: phage heads were icosahedral, 55–60 nm in diameter, and the tails were curved, striated, and up to 450 nm in length. Other properties including host range and restriction endonuclease digests of the DNA indicated that these phages, though similar, were not identical. They were, however, clearly different from φYS40 and phage φYB10 was selected for further investigation.

5.2.2 Growth Characterization and Host Range of Phage φYB10

Phage φYB10 was isolated and purified in a similar manner to φYS40. Infections were carried out at 65 °C in shake flasks on exponential phase cultures of *T. thermophilus* HB8 (2–4 × 10^8 cells ml^{-1}) at a multiplicity of 0.1. Incubation was continued until the optical density at 600nm declined. The lysates were cooled to 4 °C and centrifuged at 4000g for 15 min to remove cell debris. Titers were routinely in the range 1–5 × 10^{10} pfu ml^{-1}.

Phage φYB10 was more osmotically sensitive than φYS40, which precluded the use of PEG 6000/NaCl precipitation to prepare concentrated bacteriophage stocks. High speed centrifugation (40,000g for 3 hr) was effective in sedimenting the phage, but this was inefficient since only small volumes of lysate could be processed and viability was reduced by 90%. To overcome this phage lysates were grown on a dialyzed medium prepared by incubating 11 × concentration *Thermus* growth medium in a dialysis bag (approximate 10,000D cut-off) immersed in 10 volumes of deionized water at 4 °C for 12–16 h. The dialysate was sterilized by autoclaving and growth and infection carried out as before. Lysates were cleared of debris by low-speed centrifugation. Aliquots of the cleared supernatant were successively placed in a dialysis bag and concentrated by extraction of water with a saturated solution of PEG 6000 in *Thermus* growth medium. Large volumes of lysate could be concentrated at least 100-fold with a negligible reduction in titre. Bacteriophage stocks (10^{12}–10^{13} ml^{-1}) were prepared by filter sterilization (0.45µm pore diameter) of these concentrated lysates and stored at 4 °C. Titres decreased by approximately 20-fold per year of storage under these conditions giving a useful life as inocula for large-scale cultures of at least two years.

At 65 °C on *T. thermophilus* HB8 the latent period was determined according to Kay (1971) to be 75 min and the burst size 95–100. The thermal stability of phage φYB10 was comparable to that of φYS40 with $t_{0.5}$ of 1 hr at 80 °C in growth medium. The temperature range of φYB10 was investigated by screening for plaques on a lawn of growth of *T. thermophilus* HB8 on either agar or "Gelrite" double-layer plates at 5 °C intervals. Glass plates were required at temperatures above 70 °C and Gelrite was the preferred solidifying agent with 0.5% and 1.0% Gelrite in the upper and lower layers respectively. $CaCl_2$ and $MgCl_2$ (1.5 mM) were added to the growth medium to promote solidification without reducing plaque size significantly. At 50 °C and 55 °C no plaques were observed while at 85 °C the host was unable to to produce a bacterial lawn. Plaques were formed, however, at all intervening temperatures, including 80 °C, indicating that φYB10 is slightly more thermophilic than φYS40.

Sixty-four *Thermus* strains were selected on the basis of geographical distribution and taxonomic grouping for host range determination. Aliquots of bacteriophage suspension containing 10^1–10^8 pfu ml^{-1} were plated out on soft agar overlays with 0.1ml aliquots of exponential cultures of each prospective host. Plates were incubated overnight at 65 °C then examined for evidence of plaques. Nine strains were found to be susceptible to infection: *T. aquaticus* YT1 (ATCC 25104), *T. thermophilus* HB8 (ATCC 27634), *T. aquaticus* YVII 51-B (ATCC 25105), *T. flavus* AT62 (DSM674), *T. lacteus* (ATCC 31557), *Thermus* sp. X1 (ATCC 27978), *Thermus* sp. T2 (ATCC27737), *Thermus* sp. RQ3 (M. S. da Costa; Ribeira Quente, São Miguel, Açores), *Thermus* sp. YS041 (R. J. Sharp; Yellowstone National Park, Wyoming, USA). No general correlation was apparent between the host range and either the geographical location of the isolate or its taxonomic grouping.

5.2.3. Isolation of Phage Nucleic Acid

Initially, nucleic acids were isolated from high titer bacteriophage stocks of φYB10 by formamide dialysis (Davis *et al.*, 1980). Once the genome of φYB10 had been established as double stranded DNA (by nuclease sensitivity pattern and restriction endonuclease digestion) an alternative rapid method was developed. Since viable phages were not required for DNA isolation, rigorous methods for precipitating phage from dilute solutions were tested. Initially, culture lysates were cooled to 37 °C and treated with DNase I (1μg ml^{-1}) and RNase A (1μg ml^{-1}) to hydrolyze free nucleic acids, then reheated to 65 °C and incubated for at least 10 min to inactivate the deoxyribonuclease. Cell debris was removed by low-speed centrifugation and the supernatant recovered. Cationic detergents like cetyl pyridinium bromide (CPB) and cetyl trimethylammonium bromide precipitate polyanions such as nucleic acids from dilute solutions (Scott, 1962), but proteins also precipitate at pHs above their isoelectric points (Scott, 1955). The cleared lysates were, therefore, tested for the production of a precipitate with CPB. Lysates became cloudy immediately upon addition of 0.01 volumes of a 10% solution of CPB (dissolved by warming to 45 °C). After a few min a flocculent precipitate formed, which was pelleted by low-speed centrifugation. If the phage were precipitated they would redissolve in a high-salt solution added to the precipitate and release their DNA by osmotic lysis. Ammonium acetate (2.5M) was chosen for the high-salt solution, as nucleic acids can be precipitated directly from it with ethanol (Crouse and Amorese, 1987). To facilitate lysis of the phages, denaturing of phage-coat proteins and removal of excess CPB, redissolved precipitates were vortexed with chloroform/isoamylalcohol (24:1). After centrifugation two volumes of ethanol were added to the aqueous phase and a spoolable precipitate of DNA was formed. Residual ammonium acetate was removed by immersing the spooled DNA in 70% ethanol prior to air-drying. The φYB10 DNA was redissolved in TE buffer (10mM Tris HCl, 0.1mM EDTA; pH7.5) and its integrity confirmed by restriction endonuclease analysis. The optimized procedure allowed routine production of *Thermus* phage DNA at high purity (A^{260}/$_{280}$ 1.9–2.0) in high yield within 2–3 hours (Table III).

5.2.4. Analysis of the Genome of Bacteriophage φYB10

Restriction endonuclease analysis with 24 restriction endonucleases having recognition sequences containing at least 6 nonredundant nucleotides showed no submolar fragments, indicating a nonpermuted genome. Comparison of the fragment sizes with molecular weight standards (λHind III, 1kb DNA ladder), indicated a genome size of approximately 80kb. This was in general agreement with the 74kb determined by electron microscopy of full-length DNA visualized by spontaneous cytochrome *c* adsorption (Lang and Mitani, 1970).

The nucleotide composition of the phage genome was analyzed by

Table III. Rapid Purification of *Thermus* Phage DNA

1. Culture lysate cooled to 37 °C and 1 μg ml⁻¹ of both deoxyribonuclease I and ribonuclease A added. Incubation continued at 37 °C for 30 minutes.
2. Heated to 65 °C and maintained at that temperature for at least 10 min to inactivate DNase.
3. Cell debris cleared by centrifugation at 5000 g for 10 min. Supernatant recovered and cooled to room temperature.
4. Cetyl pyridinium bromide (0.01 volumes of a 10% solution prepared by warming CPB in water at 45 °C) added and incubated for 10 min.
5. Precipitate centrifuged at 5000 g for 5 min and the supernatant decanted. Pellet redissolved/resuspended in 0.01 volumes (of the original lysate) of 2. 5 M ammonium acetate solution.
6. Equal volume of chloroform/isoamylalcohol (24:1 v/v) added and the mixture vortexed for 1 min. Centrifuged at 10,000 g for 10 min to resolve phases.
7. Upper phase recovered and a further volume of 2.5 M ammonium acetate added to the lower phase plus interface material.
8. Mixture vortexed for 1 min, centrifuged as before and the recovered upper phases combined.
9. Recovered aqueous phase vortexed against an equal volume of chloroform/isoamylalcohol and centrifuged as above. Upper aqueous phase recovered and CHCl₃/IAA extraction repeated.
10. Two volumes of ethanol layered above the recovered aqueous phase. DNA present spooled out on a glass rod by gentle mixing of the layers.
11. DNA precipitate on glass rod carefully lowered into 70% ethanol to remove ammonium acetate.
12. Precipitate allow to air dry on a glass rod to remove ethanol before redissolving DNA in T.E. buffer. Resultant DNA solution stored frozen at −20 °C until use.

HPLC of nuclease PI hydrolysates (modified from Katayama-Fujimura *et al.*, 1984). No major modified nucleotides were observed and a G+C content of 54.7% was determined. A G+C content of 54.1% was determined from DNA melting temperature in 0.1× standard saline citrate (Owen and Hill, 1979).

For the more accurate estimation of genome size and the preparation of a restriction map, φYB10 DNA was analyzed by the pulsed field gel electrophoresis of fragments generated from single and double digests with infrequent cutting restriction endonucleases. A coherent map was established for 8 enzymes giving a total genome size for φYB10 of 81kb (Figure 3). Mapping also confirmed that the genome was linear, non-permuted and contained a directly repeated terminal redundancy of approximately 7kb.

Since *T. thermophilus* HB8 had been shown to be naturally competent for the uptake of both chromosomal DNA (Koyama *et al.*, 1986) and plasmid DNA (Mather and Fee, 1992) a number of transfection experi-

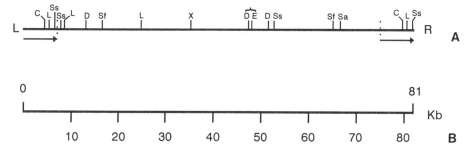

Figure 3. Restriction map of bacteriophage φYB10 DNA. (A) Restriction endonuclease designations: C, ClaI; D, DraI; E, EcoRV; L, LspI; Sa, SalI; Sf, SfiI; Ss, SspI; X, XhoI. (B) Orientation not established.

ments were carried out. Although possible differences in methylation and DNA conformation were addressed, transfection was not achieved.

6. UTILITY OF A *THERMUS* CLONING SYSTEM

Thermus appears to be the most suitable extreme thermophile for routine genetic experimentation. The simplicity of its transformation coupled with its higher maximum growth temperature means that *Thermus* is now likely to be preferred to *B. stearothermophilus*. Since *E. coli* is the most frequently used and most sophisticated mesophilic system, the availability of promoters, genes, and vectors able to function in both organisms is of considerable benefit.

Four potential areas for the use of a thermophile host–vector system are apparent (Imanaka and Aiba, 1986). These are (1) as a biocontainment system, (2) for the cloning of thermostable enzymes, (3) for the enhancement of biotechnological processes, and (4) as a tool to investigate the molecular basis of thermophily and the selection of thermostable enzyme variants. The principal interest, however, lies in this last area, the selection of catalytically active thermostable enzyme mutants *in vivo* for both academic and industrial research.

The techniques of DNA cloning and site-directed mutagenesis have made possible the systematic investigation of the biochemistry of thermophily. Total chemical synthesis of enzymes and mutant enzymes is at present impractical, and homologous enzymes from mesophiles and thermophiles are limited in number and variety. The molecular basis of thermostability has been investigated by comparing either the amino acid sequences of closely related enzymes that differ in thermostability (Matsumura *et al.*, 1984) or by the isolation of mutants with differing thermostabilities (Yutani *et al.*, 1977). Examples of the isolation of mutants with altered thermostabilities are usually of thermosensitive variants selected by the use of nonpermissive growth temperatures. Mutants with increased thermostability have been produced by chemical mutagenesis *in vitro* and cloning into *E. coli* (Makino et al., 1989) This approach, requires a convenient plate or filter assay to screen the large number of transformed colonies produced.

Site-directed mutagenesis is very highly efficient, overcoming the problem of screening large numbers of clones for potential mutants. This technique has been successfully used to engineer increased thermostability into enzymes of academic interest (T4 lysozyme) (Matsumura *et al.*, 1989) and of commercial importance (L-lactate dehydrogenase) (Kallwass *et al.*, 1992). The primary limitation to this approach is that the amino acid sequence and three-dimensional structure must be known before alterations can be made in an informed manner. Even when engineered changes induce a predicted increase in thermostability, there may be unpredicted secondary effects upon the activity of the enzyme. These include changes in the pH optimum, substrate affinity, substrate specificity (including stereospecificity), catalytic rate and cofactor binding efficiency. Maintenance of enzyme function, while retaining increased thermostability, frequently requires the introduction of additional subtle amino acid substitutions. Such changes are more difficult to define from three-dimensional models. A technique is needed to select mutants that are more thermostable but retain their other activity characteristics. An *in vivo* system based upon enzyme activity-dependent growth over a range of temperatures meets this requirement. Such a system would provide further information on the biochemistry of thermophily (complementary to that produced by protein engineering), potentially enhance the properties of commercially important enzymes, and allow the study of enzymes for which sequence data and three-dimensional structures are not available.

Selection *in vivo* in *Bacillus stearothermophilus* was effective in increasing the thermal stability kanamycin nucleotidyl transferase after random mutagenesis of the isolated gene *in vitro* and *in vivo* (Matsumura and Aiba, 1985; Liao *et al.*, 1986). This is not an isolated phenomenon; a thermostable mutant of *Bacillus pumilus* chloramphenicol acetyl transferase has been selected in *Bacillus stearothermophilus* (Turner *et al.*, 1992).

Liao (1992) showed a further use for this strategy. Loss of enzyme activity due to inclusion bodies at high levels of kanamycin nucleotidyl transferase expression in *E. coli* was greatly reduced when more thermostable mutants were cloned and expressed. It was postulated that inclusion body formation was due to temperature-dependent aggregation of a folding intermediate that was thermosensitive. Since the mutants were active *in vivo* at high temperature their folding intermediates must also be thermostable and, therefore, less susceptible to the formation of aggregates. The selection of mutants able to fold correctly and be active at high temperatures may, therefore, help to overcome the problem of insolubility of some proteins overexpressed in *E. coli*.

The simplicity and high temperature range of a *Thermus* cloning system will be useful in the *in vivo* selection of thermostable mutants and be applicable to virtually any enzyme activity. Activities not possessed by *Thermus* (antibiotic resistance, novel substrate utilization) can be selected for directly after transformation of host strains. Where a host enzyme is present but not always required (amino acid biosynthesis), an inactive mutant would be needed. Selection could then take place using growth conditions where the requirement for the activity is reinstated (on a minimal medium). Even where an enzyme activity is essential for the growth of *Thermus*, a cloned gene might be introduced and expressed alongside the chromosomal counterpart of the host. The *Thermus* gene could subsequently be inactivated by homologous recombination using transforming DNA containing the same gene interrupted by a selectable marker. Selection could then be applied to the introduced gene.

Since the growth maximum of *E. coli* and the growth minimum of *Thermus* are close, mutants with increased thermostability can be selected by successive growth temperature increments from 40–45 °C to 80–85 °C if required. After each increase in temperature has successfully generated mutants, additional selective pressure may be applied to generate secondary mutations that would optimally adapt the enzyme to the new temperature. Growing *Thermus* in conditions where maximal activity of the cloned gene has a selective advantage (at high dilution rates or limiting substrate concentrations in a chemostat, or on plates where rapidly growing colonies can be detected) would achieve this. The *in vivo* selection of thermostable mutants of thermolabile enzymes using *Thermus*, therefore, appears to have considerable potential.

REFERENCES

Becker, D. A., Glass, K. A., and Starzyk, M. J., 1986, Isolation and curing of extrachromosomal DNA in *Thermus aquaticus, Microbios*. **48:**71–79.

Berquist, P. L., Love, D. R., Croft, J. E., Streiff, M. B., Daniel, R. M., and Morgan, H. W., 1987, Genetics and potential biotechnological applications of thermophilic and extremely thermophilic microorganisms, *Biotechnol. Genetic Eng. Revs.* **5:**199–244.

Bingham, A. H. A., Bruton, C. J., and Atkinson, T., 1979, Isolation and partial characterization of four plasmids from antibiotic-resistant thermophilic bacilli, *J. Gen. Microbiol.* **114**:401–408.

Birnboim, H. C., and Doly, J., 1979, A rapid alkaline extraction procedure for screening recombinant plasmid DNA, *Nucl. Acids Res.* **7**:1513–1524.

Brock, T. D., and Freeze, H., 1969, *Thermus aquaticus* gen.n. and sp.n., a non-sporulating extreme thermophile, *J. Bacteriol.* **91**:289–297.

Crouse, J., and Amorese, D., 1987, Ethanol precipitation: ammonium acetate as an alternative to sodium acetate, *Focus* (BRL) **9**:(2)3–5.

Davis, R. W., Bostein, D., and Roth, J. R., 1980, in: *Advanced Bacterial Genetics*, Cold Spring Harbor Press, Cold Spring Harbor, New York, pp. 106.

Denman, S., Hampson, K., and Patel, B. K. C., 1991, Isolation of strains of *Thermus aquaticus* from the Australian Artesian Basin and a simple and rapid procedure for the preparation of their plasmids, *FEMS Microbiol. Lett.* **82**:73–78.

Eberhard, M. D., Vasquez, C., Valenzuela, P., Vicuna, R., and Yudelevich, A., 1981, Physical characterization of a plasmid (pTT1) isolated from *Thermus thermophilus*, *Plasmid* **6**:1–6.

Eckhardt, T., 1978, A rapid method for the identification of plasmid desoxyribonucleic acid in bacteria, *Plasmid* **1**:584–588.

Faraldo, M. M., de Pedro, M. A., and Berenguer, J., 1991, Cloning and expression in E. coli of the structural gene coding for the monomeric protein of the S-layer of *Thermus thermophilus* HB8, *J. Bacteriol.* **173**:5346–5351.

Gryczan, T. J., Contente, S., and Dubnau, D., 1978, Characterization of Staphylococcus aureus plasmids introduced by transformation into *Bacillus subtilis*, *J. Bacteriol* **134**:318–329.

Hishinuma, F., Tanaka, T., and Sakaguchi, K., 1978, Isolation of extrachromosomal deoxyribonucleic acids from extremely thermophilic bacteria, *J. Gen. Microbiol*, **104**:193–199.

Holmes, D. S., and Quigley, M., 1981, A rapid boiling method for the preparation of bacterial plasmids, *Analyt. Biochem.* **114**:193–197.

Imanaka, T., and Aiba, S., 1986, Applied genetics of aerobic thermophiles, in: *Thermophiles—General Molecular and Applied Microbiology* (T. D. Brock, ed.), Wiley–Interscience, New York, pp. 159–178.

Imanaka, T., Fujii, M., and Aiba, S., 1981, Isolation and characterization of antibiotic resistance plasmids from thermophilic bacilli and construction of deletion plasmids, *J. Bacteriol.* **146**:1091–1097.

Imanaka, T., Fujii, M., Aramori, I., and Aibi, S., 1982, Transformation of *Bacillus stearothermophilus* with plasmid DNA and characterization of shuttle vector plasmids between *Bacillus stearothermophilus* and *Bacillus subtilus*, *J. Bacteriol.* **149**:824–830.

Kallwass, H. K., Surewicz, W.,K., Paris, W., Macfarlane, E. L., Luyten, M. A., Kay, C. M., Gold, M., and Jones, J. B., 1992, Single amino acid substitutions can further increase the stability of a thermophilic L-lactate dehydrogenase, *Protein Eng.* **5**:535–541.

Katayama-Fujimura, Y., Komatsu, Y., Kuraishi, H., and Kaneko, T., 1984, Estimation of DNA base composition by high performance liquid chromatography of its nuclease P1 hydrolysate, *Agric. Biol. Chem.* **48**:3169–3172.

Kay, D., 1971, Methods for studying the infectious properties and multiplication of bacteriophage, in: *Methods in Microbiology, Volume 7A* (J. R. Norris, and D. W., Ribbons, eds.) Academic Press, London, pp. 191–262.

Koh, C-L., 1985, Detection and purification of plasmids present in *Thermus* strains from Icelandic hot springs, *MIRCEN J.* **1**:77–81.

Koyama, Y., 1992, Genetic analysis systems in *Thermus thermophilus*, in: *Thermophiles—Science and Technology*, Ice Tec, Reykjavik, pp. 81.

Koyama, Y., 1993, Genetic engineering in an extreme thermophile, *Thermus thermophilus*, in: *Thermophiles '93*, Hamilton, New Zealand.

Koyama, Y., and Furukawa, 1990, Cloning and sequence analysis of tryptophan synthetase genes of an extreme thermophile, *Thermus thermophilus* HB27: plasmid transfer from replica-plated Escherchia coli recombinant colonies to competent *T. thermophilus* cells, *J. Bacteriol.* **172**:3490–3495.

Koyama, Y., Hoshino, T, Tomizuka, N., and Furukawa, K., 1986, Genetic transformation of the extreme thermophile *Thermus thermophilus* and of other *Thermus* spp. J. Bacteriol. **166**:338–340.

Koyama, Y., Okamoto, S., and Furukawa, K., 1989, Development of host–vector systems in the extreme thermophile *Thermus thermophilus*, in: *Microbiology of Extreme Environments and It's Potential for Biotechnology* (M. S. da Costa, J. C. Duarte, and R. A. D. Williams, eds.), Elsevier, London, pp. 103–105.

Koyama, Y., Arikawa, Y., and Furukawa, K., 1990a, A plasmid vector for an extreme thermophile, *Thermus thermophilus*, FEMS Microbiol. Lett. **72**:97–102.

Koyama, Y., Okamoto, S., and Furukawa, Y., 1990b, Cloning of α- and β-galactosidase genes from an extreme thermophile, *Thermus* strain T2, and their expression in *Thermus thermophilus*, Appl. Environ. Microbiol. **56**:2251–2254.

Kröger, B., Specht, T., Lurz, R., Ulbrich, N., and Erdmann, V. A., 1988, Isolation and characterisation of plasmids from a wild type strain of the extremely thermophilic eubacterium *Thermus aquaticus*, FEMS Microbiol. Lett. **50**:61–65.

Kutter, E., Guttman, B., Mosig, G., and Rüger, W., 1990, Genomic map of bacteriophage T4, in: *Genetic Maps* (S. J. O'Brien, ed.), Cold Spring Harbor Press, Cold Spring Harbor, New York, pp. 1.24–1.51.

Lang, D., and Mitani, M., 1970, Simplified quantitiative electron microscopy of biopolymers, *Biopolymers* **9**:373–379.

Lasa, I., Castór, J. R., Fernández-Herrero, L. A., de Pedro, M. A., and Berenguer, J., 1992a, Insertional mutagenesis in the extreme thermophilic eubacteria *Thermus thermophilus* HB8, Mol. Microbiol. **6**:1555–1564.

Lasa, I., de Grado, M., de Pedro, M. A., and Berenguer, J., 1992b, Development of *Thermus–Escherichia* shuttle vectors and their use for expression of the *Clostridium thermocellum* cel A gene in *Thermus thermophilus*, J. Bacteriol. **174**:6424–6431.

Liao, H., McKenzie, T., and Hageman, R., 1986, Isolation of a thermostable enzyme variant by cloning and selection in a thermophile, *Proc. Natl. Acad. Sci. USA* **83**:576–580.

Liao, H. H., 1992, Effect of temperature on the expression of wild-type and thermostable mutants of kanamycin nucleotidyltransferase in *Escherichia coli*, Protein Expr. Purif. **2**:43–50.

Makino, Y., Negoro, S., Urabe, I., and Okada, H., 1989, Stability increasing mutants of glucose dehydrogenase from *Bacillus megaterium* IWG3, *J. Biol. Chem.* **264**:6381–6385.

Mandel, J. D., and Hershey, A. D., 1960, A fractionating column for analysis of nucleic acid, *Anal. Biochem.* **1**:66–77.

Marko, M. A., Chipperfield, R., and Birnboim, H. C., 1982, A procedure for the large-scale isolation of highly purified plasmid DNA using alkaline extraction and binding to glass powder, *Analyt. Biochem.* **121**:382–387.

Mather, M. W., and Fee, J. A., 1990, Plasmid-associated aggregation in *Thermus thermophilus* HB8, *Plasmid* **24**:45–56.

Mather, M. W., and Fee, J. A., 1992, Development of plasmid cloning vectors for *Thermus thermophilus* HB8: expression of a heterologous, plasmid-borne kanamycin nucleotidyl transferase gene, *Appl. Environ. Microbiol.* **58**:421–425.

Matsumura, M., and Aiba, S., 1985, Screening for thermostable mutant of kanamycin nucleotidyltransferase by the use of a transformation system for a thermophile, *Bacillus stearothermophilus*, *J. Biol. Chem.* **260**:15298–15303.

Matsumura, M., Katakura, Y., Imanaka, T., and Aiba, S., 1984, Enzymatic and nucleotide sequence studies of a kanamycin-inactivating enzyme encoded by a plasmid from thermophilic bacilli in comparison with that encoded by plasmid pUB110, *J. Bacteriol.* **160**:413–420.

Matsumura, M., Becktel, W. J., Levitt, M., and Matthews, B. W., 1989, Stabilization of phage T4 lysozyme by engineered disulfide bonds, *Proc. Natl. Acad. Sci. USA* **86**:6562–6566.

Munster, M. J., Munster, A. P., and Sharp, R. J., 1985, Incidence of plasmids in *Thermus* spp. isolated in Yellowstone National Park, *Appl. Environ. Microbiol.* **50**:1325–1327.

Owen, R. J., and Hill, L. R., 1979, The estimation of base compositions, base pairing and genome sizes of bacterial deoxyribonucleic acids, in: *Identification methods for microbiologists* (F. Skinner and D. W. Lovelock, eds.) Academic Press, London, pp. 271–296.

Patel, B. K. C., 1985, Extremely thermophilic bacteria in New Zealand hot springs, Ph. D. Thesis, University of Waikato.

Raven, N. D. H., 1990, On the molecular biology of the genus *Thermus*, Ph. D. Thesis, University of London.

Raven, N. D. H., and Williams, R. A. D., 1985, Isolation and partial characterization of two cryptic plasmids from an extreme thermophile, *Biochem. Soc. Trans.* **13**:214.

Raven, N. D. H., and Williams, R. A. D., 1989, in: *Microbiology of Extreme Environments and Its Potential for Biotechnology* (M. S. da Costa, J. C. Duarte and R. A. D. Williams, eds.), Elsevier, London, pp. 44–61.

Sakaki, Y., and Oshima, T., 1975, Isolation and characterization of a bacteriophage infectious to an extreme thermophile, *Thermus thermophilus* HB8, *J. Virol.* **15**:1449–1453.

Scott, J. E., 1955, Reaction of long-chain quaternary ammonium salts with acidic polysaccharides, *Chem. and Ind.* 168–169.

Scott, J. E., 1962, The precipitation of polyanions by long-chain aliphatic ammonium salts, *Biochem. J.* **84**:270–275.

Stetter, K. O., Fiala, G., Huber, G., Huber, R., and Segerer, A., 1990, Hyperthermophilic microorganisms, *FEMS Microbiol. Rev.* **75**:117–124.

Terada, I., Kwon, S-T., Miyata, Y., Matsuzawa, H., and Ohta, T., 1990, Unique precursor structure of an extracellular protease, aqualysin I, with NH_2- and COOH-terminal pro-sequences and its processing in *Escherichia coli*, *J. Biol. Chem.* **265**:6576–6581.

Touhara, N., Taguchi, H., Koyama, Y., Ohta, T., and Matsuzawa, H., 1991, Production and extracellular secretion of aqualysin I (a thermophilic subtilisin-type protease) in a host-vector system for *Thermus thermophilus*, *Appl. Environ. Microbiol.* **57**:3385–3387.

Turner, S. L., Ford, G. C., Mountain, A., and Moir, A., 1992, Selection of a thermostable variant of chloramphenicol acetyltransferase (Cat-86), *Protein Eng.* **5**:535–541.

Ulrich, J. T., McFeters, G. A., and Temple, K. L., 1972, Induction and characterization of β-galactosidase in an extreme thermophile, *J. Bacteriol.* **110**:691–698.

Van Randen, J., and Venema, G., 1984, Direct plasmid transfer from replica-plated E. coli colonies of competent B. subtilis cells, *Mol. Gen. Genet.* **195**:57–61.

Vasquez, C., Venegas, A., and Vicuna, R., 1981, Characterization and cloning of a plasmid isolated from the extreme thermophile *Thermus flavus* AT-62, *Biochem. Int.* **3**:291–299.

Vasquez, C., Villenueva, J., and Vicuna, R., 1983, Plasmid curing in *Thermus thermophilus* and *Thermus flavus*, *FEBS Letters* **158**:339–342.

Vasquez, C., González, B., and Vicuna, R., 1984, Plasmids from thermophilic bacteria, *Comp. Biochem. Physiol.* **780**:507–514.

Woese, C. R., 1987, Bacterial evolution, *Microbiol. Revs.* **51**:221–271.

Yutani, K., Ogasahara, K., Sugino, Y., and Matshushiro, A., 1977, Effect of a single amino acid subsitution on stability of conformation of a protein, *Nature* **267**:274–275.

Genes and Genetic Manipulation in *Thermus thermophilus*

7

TAIRO OSHIMA

1. GENE STRUCTURE

1.1. Molecular Cloning

Nagahari and colleagues (1980) first isolated a gene from *Thermus* by cloning a HindIII fragment of chromosomal DNA containing the leuCine genes, *leuB*, *leuC*, and *leuD* from *Thermus thermophilus* strain HB27 into *Escherichia coli*. Independently, Tanaka *et al.* (1981) cloned the same genes from strain HB8, and then subcloned the *leuB* gene which was subsequently sequenced (Kagawa *et al.*, 1984). Since then, many genes have been cloned and sequenced from species of *Thermus*. Recent examples include tRNA genes linked to the *tuf* gene (Weisshaar *et al.*, 1990), ribosomal proteins (Yakhnin *et al.*, 1990), a putative insertion sequence (Ashby and Bergquist, 1990), xylose isomerase (Dekker *et al.*, 1991), DNA ligase (Lauer *et al.*, 1991), cytochrome oxidase (Mather *et al.*, 1991), elongation factor Tu (Satoh *et al.*, 1991), and phosphofructokinase (Xu *et al.*, 1991)—all of *T. thermophilus*, and lactate dehydroqenase (Ono *et al.*, 1990), and DNA polymerase (Engelke *et al.*, 1990)—both of *T. aquaticus*.

Most of these genes have been expressed in *E. coli*. Since the %G+C content of *Thermus* DNA is unusually high, coding regions of these cloned genes are also extraordinary G+C rich (Kagawa *et al.*, 1984; Oshima, 1986). The third base of the codons is usually G or C, and the codon usage of *Thermus* genes is quite different from that of *E. coli*. For instance, G+C content of gap gene of *T. aquaticus* is 96% (Hecht *et al.*, 1989). Table I

TAIRO OSHIMA • Department of Life Science, Tokyo Institute of Technology, Nagatsuta, Yokohama, Japan.

Thermus Species, edited by Richard Sharp and Ralph Williams. Plenum Press, New York, 1995.

Table I. Comparison of Codon Usage in *T. aquaticus* (Left), *T. thermophilus* (Middle, with Parentheses), and *E. coli* (Right) *leuB* Genes[a]

	Second letter															
	T				C				A				G			
T	TTT-Phe	3	(3)	8	TCT-Ser	0	(1)	3	TAT-Tyr	0	(1)	6	TGT-Cys	0	(0)	3
	TTC-Phe	10	(11)	5	TCC-Ser	5	(8)	3	TAC-Tyr	6	(4)	5	TGC-Cys	0	(0)	3
	TTA-Leu	1	(1)	5	TCA-Ser	0	(0)	1	TAA-***	1	(1)	1	TGA-***	0	(0)	0
	TTG-Leu	2	(1)	2	TCG-Ser	5	(1)	6	TAG-***	0	(0)	0	TGG-Trp	3	(3)	2
C	CTT-Leu	5	(5)	1	CCT-Pro	0	(2)	3	CAT-His	0	(0)	4	CGT-Arg	0	(0)	7
	CTC-Leu	15	(16)	0	CCC-Pro	19	(17)	0	CAC-His	7	(6)	4	CGC-Arg	5	(9)	14
	CTA-Leu	0	(1)	0	CCA-Pro	0	(0)	7	CAA-Gln	0	(0)	7	CGA-Arg	1	(1)	0
	CTG-Leu	15	(14)	27	CCG-Pro	5	(6)	9	CAG-Gln	3	(2)	6	CGG-Arg	10	(7)	1
A	ATT-Ile	0	(0)	12	ACT-Thr	0	(0)	1	AAT-Asn	0	(0)	1	AGT-Ser	0	(0)	1
	ATC-Ile	10	(10)	15	ACC-Thr	11	(5)	8	AAC-Asn	6	(5)	12	AGC-Ser	5	(6)	5
	ATA-Ile	1	(1)	0	ACA-Thr	0	(0)	0	AAA-Lys	1	(1)	9	AGA-Arg	2	(0)	0
	ATG-Met	6	(6)	9	ACG-Thr	3	(8)	3	AAG-Lys	15	(15)	4	AGG-Arg	7	(6)	0
G	GTT-Val	1	(0)	4	GCT-Ala	0	(1)	3	GAT-Asp	1	(0)	15	GGT-Gly	0	(0)	6
	GTC-Val	8	(13)	1	GCC-Ala	35	(31)	20	GAC-Asp	14	(15)	8	GGC-Gly	15	(15)	18
	GTA-Val	1	(0)	4	GCA-Ala	0	(0)	12	GAA-Glu	1	(2)	18	GGA-Gly	3	(5)	2
	GTG-Val	23	(22)	9	GCG-Ala	7	(11)	8	GAG-Glu	31	(31)	7	GGG-Gly	17	(16)	6

[a]From Kirino and Oshima, 1991, and unpublished results.

compares codon usage of the *leuB* gene (3-isopropylmalate dehydrogenase) of *T. thermophilus* and *T. aquaticus* with that of *E. coli* (Kirino and Oshima, 1991). No significant differences were observed between the codon usage of *T. thermophilus* and *T. aquaticus,* although some minor differences were found.

For expression of a *Thermus* gene in *E. coli,* transcriptional efficiency seems to be important. Although the codon usage of the two bacteria are quite different, cloned *Thermus* genes have often been expressed with a considerable efficiency, if inserted at a suitable site within an *E. coli* plasmid vector (Croft *et al.,* 1987; Yamada *et al.,* 1990). This suggests that codon usage is not critical for translational efficiency.

In some cases, the same gene has been cloned and sequenced from two different species of *Thermus: leuB* (Kirino and Oshima, 1991) and *gap* genes (Bowen *et al.,* 1988; Hecht *et al.,* 1989) from both *T. thermophilus* and *T. aquaticus;* the mdh/sucD genes from both *T. thermophilus* (Nicholls *et al.,* 1990; the strain was mistakenly called *T. aquaticus*); and "*T. flavus*" (Nishiyama *et al.,* 1991). Comparisons of gene sequences may provide an aid to studies of the taxonomic grouping of strains and species of *Thermus* (Hartmann *et al.,* 1989; Munster *et al.,* 1986; Santos *et al.,* 1989).

The difficulty of nucleotide sequencing G+C rich DNA cloned from species of *Thermus* can be overcome by technical advances including the use of SSB protein (single strand binding protein, Chase and Williams, 1986), ^{35}S labeling (Biggins *et al.*, 1983), and modified nucleosides, dITP (Mills and Kramer, 1979) or 7-deaza-dGTP (Mizusawa *et al.*, 1986).

1.2. Gene Organization

The pioneer work was done by Bergquist and his colleagues (Croft *et al.*, 1987), whose study indicated that the *leuB* gene is linked with *leuC* and *leuD* genes in the *T. thermophilus* chromosome and comprise an operon like the leucine genes in *E. coli*. However, the order of these genes differs from *E. coli* leucine operon; they proposed that *T. thermophilus* the *leuD* and *leuB* structural genes overlap. Unfortunately the sequence data contained errors and this hypothesis seemed not to be valid when the revised sequence was examined. No evidence was found for a putative *leuD* gene within the *leuB* gene of *T. aquaticus* (Kirino and Oshima, 1991).

Linkage of ribosomal RNA genes of *T. thermophilus* has been studied by Erdmann. The 5S and 23S genes are closely linked and the thermophile chromosome carries two structural genes for each RNA species (Ulbrich *et al.*, 1984; Hartmann *et al.*, 1987). However, intercistronic sequences between 16S and 23S rRNA genes are unusually long in both operons (Hartmann and Erdmann, 1989). Thus rRNA gene organization of *T. thermophilus* differs from that of other bacteria and is rather similar to that of thermophilic archaea (Tu and Zillig, 1982). In this context, the membrane bound ATPase of *T. thermophilus* is not a member of the bacterial ATPase family, but resembles archaeal ATPase (Yokoyama *et al.*, 1990). Gene organization analysis indicated that the operon structure of *T. thermophilus* ATPase genes is similar to that of the archaea *Sulfolobus acidocaldarius*, but different from that of *E. coli* (Tsutsumi *et al.*, 1991). It seems that *T. thermophilus*, and probably other *Thermus* species, share some aspects of biochemistry in common with archaea.

Gene organization around *mdh* of "*T. flavus*" Iijima *et al.* (1986) and of *Thermus* strain B (Nicholls *et al.*, 1990) was similar, but different from that of *E. coli*. Both strains belong to the *T. thermophilus* genospecies (see Chapter 1). Likewise, the overall organization of *trp* genes of *T. thermophilus* differs from that of *E. coli*, though the tail of the *trpB* gene overlaps with the initial part of *trpA* gene as in *trpAB* genes in many eubacteria (Sato *et al.*, 1988; Koyama and Furukawa, 1990).

1.3. Promoter Sequences

A putative ribosome binding site has often been found shortly before the initiation codon of a gene cloned from *Thermus*, which is highly complementary to the 3' end of *T. thermophilus* rRNA, UCACCUCCUUU. This sequence resembles the 3' end of E. coli 16s rRNA, CACCUCCUUA. In some open reading frames of *Thermus* genomes, CTG or TTG is used as the initiation codon instead of ATG (Nishiyama *et al.*, 1991) The consensus sequence for promoters of *Thermus* chromosomes is still controversial. Promoters for 16S and 23S/5S RNA genes (Hartmann *et al.*, 1987; Struck *et al.*, 1988; Hartmann and Erdmann, 1989) are similar sequences to those of *E. coli* at around −35 and −10. By contrast, the −35 region of the 4.5S RNA promoter did not resemble the *E. coli* consensus sequence. Recently Nishiyama *et al.* (1991) determined the promoter sequence for "*T. flavus*" suc-mdh operon by using the S1 nuclease mapping technique, but did not find a consensus sequence similar to that of *E. coli* in the −35 region, though the typical −10 consensus sequence was present (Figure 1). Sequences highly homologous to *E. coli* −35 and −10 consensus sequences were found upstream of the *T. thermophilus* isocitrate dehydrogenase gene of strain HB8 (Miyazaki *et al.*, 1995) and the *trpAB* gene of strain HB27 (Koyama and Furukawa, 1990). A similar structure has been pointed out by Kushiro *et al.* (1987) for the *T. thermophilus tuf* gene. These sequences are shown in Figure 1 along with promoter-like sequences proposed by other researchers (Croft *et al.*, 1987; Sato *et al.*, 1988). However, there is no direct evidence that these sequences are really involved in transcriptional regulation.

2. TRANSFORMATION

2.1. General Transformation

When *T. thermophilus* HB27 is used as DNA donor, transformation takes place very easily and with unusually high frequency (Koyama *et al.*, 1986; Takada, 1989). Although this respect this strain is extremely interesting, little is known about its chromosomal structure. Transformation has also been observed for strains such as HB8 and "*T. flavus*", and *T. aquaticus*, suggesting that the transformation at high frequency is a general property common to *Thermus* strains. Koyama and his colleagues (Koyama *et al.*, 1986, 1990b; Koyama and Furukawa, 1990) have extensively studied genetic transformation in *Thermus*. General transformation was demonstrated by Takada (1989) (Figure 2) using amino acid auxotrophs of *T. thermophilus* as hosts. These were prepared by treatment with 100μg/ml nitrosoguanidine at 70 °C for 30 min. (e.g., strain NM6 *leuC⁻*). Transformants that restore leucine biosynthesis were obtained when mu-

GENES	SEQUENCE		SD	Init.
	-35	-10		
T-16S	CCTTGACAAAAGGAGGGGGA TTGATAGCA<u>T</u>GGCTTTT			
T-23S/5S	CCTTGACAAAGGCCATGCCTCCTTGG<u>T</u>A<u>T</u>C<u>T</u><u>T</u>CCCTTTG			
T-4,5S	TCTAG<u>C</u>CTCAGGGCTTCCATGGGTGC<u>T</u>A<u>T</u>A<u>C</u><u>T</u>ACCCGAG			
T-SCS	CGG<u>C</u>CGTTTCCACGGCCCAAGGCCTATATAA<u>T</u>GCCCTTGACCGGGGTCGCCCGGCAAGGGAGGAGGTGGGTCTTG			
T-ICDH	AGTTTACAAGGCCTCAAGCCCCGTGGTGTAG<u>T</u>GTAAAGGGGGCGATTCCGCCCCC**GGAGG**TGAACCCATG			
T-TRPAB	AG<u>TTT</u>ACCGGGAGGCCCCTCCGGG<u>T</u>AGGA<u>T</u>GGGAGT TGTCTTGGCGCGAGGCGCCTTTAGGGAGCGAAGCATG			
T-TRPEG	CCTTTAG<u>CCCC</u>TGGACAGGGCCCCCG<u>T</u>GTCCGC<u>T</u>ATCCTGAGGCCATG			
T-EFTU	GGTCAATAGCGCGGC<u>T</u>TCGGGGTGGGCCTTT<u>T</u>TA<u>G</u>GGCCCGCCCCTTCCTTGCGAGGAGGACGGAGATG			
T-LEU	GC<u>T</u>TGA<u>C</u>CCCGCAGGCCTCGAGGGCTTAC<u>CCT</u>TAGGGGCAAATG			
E-ICDH	CATATGCAAC<u>G</u>TGGTGGCAGACGAGCAAACCAG<u>T</u>AGCGCTCGAAGGAGGAGGTGAATG			
E-LAC	GCTTTACAC<u>T</u>TTA<u>T</u>GC<u>T</u>TCCGGC<u>T</u>CG<u>T</u>A<u>T</u>G<u>T</u>G<u>T</u>GGAATTGTGAGCGGA<u>T</u>AACAATTCACACAGGAAACAGCTATG			
E-CON	T<u>TGACA</u>	<u>TATGTT</u>		

Figure 1. Comparison of *Thermus* promoter sequences and promoter-like sequences with those of *E. coli.* −10 and −35 regions are underlined, and ribosome binding sequences and initiation codons are marked with bold. T-16S, *T. thermophilus* 16S rRNA; T-23S/5S, *T. thermophilus* 23S/5S rRNA, T-4,5S, *T. thermophilus* 4.5S RNA; T-SCS, *T. flavus scsB-mdh*; T-ICDH, *T. thermophilus trpAB*; isocitrate dehydrogenase; T-TRPAB, *T. thermophilus trpAB*; T-TRPEG, *T. thermophilus trpEG*; T-EFTU, *T. thermophilus tuf* T-LEU; *T. thermophilus leu*; E-ICDH, *E. coli* isocitrate dehydrogenase; E-LAC, *E. coli lac*; E-CON, *E. coli* consensus.

tant NM6 was incubated with chromosomal DNA from strain HB8 (wild type), or with the recombinant *E. coli* plasmid pTH3 containing the thermophile *leuC* gene, but not with plasmid pHB2 containing only the thermophile *leuB* gene. The transformation frequency was proportional to the DNA concentration (up to 1–10 μg/ml) and was unusually high (e.g., when cells were treated with 1–2 μg/ml of DNA in the presence of 1–2 mM $CaCl_2$ and 1–2 mM of $MgCl_2$, the frequency was 10% or more) (Figure 2). Koyama *et al.*, (1986) also was reported a very high frequency (over 10%) for transformation using a *pro*⁻ mutant derived from *T. thermophilus* HB27.

The growth phase of the recipient cells is not important in transformation because *Thermus* appears to be permanently competent. The cell surface structure of *Thermus* is similar to those of gram-negative bacteria such as *E. coli*, and it is unnecessary to treat the cell with $CaCl_2$ prior to the incubation with the donor DNA. However, $CaCl_2$ in the incubation mixture did enhance the transformation frequency. Thus a simple method is available for transformation experiments; an appropri-

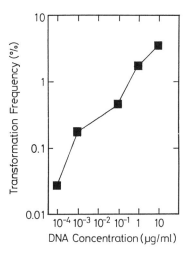

Figure 2. An example of transformation frequency. *T. thermophilus* mutant, NM31(ile⁻, val⁻), was transformed with an *E. coli* recombinant plasmid, pTH3, carrying a DNA fragment from the thermophile.

ate mixture of *T. thermophilus* cells with DNA in a culture medium (containing Mg^{2+} and Ca^{2+}) is spread on agar or gelrite (Lin and Casida, 1986) plates, and immediately incubated. Transformation also took place with a high frequency in heterologous *Thermus* systems (e.g., strain HB27) as the recipient and HB8 as the DNA donor (Koyama *et al.*, 1990a).

The restriction-modification system in the host cell has practically no effect on the transformation and the homologous recombination. Strain HB8 produces a restriction enzyme TthHB81, an isoschizomer of TaqI, and recognizes a sequence TCGA (Sato and Shinomiya, 1978). The cognate methylase modifies the adenine residue in the recognition sequence (Sato *et al.*, 1980). Chromosomal DNA from strain HB27 was susceptible to digestion with TthHB81 whereas DNA from HB8 was resistant. The thermophile plasmid pTT8 grown in HB27 was hydrolyzed by the restriction enzyme, whereas the same plasmid isolated from HB8 was not attacked by the enzyme (Takada, 1989). These results indicate that the restriction-modification system in strain HB27 differs from that of strain HB8. Though pTT8 contains three XhoI hydrolysis sites, when this plasmid DNA is propagated in *E. coli*, the plasmid isolated from either strain HB27 or HB8 was resistant, suggesting that a methylase for an isoschizomer of XhoI occurs in both strains (Takada, 1989).

2.2. Use of Thermophile Plasmids as Vectors

Plasmids occur in most strains of *Thermus* (Hishinuma *et al.*, 1978; Eberhard *et al.*, 1981; Kröger *et al.*, 1988; Vásquez *et al.*, 1981; Munster *et al.*, 1985; Mather and Fee, 1990), but most of them are cryptic, and no plasmid-borne drug resistant gene has been found. Homologous recombination takes place with unusually high frequency, and it is difficult to construct a host–vector system using a *Thermus* marker gene that complements a nutritional requirement of the host.

The 9. 7Kbp multicopy plasmid pTT8 from *T. thermophilus* HB8 (Hishinuma *et al.*, 1978) can be maintained stably in cells of *T. thermophilus* (Vasquez *et al.*, 1983). The sequence necessary for the replication of this plasmid is located in the right half of the map in Figure 3. However, no gene has been identified on this plasmid. Transformation with pTT8 either in strain HB27, or a cured strain of HB8 occurs with 0.1 to 1% frequency (Koyama *et al.*, 1986; Takada, 1989). Strain HB27 contains no plasmid and readily incorporates DNA from the environment; it is, therefore, useful as a host in *Thermus* genetic manipulation. Koyama *et al.* (1990a) found that the transformation frequency was surprisingly poor

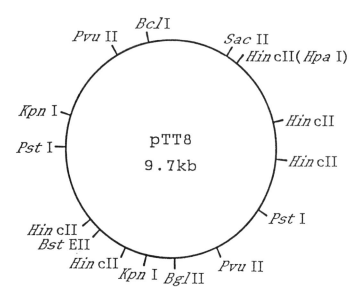

Figure 3. Physical map of pTT8.

(less than 0.001%) when a *trpB*-mutant derived from *T. thermophilus* strain HB27 was treated with chromosomal DNA from *Thermus* T2. The mutant was transformed to trp⁺ with DNA from strains HB27, HB8, and "*T. flavus*" (all isolated in Japan). The transformation was also extremely low (less than 0.001%) if the DNA was from *T. aquaticus*, which, like *Thermus* T2, is from a hot spring in the U.S.A. This observation supports the conclusion that *T. thermophilus* is a distinct species from *T. aquaticus* (Munster *et al.*, 1986; Santos *et al.*, 1989; Williams, 1989; Bateson *et al.*, 1990). Based on these findings, attempts were made to construct a *Thermus–E. coli* shuttle vector containing *trpB* gene from *Thermus* T2. A *T. thermophilus* HB27 trpB⁻ mutant was transformed with *Thermus* T2 DNA fragments ligated to an *E. coli* plasmid carrying an Amp^r gene and pTT8. The recombinant plasmid was used to transform *E. coli* to Amp^r to confirm that the plasmid was a shuttle vector. The frequency for *T. thermophilus* HB27 trp⁺ transformation was about 10⁶ per μg DNA with this shuttle vector (Koyama *et al.*, 1990a). A similar technique was used to construct a recombinant plasmid containing both α and β-galactosidase genes from *Thermus* T2 (Koyama *et al.*, 1990b). First these genes were cloned and expressed in *E. coli*, then *T. thermophilus* HB27 was transformed with the recombinant plasmid, and the β-galactosidase gene was expressed constitutively in the host.

Tellurite-resistance gene of *T. thermophilus* has been used with a plasmid (Chiong *et al.*, 1988a), and a transformation system was constructed using as a marker the cell aggregation property associated with a plasmid (Mather and Fee, 1990).

2.3. Gene Integration Experiments

An integration host–vector system to express a foreign gene in *T. thermophilus* has been constructed (Tamakoshi *et al.*, 1995; Tamakoshi, 1991). An *E. coli* recombinant plasmid pHB2, which contains the *leuB* gene from strain HB8—and its upstream and downstream flanking sequences— was partially digested with Bam H1. The products were ligated, and a plasmid which lacks the entire *leuB* gene but still contains 5′- and 3′- flanking regions was selected. This plasmid was used to transform *T. thermophilus* HB27. Since the flanking regions in the plasmid are homologous to upstream and downstream sequences of the *leuB* gene in the chromosome, a transformant lacking the entire *leuB* gene was obtained and used as the host cell in integration–expression experiments (Figure 4).

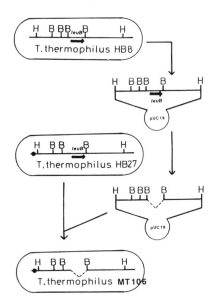

Figure 4. Construction of *leuB* deleted host cell, MT106.

Using similar procedures a vector for the integration was constructed (Figure 5). The vector is an *E. coli* plasmid containing both 3'- and 5'-flanking regions of the *Thermus leuB* gene and only one Bam H1 site that can be used to insert a gene. Since this vector is an *E. coli* plasmid, it cannot replicate in the thermophile. In preliminary experiments, the *leuB* genes from strain HB8 and a hybrid *leuB*, were integrated in the chromosome of the host cell. Since the genes are integrated in the middle of leucine operon of the host chromosome, they can be expressed in the host cell using a promoter for the leucine operon. The hybrid gene used was constructed by replacing roughly the 20–40% region from the initiation codon of the thermophile gene with that of the mesophile *Bacillus subtilis*. The hybrid gene product is a chimeric enzyme with a thermal denaturation temperature of 70 °C (Oshima, 1990). The transformation–integration in the leucine operon of the host chromosome was confirmed by Southern blotting analysis (Table II). The transformant with the hybrid gene could not grow at 76 °C or higher temperatures in a leucine deficient medium, probably because, in the absence of leucine, the maximum growth temperature of the transformant with the hybrid *leuB* gene is determined by the denaturation of the chimeric *leuB* gene product. This host–vector system can be developed to a more universal system for introducing and expressing any desired foreign gene in *T. thermophilus*.

An interesting experiment (Sen and Oriel, 1990) was the introduction of the broad host range conjugating streptococcus transposon Tn916,

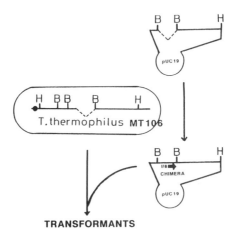

Figure 5. Integration of *leuB* gene into the chromosome of MT106.

Table II. Growth Temperatures of Transformants with
***leuB* Genes in a Leucine Deficient Medium**

Transformed with	Growth at		
	70 °C	76 °C	80 °C
None	−	−	−
Wild *leuB*	+	+	+
Hybrid *leuB*	+	−	−

carrying tetracycline resistance, into *T. thermophilus* by mating with *B. subtilis* containing Tn916 as a chromosomal insert. Tetracycline resistant thermophilic transconjugants were obtained and back transfer from *Thermus* to *B. subtilis* was also demonstrated. It may be possible to use Tn916 for stable insertion of foreign gene into chromosome of *T. thermophilus*.

3. APPLICATIONS OF *THERMUS* TRANSFORMATION

3.1. Simplified Gene Cloning

Gene cloning in *Thermus* can be simplified by using the transformation technique described above, originally invented by Randen and Venema (1984) for the cloning of *B. subtilis* genes, and modified by Koyama and Furukawa (1990) to clone *trpAB* genes from strain HB27. The first step is the selection of a *trpB* mutant of *T. thermophilus*, and preparation of *E. coli* colonies carrying DNA fragments of the thermophilic chromosome. *Trp⁻* mutant cells were plated on an agar or gelrite plate consisting of a synthetic medium without tryptophan, and the *E. coli* colonies were replica plated on the same plate and incubated at 70 °C. At this temperature the *E. coli* cells died and the plasmid DNA harbored within them was released and transformed the *Thermus*. Thus, colonies of trp⁺ transformants developed on the agar plate at 70 °C. An *E. coli* clone carrying the thermophile *trpAB* gene on a recombinant plasmid was selected by this method.

3.2. Stabilization of Enzyme Proteins

One of the major concerns in protein engineering is how to improve the thermal stability of a protein. Though many attempts have been made to stabilize several proteins so far, the evolutionary molecular engineering method, that is a combination of random mutation and screening of the mutated protein, seems to be one of the most promising ways to obtain a robust enzyme. This method does not require the elucidation of the three dimensional structure of the enzyme to be stabilized. The most remarkable example of this strategy is the stabilization of nucleotidyltransferase using a *Bacillus stearothermophilus* transformation system (Liao *et al.*, 1986). However, this method is limited to 70 °C, the maximum growth temperature of the host. It is desirable to use a genetic transformation system using an extreme thermophile to create more stable proteins.

A *T. thermophilus* transformant, MT106, which carries a hybrid *leuB* gene in the chromosome (Table II) was incubated at 76 °C for three days. Temperature insensitive mutants that could grow at 76 and 79 °C emerged from transformant MT106, which originally had a maximum growth temperature of 70 °C (Tamakoshi *et al.*, 1995; Tamakoshi, 1991). The translation product of *leuB*, 3-isopropylmalate dehydrogenase, purified from one of the temperature insensitive mutants was found to be more thermostable than the original chimeric enzyme. The mutated hybrid *leuB* gene was then cloned using the simple method described above. Nucleotide base sequence analysis of the cloned gene revealed that only one base replacement took place in the *leuB* gene: A-277 to C which replaces Ile-93 by Leu in the amino acid sequence of the enzyme. A detailed inspection on the three dimensional structure of this chimeric enzyme (Imada *et al.*, 1991; Onodera *et al.*, 1991) suggested that side chain packing is much improved by this replacement.

Other stabilized 3-isopropylmalate dehydrogenases have been isolated and analyzed by this method of "gene evolution to higher temperatures in laboratory." The gene integration system seems to be a useful tool for the study of stability and protein conformation, and for the improvement of enzymes for industrial use.

4. POLYAMINES

4.1. Unusual Polyamines

T. thermophilus and other *Thermus* species produce many unusual polyamines (Oshima, 1983, 1989). Longer polyamines such as pentaamines and hexaamines were first found in the cells of strain HB8, and so *T. thermophilus* seems to be a treasure chest for polyamine biochemists. Some of these unusual polyamines are present in organisms other than thermophiles (Oshima, 1983; Hamana *et al.*, 1990a).

Moderate thermophiles such as *B. stearothermophilus* do not produce any of the unusual polyamines found in *Thermus*, but contain spermine or tetramine(s). Mesophiles such as *E. coli* do not contain tetramine, implying a role for tetramine in life at higher temperatures. Extreme thermophiles such as *Sulfolobus* (Kneifel *et al.*, 1986; Friedman and Oshima, 1989), *Thermomicrobium* (Hamana *et al.*, 1990b), and *Thermoleophilum* (Hamana *et al.*, 1990c) also produce some of the unusual polyamines that are found in *T. thermophilus*.

Polyamine composition varies depending on growth conditions, particularly temperature. Generally longer polyamines are more abundant in cells grown at higher temperatures (Oshima, 1989). Polyamine composition differs from strain to strain (Hamana *et al.*, 1991), and can be used to distinguish *T. thermophilus* HB27 from HB8. The polyamine composition of *T. aquaticus* also differs from that of *T. thermophilus* HB8. No detectable amount of quaternary amine or longer polyamines such as caldohexamine is present in the cells of *T. aquaticus*.

Polyamines bind to DNA and stabilize it, induce conformational change in DNA and also regulate the activities of those enzymes involved in nucleic acid metabolism. The unique polyamines of thermophiles may therefore play important roles in the physiological functions of nucleic acids at high temperature.

Recently, three additional new polyamines were identified in *T. thermophilus:* an isomer of caldopentamine (1,17-diamino-4,8,13-triazaheptadecane) (Hamana *et al.*, 1990d), and two branched polyamines—tris(3-aminopropyl)amine and tetrakis-(3-aminopropyl) ammonium (Oshima *et al.*, 1987). In total, more than 11 new or rare polyamines were detected in the *Thermus* cells in addition to three normal ones: putrescine, spermidine, and spermine (Table III).

The occurrence of a tetrakisammonium compound in nature is interesting because it contains a quaternary nitrogen atom of different chemical nature from primary, secondary, and tertiary amino nitrogens. Effects of tetrakisammonium on nucleic acid conformation and metabolisms have not yet been investigated.

4.2. Effects on DNA

It is well known that the stabilization of DNA and RNA correlates with the number of amino nitrogens in the polyamine molecule used. The longer polyamines and the quaternary amines are found in *T. thermophilus*, the more effective it is in raising the melting temperatures of DNA and RNA than spermine or spermidine (Oshima *et al.*, 1989). The results suggest that at least one of the physiological roles of these unusual poly-

Table III. Names and Structures of Polyamines in *Thermus thermophilus*

Name	Chemical structure
Caldine (=norspermidine)	$NH_2(CH_2)_3NH(CH_2)_3NH_2$
Spermidine	$NH_2(CH_2)_3NH(CH_2)_4NH_2$
sym-Homospermidine	$NH_2(CH_2)_4NH(CH_2)_4NH_2$
Thermine	$NH_2(CH_2)_3NH(CH_2)_3NH(CH_2)_3NH_2$
Spermine	$NH_2(CH_2)_3NH(CH_2)_4NH(CH_2)_3NH_2$
Thermospermine	$NH_2(CH_2)_3NH(CH_2)_3NH(CH_2)_4NH_2$
Homospermine	$NH_2(CH_2)_3NH(CH_2)_4NH(CH_2)_4NH_2$
Caldopentamine	$NH_2(CH_2)_3NH(CH_2)_3NH(CH_{23}NH(CH_2)_3NH_2$
Homocaldopentamine	$NH_2(CH_2)_3NH(CH_2)_3NH(CH_2)_3NH(CH_2)_4NH_2$
Thermopentamine	$NH_2(CH_2)_3NH(CH_2)_3NH(CH_2)_4NH(CH_2)_4NH_2$
Caldohexamine	$NH_2(CH_2)_3NH(CH_2)_3NH(CH_2)_3NH(CH_2)_3NH(CH_2)_3NH_2$
Homocaldohexamine	$NH_2(CH_2)_3NH(CH_2)_3NH(CH_2)_3NH(CH_2)_3NH(CH_2)_3)_3$
Tetrakis(3-aminopropyl)ammonium	$N^+(CH_2)_3)_4$

amines is the stabilization of nucleic acids in the thermophile cell at high temperatures.

The glycosidic (C–N) linkage in purine deoxyribonucleoside in DNA is unstable in an acid environment, and an apurinic acid is formed if DNA is treated with an acid at room temperature. Depurination takes place even at neutral pH if the DNA is heated at high temperature (Lindahl and Nyberg, 1972). Kobayashi *et al.* (unpublished) found that polyamines reduce the rate of depurination of DNA at high temperature in a weakly acidic environment. Longer polyamines and quaternary ammonium compounds protected DNA more effectively than spermine or spermidine. DNA may be protected from depurination by these polyamines in the thermophile cells.

Polyamines induce the conformational change of B-DNA to Z-DNA in purine-pyrimidine alternate sequences (Rao *et al.*, 1991). Oshima *et al.* (1989) studied the effects of unusual polyamines found in *T. thermophilus* on this reaction. Longer polyamines induced Z-DNA more rapidly than spermine does. Regions of the DNA of a thermophile might be converted to Z-conformation by the unusual polyamines with the result that expression of chromosomal DNA of *Thermus* may be controlled by the cellular polyamines.

4.3. Effects on Enzymatic Activities of DNA Metabolism

Polyamines affect the catalytic activity of enzymes of nucleic acid metabolism. In some cases (DNA dependent RNA polymerase, staphylococcal nuclease and *in vitro* protein synthesis using a cell-free extract of *T. thermophilus*), polyamines are an essential component in the reaction. Polyamines may inhibit or enhance the activities of nucleases (Shishido, 1985; Conrad and Topa, 1989) Unusual polyamines affect some restriction enzymes (Kirino *et al.*, 1990), enhancing the activity of SmaI, RsaI, and NaeI when dilute, but inhibiting them when more of the enzyme was used. Some restriction sites were protected by the addition of polyamines, and larger hydrolysis fragments were obtained when DNA was treated with restriction enzymes in the presence of polyamines. In some experiments, GC rich DNA from *T. thermophilus* was used instead of *E. coli* DNA, but no meaningful difference was observed using the thermophile and mesophile DNAs as the substrate.

Hamasaki *et al.* (1995) showed three effects of unusual polyamines on calf thymus topoisomerase I:

1. polyamines activate the enzyme (practically essential for the reaction, very low activity without polyamine),
2. polyamines stabilize supercoiled or partially supercoiled DNA and inhibit the formation of fully relaxed DNA,
3. higher concentration DNA aggregates and the topoisomerase are inhibited.

The stimulation of the reaction is independent of number of nitrogen atoms in the polyamine molecule, but positively correlated with the distance between two terminal primary amino groups. For instance, 1,12-diaminododecane, which is not a natural diamine, enhances the enzymatic reaction at much lower concentrations than putrescine or cadaverine. Likewise, longer polyamines found in *Thermus* are more effective than shorter ones such as spermidine or spermine. Polyamines stabilize supercoiled DNA and the effect is proportional to the basicity of polyamine (the number of nitrogen atoms in the molecule). Pentaamines, which are the major polyamines in *T. thermophilus* cells, protect more efficiently than tetraamines. In the presence of the optimum concentration of caldopentamine (one of the most abundant polyamines of *T. thermophilus*), the final reaction products were mixture of partially relaxed DNAs with different degree of supercoiling. DNA aggregation by polyamines correlates positively with basicity and size of polyamine used. Thus, longer polyamines, such as caldopentamine and caldohexamine, inhibited the enzyme reaction at relatively low concentrations and at concentrations close to those estimated to be physiological. These results suggest that unusual polyamines found in *Thermus* regulate the higher structure of DNA and the transcription of genes. However, the study has used a eucaryote enzyme. Similar studies should be done using topoisomerase(s) of *T. thermophilus* to be conclusive.

REFERENCES

Ashby, M. K., and Bergquist, P. L., 1990, Cloning and sequence of IS1000, a putative insertion sequence from *Thermus thermophilus* HB8, *Plasmid* **24**:1–11.

Bateson, M. M., Thibault, K. J., and Ward, D. M., 1990, Comparative analysis of 16S ribosomal RNA sequences of *Thermus* species, *System. Appl. Microbiol.* **13**:8–13.

Biggins, M. D., Gibson, T. J., and Hong, G. F., 1983, Buffer gradient genNals and ^{32}S label as an aid to rapid DNA sequence determination, *Proc. Natl. Acad. Sci. USA* **80**:3963–3955.

Bowen, D., Littlechild, J. A., Fothergill, J. E., Watson, H. C., and Hall, L., 1988, Nucleotide sequence of the phosphoglycerate kinase gene from the extreme thermophile *Thermus thermophilus*, *Biochem. J.* **254**:509–517.

Chase, J. W., and Williams, K. R., 1986, Single-stranded DNA binding proteins required for DNA replication, *Ann. Rev. Biochem.* **55**:103–106.

Chiong, M., Barra, R., Gonzalez, E., and Vasquez, C., 1988a, Resistance of *Thermus* spp. to potassium tellurite, *Appl. Environ. Microbiol.* **54**:610–612.

Chiong, M. N., Gonzalez, E., Barra, R., and Vasquez, C., 1988b, Purification of biochemical characterization of tellurite–reducing activities from *Thermus thermophilus* HB8, *J. Bacteriol.* **170**:3269–3273.

Conrad, M., and Topal, M. D., 1989, DNA and spermidine provide a switch mechanism to regulate the activity of restriction enzyme NaeI, *Proc. Natl. Acad. Sci. USA* **86**:9707–9711.

Croft, J. E., Love, D. R., and Bergquist, P. L., 1987, Expression of leucine genes from an extremely thermophilic bacterium in *Escherichia coli*, *Mol. Gen. Genet.* **210**:490–497.

Dekker, K., Yamagat, H., Sakaguchi, K., and Udaka, S., 1991, Xylose(Glucose) isomerase gene from the thermophile *Thermus thermophilus:* Cloning, sequencing and comparison with other thermostable xylose isomerases. *J. Bacteriol.* **173**:3078–3083.

Eberhard, M. D., Vasquez, C., Valenzuela, P., Vicuna, R., and Yudelevich, A., 1981, Physical characterization of a plasmid (pTT1) isolated from *Thermus thermophilus*, *Plasmid* **6**:1–6.

Engelke, D. R., Krikos, A., Bruck, M. E., and Ginsburg, D., 1990, Purification of *Thermus aquaticus* DNA polymerase expressed in *Escherichia coli, Anal. Biochem.* **191**:396–400.

Friedman, S. M., and Oshima, T., 1989, Polyamines of sulfur-dependent archaebacteria and their role in protein synthesis, *J. Biochem.* **105**:1030–1033.

Hamana, K., Matsuzaki, S., Niitsu, M., and Samejima, K., 1990a, Synthesis of novel polyamines in *Paracoccus, Rhodabacter,* and *Micrococcus, FEMS Microbiol. Lett.* **67**:267–274.

Hamana, K., Matsuzaki, S., Niitsu, M., and Samejima, K., 1990b, Pentaamines and hexaamine are present in a thermophilic eubacterium *Thermomicrobium roseum, FEMS Microbiol. Lett.* **68**:31–34.

Hamana, K., Matsuzaki, S., Niitsu, M., and Samejima, K., 1990c, Two tertiary branched tetraamines, N^4-aminopropylnorspermidine and N^4-aminopropylspermidine, are the major polyamines in thermophilic eubacteria *Thermoleophilum, FEMS Microbiol. Lett.* **66**:35–38.

Hamana, K., Niitsu, M., Samejima, K., and Matsuzaki, S., 1990d, Thermopentamine, a novel linear pentaamine found in *Thermus thermophilus, FEMS Microbiol. Lett.* **68**:27–30.

Hamana, K., Niitsu, M., Samejima, K., and Matsuzaki, S., 1991, Polyamine distributions in thermophilic eubacteria belonging to *Thermus and Acidothermus, J. Biochem.* **109**:444–449.

Hamasaki, N., Kuwahara, R., Kirino, H., and Oshima, T., 1995, Effects of polyamines on relaxation of DNA by calf thymus topisomerase 1. (Submitted)

Hartmann, R. K., and Erdmann, V. A., 1989, *Thermus thermophilus* is transcribed from an isolated transcription unit, *J. Bacteriol.* **171**:2933–2941.

Hartmann, R. K., Kroger, B., Vogel, D. W., Ulbrich, N., and Erdmann, V. A., 1987, Characterization of cloned rDNA from *Thermus thermophilus* HB8, *Biochem. Internat.* **14**:267–275.

Hartmann, R. K., Wolters, J., Kroger, B., Schultze, S., Sprecht, T., and Erdmann, V. A., 1989. Does *Thermus* represent another deep eubacterial branching? *System. Appl. Microbiol.* **11**:243–249.

Hecht, R. M., Garza, A., Lee, Y-H., Miller, M. D., and Pisegna, M. A., 1989, Nucleotide sequence of the glyceraldehyde-3-phosphate dehydrogenase gene from *Thermus aquaticus* YT1, *Nucleic Acids Res.* **17**:10123–10123.

Hishinuma, F., Tanaka, T., and Sakaguchi, K., 1978, Isolation of extrachromosomal deoxyribonucleic acids from extremely thermophilic bacteria. *J. Gen. Microbiol.* **104**:193–199.

Iijima, S., Uozumi, T., and Beppu, T., 1986, Molecular cloning of *Thermus flavus* malate dehydrogenase gene, *Agric. Biol. Chem.* **50**:589–592.

Imada, K., Sato, M., Tanaka, N., Katsube, Y., Matsuura, Y., and Oshima, T., 1991, Three-dimensional structure of a highly thermostable enzyme, 3-isopropylmalate dehydrogenase of *Thermus thermophilus* at 2.2 ÅÅ resolution, *J. Mol. Biol.* **222**:725–738.

Jahn, O., Hartmann, R. K., and Erdmann, V. A., 1991, Analysis of the spc ribosomal protein operon of *Thermus aquaticus. Eur. J. Biochem.* **197**:733–740.

Kagawa, Y., Nojima, H., Nukiwa, N., Ishizuka, M., Nakajima, T., Yasuhara, T., Tanaka, T., and Oshima, T., 1984, High guanine plus cytosine content in the third letter of codons of an extreme thermophile: DNA sequence of the isopropylmalate dehydrogenase of *Thermus thermophilus, J. Biol. Chem.* **259**:1956–2960.

Kirino, H., and Oshima, T., 1991, Molecular cloning and nucleotide sequence of 3-isopropylmalate dehydrogenase gene (leuB) from an extreme thermophile *Thermus aquaticus* YT-1, *J. Biochem.* **109**:852–857.

Kirino, H., Kuwahara, R., Hamasaki, N., and Oshima, T., 1990, Effect of unusual polyamines on cleavage of DNA by restriction enzymes, *J. Biochem.* **107**:661–665.

Kneifel, H., Stetter, K. O., Andreesen, J. R., Wiegel, J., König, H., and Schoberth, S. M., 1986, Distribution of polyamines in representative species of archaebacteria, *System. Appl. Microbiol.* **7**:241–244.

Koyama, Y., and Furukawa, K., 1990, Cloning and sequence analysis of tryptophan synthetase genes of an extreme thermophile *Thermus thermophilus* HN27: Plasmid transfer from replica-plated Escherichia coli recombinant colonies to competent *T. thermophilus* cells, *J. Bacteriol.* **172**:3490–3495.

Koyama, Y., Hoshino, T., Tomizula, N., and Furukawa, K., 1986, Genetic transformation of the extreme thermophile *Thermus thermophilus* and of other *Thermus* spp., *J. Bacteriol.* **166**:338–340.

Koyama, Y., Arikawa, Y., and Furukawa, K., 1990a, A plasmid vector for an extreme thermophile *Thermus thermophilus, FEMS Microbiol. Lett.* **72**:97–102.

Koyama, Y., Okamoto, S., and Furukawa, K., 1990b, Cloning of alpha- and beta-galactosidase genes from an extreme thermophile *Thermus* strain T2 and their expression in *Thermus thermophilus* HB27, *Appl. Environ. Microbiol.* **56**:2251–2254.

Kröger, B., Specht, T., Lurz, R., Ulbrich, N., and Erdmann, V. A., 1988, Isolation and characterization of plasmids from a wild-type strain of the extremely thermophilic eubacterium *Thermus aquaticus, FEMS Microbiol. Lett.* **50**:61–65.

Kushiro, A., Shimizu, M., and Tomita, K., 1987, Molecular cloning and sequence determination of the tuf gene coding for the elongation factor Tu of *Thermus thermophilus* HB8, *Eur. J. Biochem.* **170**:93–98.

Lauer, G., Rudd, E. A., and McKay, D. L., Ally, A., Ally, D., and Backman, K. C., 1991, Cloning, nucleotide sequence and engineered expression of *Thermus thermophilus* DNA ligase, a homolog of *Escherichia coli* DNA liage, *J. Bacteriol.* **173**:5047–5053.

Liao, H., McKensie, T., and Hageman, R., 1986, Isolation of a thermostable enzyme variant by cloning and selection in a thermophile, *Proc. Natl. Acad. Sci. USA.* **83**:576–580.

Lin, C. C., and Casida, L. E., Jr., 1986, Gelrite as a gelling agent in media for the growth of thermophilic microorganisms, *Appl. Environ. Microbiol.* **47**:427–429.

Lindahl, T., and Nyberg, B., 1972, Rate of depurination of native deoxyribonucleic acid, *Biochemistry* **11**:3610–3618.

Mather, M. W., and Fee, J. A., 1990, Plasmid-associated aggregation in *Thermus thermophilus* HB8. *Plasmid* **24**:45–56.

Mather, M. W., Springer, P., and Fee, J. A., 1991, Cytochrome oxidase genes from *Thermus thermophilus*. Nucleotide sequence and analysis of the deduced primary structure of subunit 11c of cytochrome caa, *J. Biol. Chem.* **266**:5025–5035.

Mills, D. R., and Kramer, F. R., 1979, Structure-independent nucleotide sequence analysis, *Proc. Natl. Acad. Sci. USA* **76**:2232–2235.

Miyazaki, K., Eguchi, H., Yamagishi, A., Wakagi, T., and Oshima, T., 1995, Structure of isocitrate dehydrogenase gene of an extreme thermophile *Thermus thermophilus* HBB, *Appl. Environ. Microbiol.* (submitted).

Mizusawa, S., Nishimura, S., and Seela, F., 1986 Improvement of the dideoxy chain termination method of DNA sequencing by use of deoxy-7-deazaguanosine triphosphate in place of dGTP, *Nucleic Acids Res.* **14**:1319–1324.

Munster, M. J., Munster, A. P., and Sharp, R. J., 1985, Incidence of plasmids in *Thermus* spp. isolated in Yellowstone National Park, *Appl. Environ. Microbiol.* **50**:1325–1327.

Munster, M. J., Munster, A. P., Woodrow, J. R., and Sharp, R. J., 1986, Isolation and preliminary taxonomic studies of *Thermus* strains isolated from Yellowstone National Park, USA, *J. Gen. Microbiol.* **132**:1677–1683.

Nagahari, K., Koshikawa, T., and Sakaguchi, K., 1980, Cloning and expression of the leucine gene from *Thermus thermophilus* in *Escherichia coli*, *Gene* **10**:137–145.

Nicholls, D. J., Sundaram, T. K., Atkinson, T., and Minton, N. P., 1990, Cloning and nucleotide sequences of the mdh and sucD genes from *Thermus aquaticus* B, *FEMS Microbiol. Lett.* **70**:7–14.

Nishiyama, M., Horinouchi, S., and Beppu, T., 1991, Characterization of an operon encoding succinyl-CoA synthetase and malate dehydrogenase from *Thermus flavus* AT-62 and its expression in *Escherichia coli*, *Mol. Gen. Genet.* **226**:1–9.

Nureki, O., Muramatsu, T., Suzuki, K., Kohda, D., Matsuzawa, H., Ohta, T., Miyazawa, T., and Yokoyama, S., 1991, Methionyl-tRNA synthetase gene from an extreme thermophile *Thermus thermophilus* HB8, *J. Biol. Chem.* **266**:3268–3277.

Ono, M., Matsuzawa, H., and Ohta, T., 1990, Nucleotide sequence and characteristics of the gene for L-lactate dehydrogenase of *Thermus aquaticus* YT-1 and the deduced amino acid sequence of the enzyme, *J. Biochem.* **107**:21–26.

Onodera, K., Moriyama, H., Takenaka, A., Tanaka, N., Akutsu, N., Muro, M., Oshima, T., Imada, K., Sato, M., and Katsube, Y., 1991, Crystallization and preliminary X-ray studies of chimeric 3-isopropylmalate dehydrogenase between *Bacillus subtilis* and *Thermus thermophilus* HB8, *J. Biochem.* **109**:1–2.

Oshima, T., 1983, Novel polyamines in *Thermus thermophilus:* Isolation, identification and chemical synthesis, *Methods Enzymol.* **94**:401–411.

Oshima, T., 1986, The genes and genetic apparatus of extreme thermophiles, in: *Thermoohiles: General, Molecular and Applied Microbiology* (T. D. Brock, ed.), Wiley, New York, pp. 137–158.

Oshima, T., 1989, Polyamines in thermophiles, in: *The Physiology of Polyamines*, Volume 2 (U. Bachrach and Y. M. Heimer, eds.), CRC Press Inc., Boca Raton, pp. 34–45.

Oshima, T., 1990, A chimera of 3-isopropylmalate dehydrogenase with a mesophile head and a thermophile tail, in *Protein Engineering '89* (M. Ikehara, K. Titani, and T. Oshima, eds.), Center for Academic Societies Press/Springer Verlag, Tokyo, pp. 127–132.

Oshima, T., Hamasaki, N., Senshu, M., Kakinuma, K., and Kuwajima, I., 1987, A new naturally occurring polyamine containing a quaternary ammonium nitrogen, J. Biol. Chem. **262**:11979–11981.

Oshima, T., Hamasaki, N., Uzawa, T., and Friedman, M., 1989, Biochemical functions of unusual polyamines found in the cells of extreme thermophiles, in: *The Biology and Chemistry of Polyamines* (S. H. Goldemberg and I. D. Algrreanati, eds.), ICSU Press/IRL Oxford, Paris and London, pp. 1–10.

Randen, J., and Venema, G., 1984, Direct plasmid transfer from replica-plated *E. coli* colonies to competent *B. subtilis* cells, *Mol. Gen. Genet.* **195**:57–61.

Rao, M. V. R., Atreyi, M., and Saxena, S., 1991, Cooperative B-Z transition of poly (dG-dC) induced by spermine at physiological temperature in aqueous medium, *FEBS Lett.* **178**:63–65.

Santos, M. A., Williams, R. A. D., and DaCosta, M. S., 1989, Numerical taxonomy of *Thermus* isolates from hot springs in Portugal, *System Appl. Microbiol.* **12**:310–315.

Sato, S., and Shinomiya, T., 1978, An isochizomer of Taq I from *Thermus thermophilus* HB8, *J. Biochem.* **84**:1319–1321.

Sato, S., Nakazawa, K., and Shinomiya, T., 1980. A DNA methylase from *Thermus thermophilus* HB8, *J. Biochem.* **88**:737–747.

Sato, S., Nakazawa, Y., Kanaya, S., and Tanaka, T., 1988, Molecular cloning and nucleotide sequence of *Thermus thermophilus* HB8 trpE and trpG. *Biochim. Biophys. Acta* **950**:303–312.

Satoh, M., Tanaka, T., Kushiro, A., Hakoshima, T., and Tomita, K., 1991, Molecular cloning, nucleotide sequence, and expression of the tufB gene encoding elongation factor Tu from *Thermus thermophilus* HB8, *FEBS Lett.* **288**:98–100.

Sen, S., and Oriel, P., 1990, Transfer of transposon TN916 from Bacillus subtilis to *Thermus aquaticus*, *FEMS Microbiol. Lett.* **67**:131–134.

Shishido, K., 1985, Effect of spermine on cleavage of plasmid DNA by nucleases S_1 and Bal 31, *Biochem. Biophys. Acta.* **826**:147–150.

Struck, J. C. R., Toschka, H. Y., and Erdmann, V. A., 1988, Nucleotide sequence of the 4.5S RNA gene from *Thermus thermophilus* HB8, *Nucleic Acids Res.* **16**:9042.

Takada, T., 1989, Studies on an extreme thermophile *Thermus thermophilus*, host-vector system. Dissertation for Master' s Degree, Tokyo Institute of Technology.

Tamakoshi, M., 1991, Construction of an extreme thermophile host-vector system and stabilization of an enzyme. Dissertation for Master's Degree, Tokyo Institute of Technology.

Tamakoshi, M., Numata, K., Yamagishi, A., and Oshima, T., 1995, Stabilization of an enzyme using a gene integration system of an extreme thermophile *Thermus thermophilus* (Submited).

Tanaka, T., Kawano, N., and Oshima, T., 1981, Cloning of 3-isopropylmalate dehydrogenase gene of an extreme thermophile and partial purification of the gene product, *J. Biochem.* **89**:677–682.

Tomita, K., 1990, Sequences of four tRNA genes adjacent to the tuf 2 gene of *Thermus thermophilus*, *Nucl. Acids Res.* **18**:1902–1920.

Tsutsumi, S., Denda, K., Yokoyama, K., Oshima, T., Date, T., and Yoshida, M., 1991, Molecular cloning of genes encoding major two subunits of a eubacterial V-type ATPase from *Thermus thermophilus*, *Biochem. Biophys. Acta*, **1098**:13–20.

Tu, J., and Zillig, W., 1982, Organization of rRNA structural genes in the archaebacterium *Thermoplasma acidophilum*, *Nucleic Acids Res.* **10**:7231–7245.

Ulbrich, N., Kumagai, I., and Erdmann, V. A., 1984, The number of ribosomal RNA genes in *Thermus thermophilus* HB8, *Nucleic Acids Res.* **12**:2055–2060.

Vásquez, C., Venegas, A., and Vicuña, R., 1981, Characterization and cloning of a plasmid isolated from the extreme thermophile *Thermus flavus* AT-62, *Biochem. Int.* **3**:291–299.

Vásquez, C., Villanueva, J., and Vicuña, R., 1983, Plasmid curing in *Thermus thermophilus* and *Thermus flavus*, *FEBS Lett.* **158**:339–342.

Weisshaar, M., Ahmadian, R., Sprinzl, M., Satoh, M., Kushiro, A., and Tomita, K., 1990, Sequences of four tRNA genes adjacent to the tuf2 gene of *Thermus thermophilus*, *Nucleic Acids Res.* **18**:1902–1902.

Williams, R. A. D., 1989, Biochemical taxonomy of the genus Thermus, in: *Microbiology of Extreme Environments and Its Potential for Biotechnology*, (M. S. da Costa, J. C. Duarte, and R. A. D. Williams, eds.), Elsevier, London, pp. 82–97.

Xu, J., Seki, M., Denda, K., and Yoshida, M., 1991, Molecular cloning of phosphofructokinase 1 gene from a thermophilic bacterium *Thermus thermophilus*, *Biochem. Biophys. Res. Commun.* **176**:1313–1318.

Yakhnin, A. V., Vorozheykina, D. P., and Matvienko, N. I., 1990, Nucleotide sequence of the *Thermus thermophilus* HB8 rps12 and rps7 genes coding for the ribosomal proteins S12 and S7, *Nucleic Acids Res.* **18**:3659–3659.

Yamada, T., Akutsu, N., Miyazaki, K., Kakinuma, K., Yoshida, M., and Oshima, T., 1990,

Purification, catalytic properties and thermal stability of threo-3-isopropylmalate dehydrogenase coded by leu B gene from an extreme thermophile *Thermus thermophilus* strain HB8, *J. Biochem.* **108:**449–456.

Yokoyama, K., Oshima, T., Yoshida, M., 1990, *Thermus thermophilus* membrane associated ATPase indication of eubacterial V type ATPase, *J. Biol. Chem.* **265:**21946–21950.

Biotechnological Applications of *Thermus*

8

PETER L. BERGQUIST AND HUGH W. MORGAN

1. INTRODUCTION

The ability of organisms to grow at elevated temperatures has often been inferred to be of use in biotechnology. Most of the applications envisage the use of enzymes derived from thermophiles and involve the extreme thermostability of such enzymes (Kristjansson, 1990). There is now extensive literature on the purification and properties of enzymes from diverse thermophilic bacteria (Coolbear *et al.*, 1992). Thermal stability is an inherent property of such proteins, but they are also more resistant to a spectrum of chaotropic agents than corresponding mesophilic enzymes. With such advantages, it might seem surprising that few enzymes from thermophiles have been marketed, and even fewer have gained market favor. There are general and specific reasons for this. Replacing an existing enzyme in an industrial process requires that either the new enzyme functions within the operating conditions of the existing procedures, or there are other gains if the process conditions have to be modified. Increased thermal stability may not be significant. For example, it is of secondary importance in the rennin enzymes used for cheese making, but specificity for the sites of casein hydrolysis is critical. For less specific uses, *e.g.*, detergent proteases, the cost of production becomes the major barrier to market acceptance, and in thermophilic bacteria, yield is often several orders of magnitude below that of commercially used mesophilic cultures (Cowan *et al.*, 1985).

We have always held that enzymes from thermophiles will be most

PETER L. BERGQUIST • Bacterial Genetics and Microbiology, School of Biological Sciences, University of Auckland, New Zealand. HUGH W. MORGAN • Thermophile Research Unit, University of Waikato, Private Bag 3105, Hamilton, New Zealand.

Thermus Species, edited by Richard Sharp and Ralph Williams. Plenum Press, New York, 1995.

successful in new processes that exploit their intrinsic properties of stability. These applications are initially likely to be niche markets, which will be relatively insensitive to the current high cost of production of thermophilic enzymes. This chapter reviews the spectacularly successful application of a *Thermus enzyme*, DNA polymerase, to a fundamentally new process [the polymerase chain reaction (PCR)] and continuing efforts to develop processes for extracellular proteinases from *Thermus* species.

2. DNA POLYMERASES FROM *THERMUS*

Thermostable DNA polymerase (DNA pol) from *Thermus acquaticus* (Taq pol) made PCR feasible, and introduced a powerful technology that complemented recombinant DNA studies and aided in the diagnosis of inherited and infectious diseases. Most attention has been paid to the DNA pol from *T. aquaticus,* which was first purified by Chien *et al.* (1976), and cloned and expressed in *E. coli* by Lawyer *et al.* (1989). However, thermostable DNA pols from *Thermus flavus, Thermus thermophilus,* and *Thermus ruber* have also been purified (Kaledin *et al.,* 1981, 1982; Ruttiman *et al.,* 1985). The successful marriage of Taq pol and the PCR reaction has resulted in only minor attention being paid to DNA pol from other *Thermus* strains and from other thermophilic bacteria. Although other enzymes from archaea are more temperature-stable, have proofreading activity, and have been cloned (Vent™ polymerase, New England Biolabs, from *Thermococcus littoralis*), the biochemical characteristics of Taq pol have been quite adequate for the repeated exposures to the denaturing temperatures that are required for the PCR. It has been estimated that 65% of the total enzyme activity remains after 50 PCR cycles at 95 °C (Gelfand and White, 1990). The stability of the template duplex limits DNA synthesis at 94 °C or more, and for PCR applications, there is little point in searching for an enzyme with a higher temperature optimum. The extension phase of the PCR reaction is routinely performed at 72 °C, although the temperature optimum for maximum yield is reported by Wittwer and Garling (1991) to be between 75–79 °C. Enzymes with a higher temperature optimum may, however, be of value in DNA sequencing applications.

2.1. Purification of *Thermus* DNA Polymerases

Chien *et al.* (1976) reported the first isolation of DNA pol from *T. aquaticus* YT1 as a protein of 66–68,000Da and with an inferred specific activity of 2–8000 units/mgm (Gelfand and White, 1990). The assay was effectively a combination of 5′ to 3′ exonuclease activity and DNA pol activity because of the substrate used. Purification followed standard procedures developed for other DNA pols and involved DEAE-Sephadex, phospho-cellulose, and DNA-cellulose column chromatography. A final purification factor could not be derived since it was necessary to add bovine serum albumin to stabilize the enzyme after the phosphocellulose step. The DNA pol was free of alkaline phosphomonoesterase, alkaline phosphodiesterase, and single-stranded exonuclease activity. The most remarkable feature of the enzyme was its temperature optimum, reported to be 80 °C, a fact ignored for more than ten years until PCR was devised.

Kaledin *et al.* (1980) also reported the isolation of a DNA pol from *T. aquaticus* YT1, which differed in some properties from the enzyme isolated by Chien *et al.* (1976). They used a five-stage purification procedure with four-column chromatography steps after an initial ammonium sulfate precipitation. They described the alkaline pH optimum and the 70 °C temperature optimum of the enzyme. Maximum activity was observed on polydA-oligodT substrates, which were presumably partially single-stranded at that temperature. They noted that the enzyme exhibited polymerase activity over a wide range of temperatures (45–90 °C) and did not show 3' to 5' or 5' to 3' exonuclease activity although no details of results are provided. The enzyme required either Mg^{2+} or Mn^{2+} as a divalent cation for optimal activity and was unstable in dilute solution, but gelatin prevented irreversible denaturation. The molecular weight also was estimated to be between 60–62,000Da.

Lawyer *et al.* (1989) reported an alternative purification procedure for the isolation of a 94,000Da protein with a 10–20 times higher specific activity than reported previously. Taq pol is produced at low levels in the native organism, and they facilitated purification of the enzyme by cloning its gene into *E. coli* under the control of the lac promoter. From the sequence of the gene they predicted that Taq pol should have an M_r of 94,000Da.

Kaledin *et al.* (1981) reported the isolation and purification of a DNA pol from *T. flavus* using similar techniques and assays to those used for the enzyme from *T. aquaticus* YT1 isolated by Kaledin *et al.* (1980) but with the inclusion of a heparin-Sepharose chromatographic step. They achieved a 1000-fold purification with a 33% yield, and the resulting protein had to be stabilized with gelatin. *T. flavus* DNA pol differed from the Taq pol in that its activity was not inhibited by antibody to Taq pol; it had a lower temperature optimum and a higher pH optimum. The M_r was reported to be 66,000Da.

The DNA pol from *T. ruber* was purified by Kaledin *et al.* (1982) using column chromatography to give a preparation with higher specific activity than reported for the *T. aquaticus* and *T. flavus* enzymes, but the enzyme had an M_r of 70,000 Da. The enzyme was devoid of endonuclease, 3' to 5' exonuclease and 5' to 3' exonuclease activity and showed optimal activity at pH 9 and 70 °C. This enzyme showed similar characteristics to the other enzymes isolated from *Thermus sp.* by Kaledin *et al.* (1980, 1981). None of these enzymes were tested for their utility in sequencing reactions and only the *T. thermophilus* enzyme has been used in the PCR, although a commercially available preparation of the *T. flavus* DNA pol (*Replinase*, Du Pont) was briefly marketed for this purpose.

Three DNA pols have been purified from *T. thermophilus* HB8 (Ruttimann *et al.*, 1985) after phospho-cellulose and *DNA*-agarose chromatography. These three enzymes were of higher molecular weights than the *Thermus* polymerases previously described (110,000–120,000Da) and had different responses to inhibitors and different temperature stabilities. The molecular weights reported are substantially higher than another *T. thermophilus* DNA pol isolated [67,000Da (Carballeira *et al.*, 1990)], and a recent report of the purification of a *T. thermophilus* DNA pol with reverse transcriptase activity (Myers and Gelfand, 1991b). The specific activity of each of the polymerases is relatively low compared with results from *T. aquaticus* and other *T. thermophilus* polymerases, and it is possible that the enzymes contain inactivated material, although Ruttiman *et al.* (1985) argue against proteolytic modification. Gelfand and White (1990) have pointed out that DNA pols are very susceptible to proteolytic attack and that care must be taken at all stages of harvesting and purification to minimize degradation so that enzymatic preparations are not attributed to fragments of the enzyme.

A second enzyme with differing properties from those reported above was isolated by Carbelleira *et al.* (1990) from *T. thermophilus*. This enzyme was shown to be substantially pure but still had a relatively low molecular weight active fragment of 67,000Da with substantially lower specific activity compared with cloned *T. aquaticus* enzyme. The enzyme was shown to be active in PCR; it lacked a 5'–3' exonuclease and had a lower extension rate than Taq pol.

2.2. Cloning of DNA Polymerases from *Thermus*

The only DNA pol from *Thermus* species that has been cloned in *Escherichia coli* is from *T. aquaticus* YT1 (Lawyer *et al.*, 1989; Engelke *et al.*, 1990; Sagner *et al.*, 1991). Since it was known that a plasmid containing the *E. coli* DNA pol gene could not be maintained, Lawyer *et al.* (1989) cloned an epitope of Taq pol using lambda gt11 libraries and antibody selection. The lambda recombinant that expressed the epitope was used to select the bacteriophage carrying the Taq pol gene from a library made in lambda

Ch35. The few *Hin*dIII fragments in the inserted DNA were subcloned into a plasmid vector with an inducible lac promoter, and one recombinant of an 8kb *Hin*dIII fragment gave low levels of Taq pol activity on IPTG induction. The gene was localized using restriction enzyme deletions extending from the right and left of the cloning vector sequences, and the reading frame of Taq pol was determined at several points by making lac fusions. Western blot analysis of deleted derivatives gave an immunoreactive band of about 65,000Da, known to be smaller than the native polymerase (94,000Da). The complete gene was assembled by cloning a portion of the B fragment, a *Bgl*II-*Hin*dIII segment (724bp) into a plasmid vector and determining the sequence of the open reading frame, which continued in phase through the *Hin*dIII site into the A fragment. PCR amplification confirmed that the A and the B fragments are contiguous on the *T. aquaticus* genome, since specific primers amplified a predicted 86bp product. Expression of the cloned gene was only detectable using Western blots of IPTG-induced cultures, but this low level was overcome on the removal of an in-phase upstream TGA codon, which caused translational termination. Apparently the Taq pol is produced as a fusion protein.

A more direct approach was taken by Engelke *et al.* (1990) who constructed a plasmid overproducing Taq pol by using specific 5′ and 3′ primers and the PCR reaction to produce a 2.5kb fragment carrying the polymerase gene, which was cloned into an expression vector with the aid of *Eco*RI and *Bgl*II sites that had been incorporated into the primers. The fusion protein uses the tac promoter, the ribosomal binding site, and the initiator methionine of the vector. The first two amino acids encoded by the native enzyme were replaced with three amino acids from the *lacZ* alpha-fragment. Taq pol was prepared from IPTG-induced cells using a simple three-step procedure with a single ion-exchange step. The purification was markedly assisted by heat treatment at 75 °C for 60 min to denature *E. coli* proteins. In contrast to other purification procedures reported, the recombinant Taq pol was assayed using a primer extension assay using M13 as template and by observing its ability to amplify a single-copy gene fragment from human DNA. Engelke *et al.* (1990) noted the importance of careful enzyme titration in the PCR reaction.

Sagner *et al.* (1991) have reported a rapid filter-based assay for the detection of DNA polymerase activity. Activated calf thymus DNA was bound to a nitrocellulose membrane. Purified enzymes, bacterial lysates or colonies were spotted or replicated onto the filter, dried, and submerged in polymerase buffer. Incorporation of radioactive dTTP into acid-insoluble material was assayed by radioautography after incubation at 25 °C or 65 °C. A genomic library of *T. aquaticus* YT1 was constructed by partial *Sau*3A digestion and size-fractionated. Fragments of more than 3kb were ligated into the *Bam*HI-site of pUC18 and transformed into *E. coli* LE392. One colony in 2,000 showed heat-stable polymerase activity, and examination using activity gels showed that the gene product comigrated at 94,000Da with authentic Taq pol. This gene was apparently expressed from its own promoter (or promoter-like sequence) because the fragment carrying it encoded about 2kb of DNA both upstream and downstream of the open reading frame. It is not clear whether it had the upstream in-frame TGA codon that limited expression in the experiments reported by Lawyer *et al.* (1989).

2.3. Other Enzymatic Activities of Taq DNA Polymerase

Although general statements had been made by earlier workers regarding the lack of 3' to 5' exonuclease (proofreading) activity of Taq pol, the first clear demonstration was provided in a sensitive assay devised by Tindall and Kunkel (1988). The assay relies on the removal of a single mispaired base from the 3'-OH terminus of a primer consisting of the minus strand of bacteriophage M13mp2 annealed to the plus strand, but with a single strand gap of 363 nucleotides. Polymerization to fill the gap, without removal of the mismatched base, results in plaques with a distinctive color phenotype after transfection of *E. coli* host cells. Comparison with avian myeloblastosis virus (AMV) reverse transcriptase and Klenow fragment showed that Taq pol resembled AMV DNA pol in its inability to remove the mismatched base, whereas Klenow fragment excised the terminal mismatch in 88% of the plaques scored. In other studies, no 3' to 5' exonuclease activity was detected under conditions of low reaction pH, or when changes in the relative concentrations of $MgCl_2$ and dNTPs were used.

Taq pol shows sequence homology with the DNA pol domain and the 5' to 3' exonuclease domain of *E. coli* DNA pol I (Lawyer *et al.*, 1989) Longley *et al.* (1990) identified and characterized a 5' to 3' exonuclease activity associated with Taq pol that cosedimented with DNA pol, and was associated with a 92,000Da band on SDS-acrylamide gel electrophoresis. The *in situ* gel assay employed a $5'^{32}$P-labeled 15-mer oligonucleotide and a $3'^{32}$P-labeled 24-mer annealed to M13 mp2, with resolution of the reaction products on a sequencing gel. At the position of the Taq pol on the gel, polymerase activity was detected by the addition of ddTTP onto the 15-mer to give the expected 16-mer. The degradation of the 24-mer gave smaller products. Other experiments showed that the 5' exonuclease activity hydrolyzed the 5'-phosphorylated substrate twofold faster than the same substrate containing 5'-terminal hydroxylated nucleotide. When two primers that could be extended when annealed individually were both hybridized to M13 mp2 DNA to simulate a nicked substrate, chain elongation occurred by nick translation. The downstream primer was degraded to a degree that roughly equaled the amount of synthesis. Longley *et al.* (1990) also confirmed that there was no 3' to 5' exonuclease activity in Taq pol, since there was no degradation of 5'-end-labeled oligonucleotide, in agreement with the results of Tindall and Kunkel (1988).

Holland *et al.* (1991) have employed the 5' to 3' exonuclease reaction to generate a specific signal for a target fragment in PCR amplification. The probe oligonucleotide was synthesized with a 3'-phosphate to prevent extension,, and the 5' phosphate was ^{32}P-labeled. The oligonucleotide with a noncomplementary 5' tail was degraded only when it was hybridized to the specific target in the single-stranded PCR product DNA. This method employed rapid thin-layer separation of the products of exonuclease activity and provided the simultaneous generation of a target-specific signal along with target amplification in the PCR. It simplified detection of products from complex genomic mixtures, and overcame problems of primer extension from partially complementary sequences.

2.3.1. Reverse Transcriptase Activity

T. aquaticus DNA pol has reverse transcriptase activity (Jones and Foulkes, 1989) and transcribed spliced mRNA for glucose-6-phosphate dehydrogenase by the addition of a single primer at 68 °C for extension, followed by a PCR reaction with the second primer for 30 cycles of amplification at 70 °C. Submitting mRNA directly to PCR did not produce amplified DNA. Shaffer *et al.* (1990) also demonstrated the usefulness of Taq pol in combining the reverse transcription and PCR reactions into a one enzyme, one-step process that used a biotin-labeled reverse transcription primer that spanned a splice site, with recovery of the PCR product on streptavidin-coated magnetic beads for sequencing. The high reaction temperature of Taq pol, compared with other reverse transcriptases, reduces the problems caused by RNA secondary structure and leads to complete cDNA product.

Reports have appeared recently on the efficient reverse transcriptase activity of a recombinant DNA pol from *T. thermophilus* [Tth pol (Myers and Gelfand, 1991a,b)]. RNA can be copied and cDNA amplified in a two-step coupled process using the one enzyme. The *T. thermophilus* reverse transcriptase activity was 100-fold more efficient than *T. aquaticus* DNA pol, but $MnCl_2$ was found to be superior to either $MgCl_2$ or $CoCl_2$ in this reaction. These two enzymes are reported to show significant amino acid sequence similarity, and it is not clear why their abilities to utilize RNA templates are so different. The use of a thermostable reverse transcriptase coupled with

PCR amplification presumably overcomes the problem of stable RNA structures and hence decreases premature termination. It should also enhance specificity of primer binding and extension as a consequence of higher incubation temperatures.

2.3.2. Terminal Transferase Activity

Clark et al. (1997) demonstrated that *E. coli* DNA pol I (Klenow fragment) was able to add one or more nucleotides to the 3'-terminus of blunt-ended DNA in a nontemplated manner. This property was subsequently found to be common to a number of procaryotic and eucaryotic DNA pol including Taq pol (Clark, 1988). Terminal transferase activity was assayed using a synthetic double-stranded oligonucleotide that contained no T residues and thus was unable to provide template information for the incorporation of dA, which, if incorporated, must represent a nontemplate addition. Taq DNA pol could carry out blunt-end addition of any one of the four dNTP precursors when supplied in equivalent concentration, with dATP being incorporated much more efficiently than the other three dNTPs. dATP alone also could be incorporated efficiently. No addition was found for the single-stranded oligonucleotide, showing that the substrate must be double-stranded DNA. Other measurements using an oligonucleotide with a recessed 3' OH end suggested that the rate of templated nucleotide addition for DNA pols and reverse transcriptase is up to several hundred times faster than nontemplated addition of dA. This nontemplated addition of dA to PCR products has been exploited by several groups for the cloning of PCR products. Restriction enzyme sites are frequently included within PCR primers, and the required fragment is then cut with the appropriate restriction endonuclease and ligated into a suitably cut vector. This procedure is expensive, since additional bases have to be included in the primers, and the process is often inefficient, as many restriction enzymes do not cleave when their recognition sequences are located within a few base pairs of the ends of the PCR fragment (Kaufman and Evans, 1990). One approach has been to remove the A residue using Klenow DNA pol (Hemsley *et al.*, 1989) whereas another has been to use a specially constructed vector that carries a protruding 3' thymidine residue after cleavage with the restriction enzyme *Hph*I in a polylinker region which contains two flanking copies of the *Hph*I recognition sequence (Mead *et al.*, 1991). An alternative procedure, which does not require a specialized vector without additional *Hph*I sites used the common cloning vector Bluescript (Stratagene, La Jolla, California). After digestion with *Eco*RV in the polylinker region to give a blunt-ended cleavage site, the terminal transferase activity of Taq pol with TTP as the only deoxynucleotide present was used to add a single T residue at each of the 3′ ends to give vector and PCR products with complementary one-base overhang (Marchuk *et al.*, 1991). The vector is unable to self-ligate or to undergo concatamerization, and the standard blue–white discrimination using X-Gal allows the identification of vector plasmids with inserts, using the standard *E. coli* lac Z alpha-complementation assay. Holton and Graham (1991) have used the same vector and terminal transferase to add a single T residue to the *Eco*RV site by using dideoxy TTP. The 3′-overhanging T residue lacks a 3′ hydroxyl group and cannot form a phosphodiester bond, whereas the other strand carries a 5′ phosphate that can ligate to the 3′-hydroxyl group from the PCR product. Presumably the nick that is left on each strand is repaired *in vivo* with the replication of the vector plus insert in the recipient *E. coli* strain.

2.4. Characteristics of *Thermus* DNA Polymerases

Taq pol is the only enzyme for which substantial data is available. Apart from studies of fidelity and misincorporation of nucleotides, most of the basic biochemical characteristics have not been published in a systematic fashion. Two reviews (Gelfand, 1989; Gelfand and White, 1990) describe a number of enzymatic characteristics of Taq pol, but most references are either personal communications or papers in preparation.

The thermal stability of Taq pol is 40 min at 95 °C and it has been estimated that about 65% of the enzyme is still active after 50 cycles of PCR with a 95 °C for 20 sec denaturation step (Gelfand and White (1990). The temperature optimum is 75–80 °C, with the enzyme having low activity (but high stability) above 90 °C and below 30 °C (Gelfand and White, 1990). DNA synthesis at high temperatures is obviously limited by the denaturation and dissociation of the primer and template.

Taq pol is sensitive to the concentration of magnesium ions in the reaction mixture, but the optimal concentration appears to depend on both the dNTP concentration and on the particular primers used (Saiki, 1989). The dNTP's bind magnesium, so that the amount of free Mg^{2+} in the PCR reaction will depend on the dNTP concentration. $MgCl_2$ concentrations from 1.5–3.0mM are frequently used, but the optimal concentration usually has to be determined empirically using fixed concentrations of all four dNTPs. Potassium chloride, gelatin, dimethylsulphoxide, detergents, urea, and formamide have been claimed to improve or inhibit the amount of product found in the PCR (Gelfand, 1989; Gelfand and White, 1990).

2.5 Fidelity of DNA Synthesis with *Thermus* DNA Polymerase

The lack of a "proofreading" 3' to 5' exonuclease activity in Taq pol has caused concern regarding the fidelity of DNA replication using this enzyme. Mendelmann *et al.* (1989, 1990) have shown that other DNA pols lacking proofreading activity misincorporate deoxynucleotide triphosphates as a function of the dNTP concentration in the reaction. Several methods have been used to characterize misincorporation by Taq pol. The simplest is sequencing of cloned PCR fragments, followed by reporting the number of errors per target sequence (Saiki *et al.*, 1988). A more elaborate procedure has used denaturing gradient electrophoresis to isolate mutant heteroduplexes from a single synthetic cycle (Keohavong and Thilly, 1989); another is to use the highly sensitive M13 mp2 system that allows detection of mutant enzymes based on a color-assay with the chromogenic substrate X-Gal. Interpreting these different types of assays is difficult as has been discussed by Bloch (1991). Tindall and Kunkel (1988) reported that Taq pol introduces single-base substitution errors at a rate of 1 in 9000 nucleotides polymerized and frameshift mutations at a frequency of 1 in 41,000 nucleotides polymerized. Taq pol was claimed to be 4–8 times more error-prone than Klenow DNA pol assayed under similar conditions, although it should be noted that high dNTP concentrations were used, which should have facilitated misincorporation. However, Eckert and Kunkel (1990) have shown that low dNTP concentrations and $MgCl_2$ in only a slight excess over the total dNTP concentrations should be used when high fidelity synthesis is required, for example, for the distinction of individual alleles in heterozygotes or for the expression of products of PCR-amplified genes. Gelfand and White (1990) point out that a significant increase in the fidelity of DNA amplification using Taq pol has been reported using lowered dNTP and $MgCl_2$ concentrations (Goodenow *et al.*, 1989; Fuchareon *et al.*, 1989; Reiss *et al.*, 1990).

The effect of primer-template mismatches have been analyzed by Kwok *et al.* (1990) using a PCR fragment from human immunodeficiency virus as a model template. Mismatches in the primer and the template that were analyzed included all possible 3' mismatches between the two components of the duplex. Some mismatches at the 3' terminal position appeared to amplify as efficiently as the nonmismatched primer-template combination, whereas A:G, G:A, and C:C mismatches markedly reduced the yield of product. A or T mismatched with any of the other three bases did not have a significant effect on the efficiency of the PCR and two T's at the 3' end of the primer had a relatively minor effect on efficiency. Hence the presence of two T's enabled amplification irrespective of the other nucleotides involved. These studies, while different in some details from similar experiments with 3′ terminal mismatched primers reported by Newton *et al.* (1989), are important in relation to the design of primers that will amplify sequences despite mismatches and for achieving allele-specific amplification where only one particular target is required to be amplified.

2.6. Applications of *Thermus aquaticus* Polymerase Involving the PCR

This area has developed exponentially with a growing awareness of the potential uses of the PCR reaction in applications such as specific mutagenesis, amplification, and cloning of particular gene products or related genes using consensus primers, and DNA sequencing. With auto-mated sequencing, a thermostable polymerase has obvious advantages at high temperature because its processivity is not interrupted by secondary structure, which is melted at the assay temperature. Papers published recently reviewed details of the biochemistry and molecular biology of the reactions involved (Innis *et al.*, 1990; Bloch, 1991; Erhlich *et al.*, 1991). Of the many academic, diagnostic, and forensic applications of PCR, only two will be discussed here. PCR techniques using Taq pol have allowed systematic acquisition of sequence information in a known direction from a point at which the sequence is already determined (genome walking), and it has allowed both random and directed mutagenesis to be carried out on known coding sequences. Only a few of the possible applications in these areas are mentioned in this study, but they indicate the importance of thermostable DNA pol for rapid progress in genome mapping and functional studies of expressed proteins.

One of the first uses of PCR was to isolate unknown upstream or downstream sequences by designing primers that faced away from the known sequences (inverse PCR, Ochman *et al.*, 1988; Triglia *et al.*, 1988). This technique was limited by the necessity for ligation and recleavage by conveniently located restriction endonuclease sites. A simpler approach was reported by Shyamala and Ferro-Luzzi Ames (1989), which involved ligation of genomic fragments resulting from specific restriction nuclease digestion with M13 vector cut with the same nuclease. The reverse sequencing primer was used as one primer, and a primer facing out from the known sequence was used to amplify the intervening DNA, so that a primer was required for only one side of the target sequence. An *in vivo* footprinting method for examining DNA binding proteins employed a synthetic adaptor that could be ligated to a nicked or restriction enzyme-cleaved site, with PCR extension followed by amplification and the addition of internal sequencing primers (Mueller and Wold, 1989) This single-sided PCR technique was further exploited by Fors *et al.* (1990) to clone the shark PO promoter. Systematic construction of adaptors, which included the forward and reverse sequencing primers and which were capable of ligation to a variety of specific restriction enzyme cleavage sites, was reported by Copley *et al.* (1991). These adaptors allowed the construction of restriction enzyme maps after the isolation and sizing of fragments produced from specific adaptors, as well as allowing determination of intervening DNA sequences in genomic walking studies of *Chlamydia trachomatis*, an obligate intracellular bacterium. We have carried out similar constructions involving adaptors that were constructed with both 5′ and 3′ overhangs to allow ligation to genomic DNA cut with 14 different enzymes, with the objective of linking the PCR reaction to automated sequencing procedures (Bergquist and Saul, unpublished). For reasons not currently understood, the maximum length of unknown sequence that can be amplified is 2–3kb (see also Copley *et al.*, 1991). The fact that these single-sided PCR techniques can allow directionality to be achieved in genomic walking means that it should be possible to systematically walk away from known sequences in a given direction while simultaneously accumulating sequence information for the synthesis of new primers to continue the walk. Initially, site-directed mutagenesis was limited to introducing mismatches into the primer sequence or by supplying a 5′ terminal addition to the primer (Scharf *et al.*, 1986). However, it was recognized that if a gene was amplified in two sections, it was possible to introduce a mutation by including it in the primers for amplification by providing for overlap at the site of mutation (Higuchi *et al.*, 1988). Two PCR reactions are performed, the products are denatured, reannealed, and extended using Taq pol. This basic protocol can be modified to allow the construction of deletions or fusions (Ho *et al.*, 1989; Vallette *et al.*, 1989; Yon and Fried, 1989; Horton *et al.*, 1989). A more complicated method involves the introduction of specific sequences for restriction enzymes that hydrolyze at a short distance from their recognition sites (Tomic *et al.*, 1990).

A modification of overlap extension mutagenesis requires only a single new primer per mutational step (Ito *et al.*, 1991). This procedure makes use of primers that flank the multiple cloning site of the Bluescript vector (primers I and II). A third primer (III) is complementary to the polylinker sequence; it is between one of the flanking primers and the end of the insert, cloned in Bluescript. This primer has a mismatched base that destroys a restriction site. Primer IV carries the specific mutation in the gene of interest. Two PCR reactions are performed using primers I and IV plus II and III, and the amplified products are mixed, denatured, and reannealed. The resulting products are reamplified using primers I and II. The product is digested with the appropriate two restriction enzymes and recloned into the vector. Only the DNA fragment containing the mutation is able to be recloned because the other sequence from the reaction lacks one of the specific enzyme sites.

3. EXTRACELLULAR PROTEINASES FROM *THERMUS*

The genus *Thermus* is phenotypically diverse and several genospecies have been suggested (see Chapter I). *Thermus* was first isolated on a peptone medium, and most strains have been obtained on derivatives of the original medium formulation. Despite this, not all strains produce extracellular proteinase, and the properties of proteinases produced from different strains reflect the phenotypic diversity. All are alkaline serine proteinases with molecular weights in the range 23–34 kDa and an isoelectric point between 8.5–8.9 (Coolbear *et al.*, 1992), but vary in their response to metal ion chelators, thermostability, and rates of hydrolysis of chromogenic peptides (Cowan *et al.*, 1987a). The type strain *T. aquaticus* YT1 produces two distinct proteinases, aqualysin I—the typical alkaline serine proteinase produced in mid-log to early stationary phase of growth, and aqualysin II—a neutral proteinase produced in late stationary phase of growth. Aqualysin I has a strong amino and sequence homology with alkaline serine proteinases that have been purified from other *Thermus* strains, and to a lesser extent, with the subtilisin group of proteins.

3.1. Stability of *Thermus* Extracellular Proteinases

Expectedly, proteinases from *Thermus* strains are thermostable (Table I). Although there are obvious differences in the thermostabilities of *Thermus enzymes*, some care must be exercised in making comparisons. Stability is affected differently by available Ca^{2+}, ionic concentration, presence of surfactants, and protein–concentration-related autolysis. The stability of the enzyme can be dramatically affected by some of these factors, e.g., the half-life of the *Thermus* Rt41A proteinase at 75 °C is reduced from >254 hrs to 2.5 hrs to 2.9 min as the calcium concentration is reduced from 5 to 10mM to no free calcium, and this is due to both thermal destabilization and increasing autolysis. The only chelator-resistant *Thermus* proteinase that has been investigated in detail is that from strain Tok3. This enzyme retains full activity and thermostability after EDTA treatment but it was unstable at low ionic strength, with 0.4 M NaCl required for full stability (Saravani *et al.*, 1989).

The resistance of *Thermus* proteinases to other denaturing agents is equally impressive (Table I), and this is likely to be a property common to all thermostable proteins. In addition, proteinase from *Thermus* strain

Table I. Stability of Some *Thermus* sp. Proteinases

Organism	Proteinase	Thermostability T½ at 80 °C (mins.)	% activity remaining following exposure to denaturing agents[a]				References
			8M Urea	6M Guanidin HCl	1% SDS	EDTA 10mM	
Thermus aquaticus YT1	Aqualysin I	480	<1	<1	<1	50	Matsuzawa *et al.*, 1988
Thermus T351	Caldolysin	1800	<1	<1	3	32	Cowan *et al.*, 1982
Thermus Rt4.1AP	PreTaq	810	62	5	73	<1	Peek *et al.*, 1992
Thermus Tok3	Caldolase	840	31	20	41	97	Saravani *et al.*, 1989

[a]% activity remaining follows 1 hr. exposure to denaturing agent at 70 °C. At room temperature, all proteinases retain all activity when exposed to denaturing agents for 1 hr.

Rt41A was readily immobilized onto glass beads with no loss of enzyme activity when low enzyme concentrations were used (Peek *et al.*, 1992a). When hydrolyzing mesophilic protein substrates the specific activity of thermophilic proteinases is up to an order of magnitude greater than that of mesophilic proteinases (Cowan *et al.*, 1987b). Presumably this high specific activity of thermophilic enzymes could be even further enhanced if the destabilizing effect of other chaotropic agents on the substrate were combined with those of heat denaturation. The hydrolysis of oxidized insulin B chain by proteinases from *Thermus* strains Rt41A and *Thermus aquaticus* YT-1 gave almost identical cleavage patterns, and were similar to that of proteinase K (Peek *et al.*, 1992b). Thus the enzymes have a broad substrate specificity (Cowan *et al.*, 1987b), particularly on mesophilic substrates at high temperature and pH.

3.2. Industrial Potential for *Thermus* Proteinases

Although the stability and activity of *Thermus* proteinases have potential for industrial exploitation, this has not yet been exploited. In the biggest market for thermostable proteinases, the detergent industry, the sequestering agents added to formulations to remove calcium would inhibit the use of many *Thermus* proteinases, and they do not possess the required specificity for rennet production, the second largest market for proteinases. Ng and Kenealy (1986) have critically discussed properties required for a thermostable enzyme to be of industrial use and the advantages of using thermostable enzymes from mesophilic organisms. They consider it is unlikely that thermophilic proteinases will find application in current bulk market applications. Cowan *et al.* (1985) identified possible applications where thermophilic proteinases possessed appropriate properties, but their use was ruled out by expense, such as the growth costs for the producing organism, costs associated with purification and loss of enzyme during purification, and the amount of enzyme produced by the culture concerned.

Thermus can be grown on a completely defined medium with glucose, acetate or, more commonly, monosodium glutamate as sole carbon source. Growth rates of 0.2 hr^{-1} have been reported in pH-controled culture on glutamate medium. Cell yields were up to 3g dry weight per liter (Janssen *et al.*, 1991). Although the organism can be grown cheaply on a laboratory scale, the cost of scaling up to large fermenter volumes has not been investigated. For a thermophilic aerobe, effective aeration at high cell mass is a likely problem. *Thermus* strain Rt41A proteinase has been purified from culture supernatant to apparent homogeneity by a simple three-step process, resulting in over 20% recovery with 67-fold purification (Peek *et al.*, 1992b). This high yield is to be expected with thermostable enzymes because losses due to denaturation are minimal. Thus, in *Thermus*, the major hurdle to producing proteinase economically is overcoming the low level of production, which is commonly one or two orders of magnitude lower than for mesophiles (Cowan *et al.*, 1985). In both *Thermus* strains YT-1 and Rt41A, proteinase production is directly linked to metabolic activity and growth (Janssen *et al.*, 1991; Matsuzawa *et al.*, 1983). Substrate depletion halts growth and proteinase production immediately, and proteolytic activity of the culture declines rapidly. The combination of these effects can dramatically alter the production of enzyme; the same strain assayed at slightly different times of culture may show quite marked differences of activity. Unless referred to a time course of production for each culture, comparisons of enzyme activity are of little value. Under conditions of slow growth, the rate of decay of proteinase activity can be less than the rate of enzyme production, so that the culture has barely detectable activity (Kanasawud *et al.*, 1989). Published values of proteinase production by strains of *Thermus* should be interpreted with caution (Cowan and Daniel, 1982; Cowan *et al.*, 1985, Jones *et al.*, 1988; Kanasawad *et al.*, 1989). What is normally measured is proteinase yield, the resultant of production less loss under fermenter conditions. In a detailed examination of proteinase production by *Thermus* strain Rt41A, Janssen *et al.* (1991) showed that calcium-binding medium constituents decreased the stability of the enzyme. Phosphate was particularly important in this process: a medium containing anabolic amounts of inorganic phosphate increased the half-life of the proteinase in the fermenter from 6 to over 90 hrs, and proteinase yield was increased up to 10-fold. Even proteinases apparently insensitive to chelators are unstable in culture broth (Matsuzawa *et al.*, 1983) so that similar approaches may result in improved productivity. While such gains in yield are significant and might be further increased with appropriate fermenter design, production of proteinase is still directly linked to culture growth and remains low compared with the *Bacillus* cultures used industrially. Even under optimum conditions, the yield of proteinase in *Thermus* strain Rt41A was less than 75 mg 1^{-1}, and production for commercial use can seemingly only be met by cloning and overexpressing the gene in a suitable host. Terada *et al.* (1990) have successfully cloned a 1.1kb fragment that contains the sequence for aqualysin I from *T. aquaticus* YT-1 into *E. coli*, but the mature enzyme was either not expressed, or was lethal to the host cell. The enzyme was produced with three domains; an aminoterminal leader necessary for inner membrane transport; the mature enzyme sequence and; a carboxy-terminal domain thought to be required for extracellular secretion. Partially processed enzyme accumulated in the *E. coli* membrane, and exposure to 65 °C was needed to release active enzyme. The overall productivity was low, possibly because loading the membrane of the *E. coli* recombinant with enzyme protein reduced its growth rate. Lee *et al.* (1992) have indicated that the amino terminal leader may have a chaperonin-like function necessary for the correct folding of the proteinase domain. Deletion of the carboxy-terminal prosequence did not affect the production of active aqualysin I in *E. coli*, indicating that this component was not essential for formation of mature proteinase. A role for the C-terminal prosequence in secretion of the proteinase across the outer membrane of *T. aquaticus* is doubtful and other functions are being investigated.

Recently, aqualysin I has been expressed and secreted in *T. thermophilus* using a *Thermus–E. coli* shuttle vector. The entire aqualysin gene, which had been cloned as two separate plasmids in *E. coli* (Terada *et al.*, 1990), were combined on shuttle vector pYK105 (Touhara *et al.*, 1991). *T. thermophilus* HB27 was transformed; it produced aqualysin during growth and in stationary phase. However, final yield of proteinase was only equivalent to that of *T. aquaticus* YT-1. The explanation for increased activity in stationary phase was due to a less efficient secretion by strain HB27. Overproduction of the enzyme has still not been achieved, though it might be possible to improve the promoter for the aqualysin I gene and/or increase the copy number of the expression plasmid.

Opportunities for use of the enzyme are limited, in the meantime, to high cost niche markets. Although the proteinase from *Thermus* strain Rt41A performed credibly as a cleaning agent for specialty ultrafiltration membranes, chemical methods gave superior clearing of soil and higher flux levels and were likely to be preferred, even though they reduced membrane life. The stability of the enzymes to organic solvents might be of use in peptide synthesis. Peek *et al.* (1992a) have demonstrated the efficiency of immobilization of *Thermus* proteinases and the use of the immobilized enzyme for peptide synthesis with different solvents. Thermolysin, a thermostable proteinase from *Bacillus thermoproteolyticus*, is used to produce aspartame by peptide synthesis, and since replacement of this enzyme would offer no advantages, the requirement for other peptides needs to be defined.

A niche market that is currently under development is the application of *Thermus* strain Rt41A proteinase to prepare DNA in crude cell lysates for PCR amplification. The requirements for this application are quite specific. The thermophilic proteinase must have good proteolytic activity at high temperature where native cellular proteins will rapidly denature. A precursor step to PCR amplifications could include an incubation of the cell material with the proteinase at 94 °C for periods up to 1 hr (McHale *et al.*, 1991; Fung and Fung, 1991). During this period, the proteinase itself undergoes autolytic digestion and denaturation so that the PCR amplification protocol can follow directly, with no loss of Taq polymerase activity due to proteolysis. Since the *Thermus* Rt41A proteinase is stable indefinitely at 70–75 °C in the presence of 5 mM calcium, longer periods of hydrolysis can be used if required. The pronounced chelator sensitivity of the proteinase can also be used to completely remove any residual proteinase activity. Thus, reducing the free calcium by chelator addition while raising the temperature to 94 °C, will remove proteinase in less than 30 sec. The proteinase, marketed as PreTaq, has also been used in the preparation of genomic fragments for pulse-field electrophoresis or high molecular weight chromosomal DNA from bacteria and has shown to have advantages over the use of proteinase K (Borges and Bergquist, 1992; McHale *et al.*, 1991).

The successful cloning and overexpression of this and other *Thermus* enzymes is likely to be a necessary prelude to their success in larger scale applications.

REFERENCES

Bloch, W., 1991, A biochemical perspective of the polymerase chain reaction, *Biochem.* **30:**2735–2747.

Borges, K. M., and Bergquist, P. L., 1992, A rapid method for preparation of bacterial chromosomal DNA in agarose plugs using *Thermus* Rt41A proteinase, Biotechniques **12:**222–223.

Carballeira, N., Nazabal, M., Brito, J., and Garcia, O., 1990, Purification of a thermostable DNA polymerase from *Thermus thermophilus* HB8, useful in the polymerase chain reaction, Biotechniques **9**:275–281.

Chien, A., Edgar, D. B., and Trela, J. M., 1976, Deoxyribonucleic acid polymerase from the extreme thermophile *Thermus aquaticus, J. Bacteriol.* **127**:1550–1557

Clark, J. M., 1988, Novel non-templated nucleotide addition reactions catalyzed by procaryotic and eucaryotic DNA polymerases, *Nucl. Acids Res.* **16**:9677–9686.

Clark, J. M., Joyce, C. M., and Beardsley, G. P., 1987, Novel blunt-end addition reactions catalysed by DNA polymerase 1 of *Escherichia coli, J. Mol. Biol.* **198**:123–127.

Coolbear, T., Daniel, R. M., and Morgan, H. W., 1992, The enzymes from extreme thermophiles: bacterial sources, thermostabilities and industrial relevance, *Adv. Biochem. Eng. Biotechnol.* **45**:57–98.

Copley, C. G., Boot, C., Blundell, K., and McPheat, W. L., 1991, Unknown sequence amplification: application to *in vitro* genome walking in *Chlamydia trachomatis* L2, *Bio-Technol.* **9**:74–79.

Cowan, D. A., and Daniel, R. M., 1982, Purification and properties of an extracellular protease (caldolysin) from an extreme thermopile, *Biochem. Biophys. Acta.* **705**:293–305.

Cowan, D. A., Daniel, R. M., and Morgan, H. W., 1985, Thermophilic proteases: properties and potential applications, *Trends Biotechnol.* **3**:68–72.

Cowan, D. A., Daniel, R. M., and Morgan, H. W., 1987a, A comparison of extracellular serine proteases from four strains of *Thermus aquaticus, FEMS Microbiol. Lett.* **43**:155–159.

Cowan, D. A. . Daniel, R. M., and Morgan, H. W., 1987b, The specific activities of mesophilic and thermophilic proteinases, *Internat. J. Biochem.* **19**:741–743.

Eckert, K. A., and Kunkel, T. A., 1990, High fidelity DNA synthesis by the *Thermus aquaticus* DNA polymerase, *Nucl. Acids Res.* **18**:3739–3744.

Erhlich, H. A., Gelfand, D., and Sinsky J. J., 1991, Recent advances in the polymerase chain reaction, *Science* **252**:1643–1651.

Engelke, D. R., Krikos, A., Bruck, M. E., and Ginsburg, D., 1990, Purification of *Thermus aquaticus* DNA polymerase expressed in *Escherichia coli, Analyt. Biochem.* **191**:396–400.

Fors, L., Scavedra, R. A., and Hood, L., 1990, Cloning of the shark PO promoter using a genomic walking technique based on the polymerase chain reaction, *Nucl. Acids Res.* **18**:2793–2799.

Fuchareon, S., Fuchareon, G., Fuchareon, P., and Fukumaki, Y. A., 1989, Novel active mutation in the β-thalassemia gene of a Thai. Identification by direct cloning of the entire β-globin gene amplified using polymerase chain reaction, *J. Biol. Chem.* **1264**:7780–7783.

Fung, M-C., and Fung, K. Y-M. 1 1991, PCR amplification of mRNA directly from a crude cell lysate prepared by thermophilic protease digestion, *Nucl. Acids Res.* **19**:4300.

Gelfand, D., 1989, Taq DNA polymerase, in: *PCR Technology: Principles and Applications for DNA Application* (H. R. Erhlich, ed.), Stackton Press, New York, pp. 17–22.

Gelfand, D. H., and White, J. J., 1990, Thermostable DNA polymerases, in: *PCR Protocols: A Guide to Methods and Applications* (M. A. Innis, D. H. Gelfand, J. J. Sninsky, and T. J. White, eds.), Academic Press, San Diego, pp. 129–141.

Goodenow, M., Hunt, T., Savrin, W., Kwok, S., Sninsky, J., and Wain-Hobson, S., 1989, HIV1 isolates are rapidly evolving quasispecies: evidence for viral mixtures and preferred nucleotide substitutions, *Jour. AIDS* **2**:344–352.

Hemsley, A., Tong, M. D., Cartopussi, G., Arnheim, N., and Gallas, D. J. 1 1989, A simple method for site-directed mutagenesis using the polymerase chain reaction, *Nucl. Acids Res.* **17**:6545–6551.

Higuchi, R., Krummel, B., and Saiki, R. K., 1988, A general method of *in vitro* preparation and specific mutagenesis of DNA fragments: study of protein and DNA interactions, *Nucl. Acids Res.* **16:**7351–7367.

Ho, S. N., Hunt, H. D., Harton, R. M., Pullen, J. K., and Pease, L. R., 1989, Site-directed mutagenesis by overlap extension using the polymerase chain reaction, *Gene* **77:**5159.

Holton, T. A., and Graham, M. W., 1991, A simple and efficient method for direct cloning of PCR products using ddT-tailed vectors, *Nucl. Acids Res.* **19:**1156.

Holland, P. M., Abramson, R. D., Watson, R., and Gelfand, D. H., 1991, Detection of specific polymerase chain reaction product by utilizing the 5′ ~ 3′ exonuclease activity of *Thermus aquaticus* DNA polymerase, *Proc. Nat. Acad. Sci. USA* **88:**7276–7280.

Horton, R. M., Hunt, H. D., Ho, S. N., Pullen, J. K., and Pease, L. R., 1989, Engineering hybrid genes without the use of restriction enzymes: gene splicing by overlap extension, *Gene* **77:**61–68.

Innis, M. A., Gelfand, D. H., Sninsky, J. J., and White, T. J., 1990, (Eds.), PCR Protocols: A Guide to Methods and Applications. Academic Press, San Diego.

Ito, W., Ishiguro, H., and Kurosawa, Y., 1991, A general method for introducing a series of mutations into cloned DNA using the polymerase chain reaction, *Gene* **182:**67–70.

Janssen, P. H., Morgan, H. W., and Daniel, R. M., 1991, Effects of medium composition on extracellular proteinase stability and yield in batch cultures of a *Thermus sp, Appl. Microbiol. Biotechnol.* **34:**789–793.

Jones, C. W., Morgan, H. W., and Daniel, R. M., 1988, Aspects of protease production by *Thermus* strain OK6 and other New Zealand isolates, *J. Gen. Microbiol.* **134:**191–198.

Jones, M. D., and Foulkes, N. S., 1989, Reverse transcription of mRNA by *Thermus aquaticus* DNA polymerase, *Nucl. Acids Res.* **17:**8387–8388.

Kaledin, A. S., Slyusarenko, A. G., and Gorodetskii, S. I., 1980, Isolation and properties of DNA polymerase from extremely thermophilic bacterium *Thermus aquaticus* YT1, *Biokhimiya* **45:**644–651.

Kaledin, A .S., Slyusarenko, A. G., and Gorodetskii, S. I., 1981, Isolation and properties of DNA polymerase from the extremely thermophilic bacterium *Thermus flavus, Biokhimiya* **46:**1576–1584.

Kaledin, A. S., Slyusarenko, A. G., and Gorodetskii, S. I., 1982, Isolation and properties of DNA polymerase from the extremely thermophilic bacterium *Thermus ruber, Biokhimiya,* **47:**1785–1791.

Kanasawud, P., Hjorleifsdottir, S., Holst, O., and Mattiason, B., 1989, Studies on immobilization of the thermophilic bacterium *Thermus aquaticus* YT-1 by entrapment in various matrices, *Appl. Microbiol. Biotechnol.* **31:**228–233.

Kaufmann, D. L., and Evans, G. A., 1990, Restriction endonuclease cleavage at the termini of PCR products, *Biotechniques* **9:**304–306.

Keohavang, P., and Thilly, W. G., 1989, Fidelity of DNA polymerases in DNA amplification, *Proc. Natl. Acad. Sci.* **86:**9253–9257.

Kristjansson, J. K., 1990, Thermophilic organisms as sources of thermostable enzymes., TIBTECH. **7:**349–353.

Kwok, S., Kellogg, D. E., McKinney, N., Spasic, D., Goda, L., Levenson, C., and Sninsky, J. J., 1990, Effects of primer-template mismatches on the polymerase chain reaction: human immunodeficiency virus type 1 model studies, *Nucl. Acids Res.* **18:**999–1005.

Lawyer, F. C., Stoffel, S., Saiki, R. K. Myambo K., Drummond, R., and Gelfand, D. H., 1989, Isolation, characterization and expression in *Escherichia coli* of the DNA polymerase gene from *Thermus aquaticus, J. Biol. Chem.* **264:**6427–6437.

Lee, C-Y., Ohta, T., and Matsuzawa, H., 1992, A non-covalent NH2-terminal pro-region aids the production of active aqualysin I (a thermophilic protease) without the COOH-terminal pro-sequence in *Escherichia coli, FEMS Microbiol. Letts.* **92:**73–78.

Longley, M. J., Bennett, S. E., and Mosbaugh, D. W., 1990, Characterization of the 5′ to 3′ exonuclease associated with *Thermus aquaticus* DNA polymerase, *Nucl. Acids Res.* **18:**7317–7322.

Marchuk, D., Drumm, M., Saulino, A., and Collins, F. S., 1991, Construction of T-vectors, a rapid and general system for direct cloning of unmodified PCR products, *Nucleic Acids Res.* **19:**1154.

Matsuzawa, H., Hamaoki, N., and Ohta, T., 1983, Production of thermophilic extracellular proteases (Aqualysins I and II) by *Thermus aquaticus YT-1*, an extreme thermophile, *Agric. Biol. Chem.* **47:**25–28.

Matsuzawa, H., Tokugawa, K., Hamaoki, M., Mizoguchi, M., Taguchi, H. Terada, I., Kwon, S-K., and Ohta, T., 1988, Purification and characterization of aqualysin I (a thermophilic alkaline serine protease) produced by *Thermus aquaticus* YT-1, *Eur. J. Biochem.* **171:**441–447.

McHale, R. H., Stapleton, P., and Bergquist, P. L., 1991, A method for the rapid preparation of blood and cervical samples for PCR. *Biotechniques* **10:**20–21.

Mead, D. A., Pey, N. K., Herrnstadt, C., Marcil, R. A., and Smith, L. M., 1991, A universal method for the direct cloning of PCR amplified nucleic acid, *BioTechnol.* **2:**657–663.

Mendelman, L. V., Boosalis, M. S., Petruska, J., and Goodman, M. F., 1989, Nearest neighbour influences on DNA polymerase extension fidelity, *J. Biol. Chem.* **264:**14415–14423.

Mendelman, L. V., Petruska, J., and Goodman M. F., 1990, Base mispair extension kinetics. Comparison of DNA polymerase and reverse transcriptase, *J. Biol. Chem.* **265:**2338–2346.

Mueller, P. R., and Wold, B., 1989, *In vivo* footprinting of a muscle-specific enhancer by ligation mediated PCR, *Science* **246:**780–786.

Myers, T. W., and Gelfand, D. H., 1991a, Reverse transcription and DNA amplification by a DNA polymerase from *Thermus thermophilus*, *FASEB Jour.* **5:**A1552.

Myers, T. W., and Gelfand, D. H., 1991b, Reverse transcription and DNA amplifilcation by a *Thermus thermophilus* DNA polymerase, *Biochem.* **38:**7661–7666.

Newton, C. R., Graham, A., Heptinstall, L. E., Powell, S. J., Summers, C., Kalshcher, N., Smith, J. C., and Markham, A. F., 1989, Analysis of any point mutation in DNA. The amplification refractory mutation system, *Nucl. Acids Res.* **17:**2503–2516.

Ng, T. K., and Kenealy, W. R., 1986, Industrial applications of thermostable enzymes, in: *Thermophiles: General, Molecular, and Applied Microbiology.* (T. D. Brock, ed.). Wiley, New York, pp. 197–215.

Ochman, H., Gerber, A. S., and Hartl, D. L., 1988, Genetic application of an inverse polymerase chain reaction, *Genetics* **120:**621–623.

Peek, K., Wilson, S-A., Prescott, M., and Daniel, R. M., 1992a, Some characteristics of a serine proteinase isolated from an extreme thermophile for use in kinetically controlled peptide bond synthesis, *Anns. New York Acad. Sci.* **672:**471–477.

Peek, K., Daniel, R. M., Monk, C., Parker, L., and Coolbear, T., 1992b, Purification and characterization of a thermostable proteinase isolated from a *Thermus sp.* strain Rt41A, *Eur. J. Biochem.* **207:**1035–1044.

Reiss, J., Krawczak, M., Schloesser, M., Wagner, M., and Cooper, D. N., 1990, The effect of replication errors on the mismatch analysis of PCR-amplified DNA, *Nucl. Acids Res.* **18:**973–978.

Ruttiman, C., Coloras, M., Zaldivar, J., and Vicuna, R., 1985, DNA polymerases from the extremely thermophilic bacterium *Thermus thermophilus* HB8, *Eur. J. Biochem,* **149:**41–46.

Sagner, G., Ruger, R., and Kessler, C., 1991, Rapid filter assay for the detection of DNA polymerase activity: direct identification of the gene for the DNA polymerase from *Thermus aquaticus* DNA polymerase expressed in *Escherichia coli*, *Gene* **97:**119–123.

Saiki, R. K., Scharf, S., Faloona, F., Mullis, K. B., Horn, G. T., Erlich, H. A., and Arhein, N., 1988, Primer directed enzymatic amplification of DNA with a thermostable DNA polymerase, *Science* **239:**487–491.

Saravani, G. A., Cowan, D. A., Daniel, R. M., and Morgan, H. W., 1989, Caldolase, a chelator-insensitive extracellular serine proteinase from a *Thermus* spp, *Biochem. J.* **262:**409–416.

Scharf, S. J., Horn, G. T., Erlich, H. A., 1986, Direct cloning and sequence analysis of enzymatically amplified genomic sequences, *Science* **233:**1076–1078.

Shaffer, A. L., Wojnar, W., and Nelson, W., 1990, Amplification, detection and automated sequencing of Gibbon inteleukin-2 mRNA by *Thermus aquaticus* DNA polymerase reverse transcription and polymerase chain reaction, *Anal. Biochem.* **190:**292–296.

Shymala, V., and Ferro-Luzzi Ames, G., 1989, Genome walking by single-specific-primer polymerase chain reaction: SSP-PCR, *Gene* **84:**1–8.

Spanos A., Sedgwick, S. G., Yarraton, G. T., Huebscher, U., and Banks, G. R., 1981, Detection of the catalytic activity of DNA polymerase and their associated exonuclease following SDS polyacrylamide electrophoresis, *Nucl. Acids Res.* **9:**1825–1839.

Terada, I., Kwon, S-T., Miyata, Y., Matsuzawa, H., and Ohta, T., 1990, Unique precursor structure of an extracellular protease, aqualysin I, with NH2- and COOH-terminal prosequences and its processing in *Escherichia coli*, *J. Biol. Chem.* **265:**6576–6581.

Tindall, K. R., and Kunkel, T. A., 1988, Fidelity of DNA synthesis by the *Thermus aquaticus* DNA polymerase, *Biochem.* **27:**6008–6013.

Tomic, M., Sunjevaric, I., Savtchenko, E. S., and Blumenberg, M. A., 1990, Rapid and simple method for introducing specific mutations into any position of DNA leaving all other positions unaltered, *Nucl. Acids Res.* **18:**1656.

Touhara, N., Taguchi, H., Koyama, Y., Ohta, T., and Matsuzawa, H., 1991, Production and extracellular secretion of Aqualysin I (a thermophilic subtilisin-type protease) in a hostvector system for *Thermus thermophilus*, *Appl. Environ. Microbiol.* **57:**3385–3387.

Triglia, T., Peterson, M. G., and Kemp, D. J., 1988, A procedure for *in vitro* amplification of DNA segments that lie outside the boundaries of known sequences, *Nucl. Acids Res.* **16:**8186.

Vallette, F., Mege, E., Reiss, A., and Adnesik, M., 1989, Construction of mutant and chimeric genes using the polymerase chain reaction, *Nucl. Acids Res.* **17:**723–733.

Wittwer, C. T., and Garling, D. J., 1991, Rapid cycle DNA amplification: time and temperature optimization, *Biotechniques* **10:**76–84.

Yon, J., and Fried, M., 1989, Precise gene fusion by PCR, *Nucl. Acids Res.* **17:**4895.

Index

DATE DUE